CURRENT CLINICAL PATHOLOGY

ANTONIO GIORDANO, MD, PHD

SERIES EDITOR

For further volumes:
http://www.springer.com/series/7632

Antonio Russo • Stefano Iacobelli • Juan Iovanna
Editors

Diagnostic, Prognostic and Therapeutic Value of Gene Signatures

Humana Press

Editors
Antonio Russo, MD, PhD
Section of Medical Oncology
Department of Surgical and Oncological Sciences
University of Palermo
Palermo, Italy
antonio.russo@usa.net

Stefano Iacobelli, MD
National Consortium for Research on Cancer (CINBO)
University D'Annunzio
Chieti 66013, Italy
iacobell@unich.it

Juan Iovanna, MD, PhD
INSERM U624
Marseille 13288, France
iovanna@marseille.inserm.fr

ISBN 978-1-4939-6259-4 ISBN 978-1-61779-358-5 (eBook)
DOI 10.1007/978-1-61779-358-5
Springer New York Dordrecht Heidelberg London

Printed on acid-free paper

Humana Press is part of Springer Science+Business Media (www.springer.com)

Preface

Gene expression studies have revealed diagnostic profiles and upregulation of specific pathways in many solid tumors. Some gene-expression signatures are already used as predictors of relapse in early breast cancer patients. The explosion of new information in gene expression profiling could potentially lead to the development of tailored treatments in many solid tumors. In addition, many studies are ongoing to validate these signatures also in predicting response to hormonal, chemotherapeutic, and targeted agents in breast cancer as well as in other tumors.

This book has been carried out with the aim of providing readers a useful and comprehensive resource about the range of applications of microarray technology on oncological diseases.

The book is principally addressed to resident and fellow physicians, medical oncologists, molecular biologists, biotechnologists, and those who study oncological diseases. The chapters have been written by leading international researchers on these topics who have prepared their manuscripts according to current literature and field experience with microarray technology.

Palermo, Italy	Antonio Russo
Chieti, Italy	Stefano Iacobelli
Marseille, France	Juan Iovanna

Contents

Contributors

Giuseppe Badalamenti, MD, PhD Section of Medical Oncology, Department of Surgical and Oncological Sciences, University of Palermo, Palermo, Italy

Claudio Bellevicine, MD Scienze Biomorfologiche e Funzionali, Universita degli Studi di Napoli Federico II, Naples, Italy

Ezequiel Luis Calvo, MD, PhD Centre de Recherche du CHUL (CHUQ), Quebec, PQ, Canada

Lidia Corsini, PhD Section of Medical Oncology, Department of Surgical and Oncological Sciences, University of Palermo, Palermo, Italy

Bernard Czernobilsky, MD Patho-Lab Diagnostics, Ness Ziona, Israel

Jean Charles Dagorn, PhD INSERM U624, Marseille, France

Maria Debiec-Rychter, MD, PhD Department of Human Genetics, Catholic University of Leuven, Leuven, Belgium

Nelson Dusetti, PhD INSERM U624, Marseille, France

Mario Falchetti, PhD Department of Molecular Medicine, Policlinico Umberto I, Sapienza University of Rome, Rome, Italy

Daniele Fanale, PhD Section of Medical Oncology, Department of Surgical and Oncological Sciences, University of Palermo, Palermo, Italy

Pierre Ferrier, PhD Centre d'Immunologie de Marseille-Luminy, Université Aix-Marseille, France

CNRS, UMR6102, Marseille, France

Inserm, U631, Marseille, France

Anna Maria Frezza, MD Department of Medical Oncology, University Campus Bio-Medico, Rome, Italy

Gaorav P. Gupta, MD, PhD Department of Radiation Oncology, Memorial Sloan-Kettering Cancer Center, New York, NY, USA

Stefano Iacobelli, MD National Consortium for Research on Cancer (CINBO), University D'Annunzio, Chieti, Italy

Juan Iovanna, MD, PhD INSERM U624, Marseille, France

Jinsong Jia, MD, PhD Centre d'Immunologie de Marseille-Luminy, Université Aix-Marseille, France

CNRS, UMR6102, Marseille, France

Inserm, U631, Marseille, France

Institute of Hematology, People's Hospital, Beijing University, Beijing, People's Republic of China

Laura La Paglia, PhD Section of Medical Oncology, Department of Surgical and Oncological Sciences, University of Palermo, Palermo, Italy

Leonor Leider-Trejo, MD Institute of Pathology, Sourasky Medical Center, Tel Aviv, Israel

Alessandro Lugli, MD Institute of Pathology, University of Bern, Bern, Switzerland

Umberto Malapelle, MD, BS Scienze Biomorfologiche e Funzionali, Universita degli Studi di Napoli Federico II, Naples, Italy

John W.M. Martens, PhD Department of Medical Oncology, Daniel den Hoed Cancer Center, Erasmus Medical Center, Rotterdam, The Netherlands

Gabriella Nesi, MD, PhD Department of Critical Care Medicine, Division of Pathological Anatomy, University of Florence, Florence, Italy

Laura Ottini, MD, PhD Department of Molecular Medicine, Policlinico Umberto I, Sapienza University of Rome, Rome, Italy

Aleksandra Pekowska, PhD Centre d'Immunologie de Marseille-Luminy, Université Aix-Marseille, France

CNRS, UMR6102, Marseille, France

Inserm, U631, Marseille, France

Joanna Przybył, MSc Department of Molecular Biology, Maria Sklodowska-Curie Memorial Cancer Center and Institute of Oncology, Warsaw, Poland

Lajos Pustzai, MD, DPhil Department of Breast Medical Oncology, MD Anderson Cancer Center, University of Texas, Houston, TX, USA

Stefania Redana, MD Department of Medical Oncology I, Institute for Cancer Research and Treatment, Fondazione Piemontese per L'Oncologia, Candiolo, Italy

Sergio Rizzo, MD, PhD Section of Medical Oncology, Department of Surgical and Oncological Sciences, University of Palermo, Palermo, Italy

Delvys Rodriguez-Abreu, MD Department of Medical Oncology, Universitario Insular de Gran Canaria, Las Palmas de Gran Canaria, Spain

Christian Diego Rolfo, MD, PhD Medical Oncology, Oncology Unit, Clinica Rotger, Palma de Mallorca, Spain

Rafael Rosell, PhD Department of Medical Oncology, Germans Trias I Pujol, Catalan Institute of Oncology, Badalona, Spain

Antonio Russo, MD, PhD Section of Medical Oncology, Department of Surgical and Oncological Sciences, University of Palermo, Palermo, Italy

Piotr Rutkowski, MD, PhD Department of Soft Tissue/Bone Sarcoma and Melanoma, Maria Sklodowska-Curie Memorial Cancer Center and Institute of Oncology, Warsaw, Poland

Daniele Santini, MD, PhD Department of Medical Oncology, University Campus Bio-Medico, Rome, Italy

Gaia Schiavon, MD Department of Medical Oncology, Daniel den Hoed Cancer Center, Erasmus Medical Center, Rotterdam, The Netherlands

Marcel Smid, BSc Department of Medical Oncology, Daniel den Hoed Cancer Center, Erasmus Medical Center, Rotterdam, The Netherlands

Salvatore Spicuglia, PhD Centre d'Immunologie de Marseille-Luminy, Université Aix-Marseille, France

CNRS, UMR6102, Marseille, France

Inserm, U631, Marseille, France

Miquel Taron, PhD Department of Medical Oncology, Germans Trias I Pujol, Catalan Institute of Oncology, Badalona, Spain

Giuseppe Tonini, MD, PhD Department of Medical Oncology, University Campus Bio-Medico, Rome, Italy

Giancarlo Troncone, MD, PhD Scienze Biomorfologiche e Funzionali, Universita degli Studi di Napoli Federico II, Naples, Italy

Bruno Vincenzi, MD, PhD Department of Medical Oncology, University Campus Bio-Medico, Rome, Italy

Jia Wei, MD Medical School of Nanjing University and Clinical Cancer Institute of Nanjing University, The Comprehensive Cancer Centre of Drum Tower Hospital, Nanjing, China

Joaquin Zacarias-Cabeza, PhD Centre d'Immunologie de Marseille-Luminy, Université Aix-Marseille, France

CNRS, UMR6102, Marseille, France

Inserm, U631, Marseille, France

Inti Zlobec, PhD Institute of Pathology, University of Bern, Bern, Switzerland

Chapter 1
Gene Signatures and Soft Tissue Sarcomas: Status of Art and Perspectives

Bruno Vincenzi, Anna Maria Frezza, Daniele Santini, and Giuseppe Tonini

Introduction

Soft tissue sarcomas represent an extremely heterogeneous group of tumors comprising more than 70 different histotypes, as described in the current World Health Organization classification from 2002 [1]. To date, the diagnosis of soft tissue sarcomas is still challenging and based on morphological appearance; moreover, some conventional names included in the classification results pathobiologically inaccurate (e.g., synovial sarcoma) although they denote well-defined entities mutually understood by both pathologists and clinicians.

Some well-known inherited syndromes are associated with the onset of soft tissue sarcoma, underlining the importance of genetic alterations in the pathogenesis of this aggressive disease. For example, hereditary retinoblastoma syndrome [2], a rare genetic disorder deriving from an RB gene mutation, is known to be associated with a higher risk of developing osteosarcoma [3]; type 1 neurofibromatosis [4], due to the mutation of NF1 gene on chromosome 17, bring to the onset of malignant peripheral nerve sheet tumors (MPNSTs) and gastrointestinal stromal tumors (GISTs), often multiples and arising from the small bowel [5]. At the same time, Li–Fraumeni syndrome (deriving from a germ line p53 mutation) [6], Gardner syndrome [7] (due to an APC mutation), and Beckwith–Wiedemann syndrome, the pathogenesis of which is still not completely understood [8], are associated with a higher incidence of soft tissue and bone sarcoma.

Despite these data, the majority of soft tissue sarcomas arise sporadically, but are often characterized by molecular aberrations. Genetically, sarcomas can be broadly divided into two groups: the first group, usually characteristic of older patients, comprise those with a complex karyotype bringing nonspecific genetic alterations, such as genetic deletions and amplifications, nonbalanced translocations, and changes in chromosome number. These aberrations are usually not helpful for diagnosis, but they are often prognostically relevants. The second group is made up of sarcomas characterized by simple karyotypes, which create specific genetic alterations usually represented by chromosomal translocations or oncogenic mutations (e.g., the KIT mutation in GISTs). These kinds of sarcomas often arise during childhood or adolescence and the detection of their tumor-specific genetic alterations by cytogenetic or molecular genetic techniques are extremely diagnostically useful.

B. Vincenzi (✉) • A.M. Frezza • D. Santini • G. Tonini
Department of Medical Oncology, University Campus Bio-Medico, Rome, Italy
e-mail: b.vincenzi@unicampus.it

A. Russo et al. (eds.), *Diagnostic, Prognostic and Therapeutic Value of Gene Signatures*, Current Clinical Pathology, DOI 10.1007/978-1-61779-358-5_1, © Springer Science+Business Media, LLC 2012

Soft Tissue Sarcoma with Complex Genomic Profiles

Soft tissue sarcomas with complex genomic profiles represent almost 50% of all soft tissue sarcomas, and they are mainly represented by sarcoma with spindle cell/pleomorphic differentiation. These kinds of sarcomas can harbor a wide variety of genetic alterations, such as gain or loss of chromosomes or chromosome regions and amplifications; these aberrations have been proven to play a key role in tumor progression, metastatic dissemination, and they often correlate with prognosis and treatment sensitivity [9].

Leiomyosarcoma

A subtype of sarcomas, characterized by a spindle cell differentiation, account for 8–10% of all adult soft tissue sarcomas [10]. They usually show complex karyotypic alterations comprising mainly chromosomal imbalances and amplifications [11]. Recent studies have proven that some of these modifications are associated with unfavorable prognosis (p16INK4 inactivation RASSF1A hypermethylation) [12, 13], and with a higher metastatic risk (upregulation of hypoxia inducible factor A); moreover, the identification of specific pathway disregulation has brought about the development of an interesting therapeutic implication: for instance, the upregulation of PI3K–AKT–mTOR pathway identified in leiomyosarcoma (LMS) could justify the partial activity of everolimus (RAD001, mTOR inhibitor), as recently proven in different clinical trials.

Undifferentiated High-Grade Spindle/Pleomorphic Sarcoma

Undifferentiated high-grade spindle/pleomorphic sarcoma (UPS) is the term currently used to define high-grade soft tissue sarcoma, which have failed to show any specific line of differentiation (instead of malignant fibrous histiocytomas). It is a diagnosis of elimination comprising approximately 5–7% [1] of adult sarcomas. UPS are characterized by a complex cytogenetic rearrangement involving 30–35% of the genome, and they have been proven to share many of the aberrations described for LMSs. Among the most common genomic imbalances found in UPS are the loss of chromosome 13q, RASSF1A hypermethylation, and the upregulation of several hypoxia-related genes. A recent study has evaluated the involvement of gene TRIO (coding for guanidine nucleotide exchange factor), usually upregulated in soft tissue sarcoma. TRIO, through the activation of Rho GTPase-mediated signaling pathway, plays an important part in the control of apoptosis, cytoskeleton organization, cell-to-cell adhesions, and many other basic cell functions, and it seems to be involved in sarcoma progression [14]. Ezrin (villin 2) is a protein that works as a link between cell membrane and actin cytoskeleton and its overexpression has been found to correlate with a higher metastatic potential and reduced survival in many different tumors, as also seen in UPS.

Myxofibrosarcoma

For a long time, *myxofibrosarcoma* have been included in malignant fibrous histiocytomas and for this reason, the data concerning the genetics of this kind of sarcoma are still poor.

Myxofibrosarcomas are usually characterized by a complex karyotype that can be identified both in low grade and in high-grade subtypes. Recent studies have underlined how cytogenetic aberration in myxofibrosarcoma can be associated with an increased risk of local recurrence. Moreover, a better understanding of the myxofibrosarcoma genetic profile can help make a distinction between low-grade myxofibrosarcoma and cellular myxoma (not always easy on the basis of morphology) and may allow a correct identification of those myxofibrosarcomas previously misclassified as UPS.

Pleomorphic Liposarcomas

Among liposarcoma, *pleomorphic liposarcomas* are known to be characterized by a complex genomic profile. In fact, to date, no specific genetic alterations have been identified, although some frequent events have been pinpointed, such as the disregulation of RB1 pathway or the loss of neurofibromatosis type 1 (NF1) gene [15]. A better understanding of pleomorphic sarcoma genetic profiling could help to accurately distinguish between this kind of liposarcoma and one which is dedifferentiated; despite their morphological similarities, they show different chromosomal imbalances. According to a recent study, a correct distinction could be done through the evaluation of 15 differentially expressed genes, mainly located in the 12p13–p15 region [16]; however, further data are necessary to confirm these preliminary results.

Malignant Peripheral Nerve Sheet Tumors

MPNSTs originate from peripheral nerves or from extraneural soft tissue showing nerve sheath differentiation. They are characterized by complex karyotypes and frequently show chromosomal losses, which have been proven to be more frequent than chromosomal gains [17], numerous breakpoints involving different chromosomal regions, ring chromosomes, trisomy, and rearrangements, while amplification are rare. Of interest, more than 50% of MPNSTs occur in the setting of neurofibromatosis 1, one of the most common mendelian disorders caused by heterozygous mutations of the NF1 gene; the NF1-associated MPNSTs show a higher rate of genetic complex rearrangements when compared with the sporadic forms. Finally, complex karyotypes are generally not observed in benign peripheral nerve sheath tumors (e.g., schwannoma) and this may help in the differentiation of benign and malignant forms.

Other subtypes of soft tissue sarcomas, such as angiosarcoma and pleomorphic rhabdomyosarcoma, can be included in this wide group, but their characteristics are still poorly known and further studies are needed to define their genetic signatures.

Soft Tissue Sarcoma with Simple Genomic Profiles

Nearly 30% of all soft tissue sarcomas show a simple genomic profile and are characterized by specific chromosomic aberrations. These genetic abnormalities aid in the diagnosis of sarcoma, can assume a prognostic or predictive significance, and could also represent a target for the development of new biological drugs. Soft tissue sarcomas with simple karyotypes can be further divided in two groups on the base of the presence of reciprocal translocations involving or not involving the Ewing sarcoma breakpoint region 1 (EWSR1).

Table 1.1 Soft tissue sarcoma associated with EWSR1 translocation

Tumor	Translocation	Fusion product
Ewing sarcoma/PNET	t(11;22)(q24;q12)	EWSR1–FLI1
	t(21;22)(q22;q12)	EWSR1–ERG
	t(20;22)(q13;q12)	EWSR1–NFATC2
	t(2;22)(q33;q12)	EWSR1–FEV
	t(7;22)(p22;q12)	EWSR1–ETV1
	t(17;22)(q12;q12)	EWSR1–E1AF
	t(2;22)(q31;q12)	EWSR1–SP3
	t(1;22)(p36.1;q12)	EWSR1–ZNF278
	t(6;22)(p21;q12)	EWSR1–POU5FI
Clear cell sarcoma	t(12;22)(q13;q12)	EWSR1–ATF1
	t(2;22)(q34;q12)	EWSR1–CREB1
Angiomatoid fibrous histiocytoma	t(12;22)(q13;q12)	EWSR1–ATF1
	t(2;22)(q34;q12)	EWSR1–CREB1
Desmoplastic small round cell tumor	t(11;22)(p13;q12)	EWSR1–WT1
	t(21;22)(q22;q12)	EWSR1–ERG
Extraskeletal myxoid chondrosarcoma	t(9;22)(q22;q12)	EWSR1–NR4A3
Myxoid/round cell liposarcoma	t(12;22)(q13;q12)	EWSR1–DDIT3

EWSR1 Ewing sarcoma breakpoint region 1

Soft Tissue Sarcoma Associated with EWSR1 Translocation

The EWSR1 (also known as EWS) is the most commonly involved gene in sarcoma translocations, and it seems to play an important role in the pathogenesis of Ewing's sarcoma/peripheral neuroectodermal tumor, clear cell sarcoma, angiomatoid fibrous histiocytoma, desmoplastic small round cell tumor, extraskeletal myxoid chondrosarcoma and myxoid/round cell liposarcoma (Table 1.1). The translocations involving the EWSR1 usually result in the fusion of the N-terminal transcription-activating domain of EWSR1 and the C-terminal DNA-binding domain of the fusion partner, which brings a generation of a novel constitutively activated transcription factor.

Because of the difficulties often recognized in the diagnostic phase of sarcoma work-up, the identification of EWSR1 rearrangement through fluorescence in situ hybridization (FISH) is considered an extremely useful confirmatory diagnostic tool.

Ewing Sarcoma Family Tumors

Ewing's sarcomas together with primitive neuroectodermal tumors (PNETs) constitute the *Ewing sarcoma family tumors* (ESFTs) and share a similar biological behavior and therapeutical approach. Almost 80% of ESFTs harbor a recurring translocation, t(11;22) (q24;q12), that juxtaposes the FLI1 and EWSR1 genes encoding a chimeric RNA and protein; the remaining part of ESFTs are characterized by translocations always involving EWSR1 together with different transcription factors, such as ERG (9–14%) or other rarer variants (1–5%). On the basis of this data, ESFTs are considered the prototype of tumors with fusion genes involving the TET gene family, where TET was named from the initials of TLS/FUS, EWSR1, and TAFII68 [18]. The proteins derived from the TET gene family include an RNA-recognizing motif made up of 84 amino acids which is thought to bind RNA and participate in the transcription process. The evaluation of karyotype in order to identify the t(11;22) translocation must be considered today as necessary in the initial work-up of an ESFT; moreover, where the histopathological findings strongly suggest a ESFTs diagnosis with a normal karyotype, the recourse to the FISH, using a break-apart probe targeting the EWSR1 gene, must be considered. Of interest, the detection of the precise translocation in ESFTs, through the use of a reverse transcription polymerase chain reaction (RT-PCR), has been proven to hold an important prognostic relevance; in fact, the EWSR1/FLI1 translocation seems to be associated with a better outcome when compared with the alternative fusions [19].

Soft Tissue Clear Cell Sarcomas

Soft tissue clear cell sarcomas (ST-CCS), also called melanoma of soft parts, represent a well-defined subtype of soft tissue sarcoma arising mainly from adolescents and young people's aponeuroses of distal extremities. From a pathological viewpoint, they are constituted by nests and cords of pale cells, sharing many differentiation features with melanocite, including the presence of melanin and the immunohistochemical positivity for S100, HMB45, MART1, and MITF-1 in almost all cases, with the exception of the gastrointestinal tract forms which are usually positive only for S100 [20, 21]. ST-CCS are characterized by a specific translocation t(12;22)(q13;q12), fusing the EWSR1 gene with the activating transcription factor-1 gene (ATF1) [22]. Recently, in the cases of ST-CCS deriving from the gastrointestinal tract, a different variant has been identified, involving EWSR1 and cAMP-responsive element-binding protein (CREB1). In both translocations, EWSR1's partners are represented by a leucine zipper superfamily of transcription factors: the translation of the chimeric transcripts leads to a proteinaceous product in which the basic leucine zipper domain is retained.

As for pathogenesis, the upregulation of MITF gene (a master regulator of melanocyte differentiation) has been found in ST-CCS and the assessment of multiple histone deacetylase inhibitors activity in the suppression of MITF expression is today under evaluation. To date, it is still unclear if MITF upregulation derives from the EWSR1–ATF1 binding of MITF gene promoter site or it is a preexisting characteristic of ST-CCS [22]. In order to achieve a correct diagnosis, the possibility of a malignant melanoma must be ruled out. The clinical history and the absence of junctional activity can be useful in this distinction; however, metastatic melanoma can exhibit complete morphologic overlap with ST-CCS, and in these cases the use of genetic analysis represents the only way to distinguish them.

Angiomatoid Fibrous Histicytoma

Angiomatoid fibrous histicytoma is a mesenchymal lesion of intermediate malignancy, rarely metastatizing and usually occurring in the superficial soft parts of the limbs. Of interest, despite a completely different morphologic aspect and biological behavior, angiomatoid fibrous histicytoma shares the same translocations detected in ST-CCS (EWSR1–CREB1 fusion, and more rarely, of a EWSR1–ATF1 fusion) [23]. This observation leads to the supposition of preexisting differentiation programs in the two progenitor cells.

Desmoplastic Small Round Cell Tumors

As ESFTs and ST-CCS, also *desmoplastic small round cell tumors,* usually harbor a t(11;22) translocation which brings to the fusion of EWSR1 gene, on chromosome 22, with the WT1 gene, on chromosome 11 [24]. The identification of the WT1 as a partner of EWSR1 throughout the DNA amplification at the breakpoint allows a differential diagnosis between these two sarcoma subtypes.

Moreover, WT1 has been proven to act as a transcription factor activated by EWSR1, which leads to the upregulation of oncogenic factors, such as the platelet-derived growth factor (PDGF). This mechanism is thought to play a key role in desmoplastic small round cell tumors which are histologically characterized by an abundant collagenous stroma, including cord and nest of small round cells; furthermore, this could represent a rationale for experimentation with PDGF inhibitors, such as imatinib [25], in the treatment of this complex disease.

Extraskeletal Myxoid Chondrosarcoma

Extraskeletal myxoid chondrosarcoma is a histotype of sarcoma with a yet undefined origin, involving mainly deep soft tissue of limbs and limb girdles. The translocation t(9;22)(q22;q12), fusing EWSR1 with CHN (also known as NR4A3), is found in almost 70% of extraskeletal myxoid chondrosarcomas [26], even if additional fusion partners to CHN have been subsequently identified (TAF2N, TCF12, and TFG) [27, 28]. Of interest, the translocation characterizing extraskeletal myxoid chondrosarcoma has been found to be absent in the skeletal form, confirming that these two forms must be considered different entities [29]. Recent findings suggest that the EWSR1/CHN fusion protein could activate the PPARG nuclear receptor gene. Aside from the pathogenetic implications, these data could suggest the identification of a new potential therapeutic target [30].

Myxoid Liposarcoma

Myxoid liposarcoma, also called hypercellular liposarcoma because of the increase in the cellularity so that individual tumor cells lie in direct continuity with each other without matrix interposition, accounts for 30–35% of all liposarcoma and arise mainly in limbs [31]. Myxoid liposarcomas harbor two different translocations: the first, t(12;16), fuses the DDIT3 gene on 12q13 with the FUS gene on 16p11 [32], the second, t(12;22), fuses DDIT3 with EWSR1 on 22q12 [33]. The DNA-damage-inducible transcript 3 (DDIT3) gene is a member of the CCAAT/enhancer-binding protein family which is physiologically involved in adipocyte differentiation and exerts an antiproliferation activity; this gene is lost in the chimeric product deriving from the translocations.

Soft Tissue Sarcoma with Non-EWS Translocations

Apart from soft tissue sarcomas associated with EWSR1 translocations, many other histological subtypes of sarcomas are characterized by specific translocations, not involving the Ewing sarcoma region, that have been proven to play a key part in the pathogenesis, by disrupting basic cellular functions, such as proliferation, growth, and survival, and to codify for important possible new therapeutic targets. In this group, we find synovial sarcoma, alveolar rhabdomyosarcoma, alveolar soft part sarcoma, dermatofibrosarcoma protuberans, low-grade fibromyxoid sarcoma, and inflammatory myofibroblastic tumor (Table 1.2).

Synovial Sarcoma

Despite the improper name, *synovial sarcoma* does not arise from synovium but it can occur in almost any anatomical location, usually derived from connective periarticular tissue. It represents almost 10% of all sarcomas and typically affects patients between 15 and 35 years of age. Synovial

Table 1.2 Soft tissue sarcoma associated with non-EWSR1 translocation

Tumor	Translocation	Fusion product
Synovial sarcoma	t(X;18)(p11;q11)	SS18–SSX1 (SYT–SSX1)
		SS18–SSX2 (SYT–SSX2)
Alveolar rhabdomyosarcoma	t(2;13)(q35;q14)	PAX3–FOXO1 (PAX3–FKHR)
	t(1;13)(p36;q14),	PAX7–FOXO1 (PAX7–FKHR)
Alveolar soft part sarcoma	t(X;17)(p11;q25)	ASPSCR1–TFE3 (ASPL–TFE3)
Inflammatory myofibroblastic tumor	Translocations involving 2p23 locus	ALK activation
Dermatofibrosarcoma protuberans	Ring chromosome 17 or 22	COL1A1–PDGFB
	t(17;22)(q21;q13)	COL1A1–PDGFB
Low-grade fibromyxoid sarcomas	t(7;16)(q34; p11)	FUS–CREB3L2
	t(11;16)(p11;p11)	FUS–CREB3L1

EWSR1 Ewing sarcoma breakpoint region 1

sarcomas harbor a specific translocation, t(X;18)(p11.2:q11.2) [34], in which eight amino acids at the carboxy terminal of the SYT (SS18) gene on chromosome 18 are replaced with 78 amino acids of the carboxy terminal of the SSX gene on the X chromosome. SSX recognize five different variants: those involved in the translocation of synovial sarcoma are SSX1, detected in two thirds of all cases, SSX2, and more rarely SSX4 [35]. Three different subtypes of synovial sarcomas can be identified, characterized by different genetic signatures: the monophasic variant made up of vimentin-expressing spindle cells, usually carrying the SS18–SSX2 translocation, the biphasic variant comprises a mixture of vimentin-expressing spindle cells and keratin expressing glandular epithelial cells harboring the SS18–SSX1 or SS18–SSX2 translocation, and a poorly differentiated representing 20% of synovial sarcoma cases [36]. To date, the exact function of the genes identified in synovial sarcoma is still unclear and other events might be required for sarcomagenesis. The diagnosis of synovial sarcoma is frequently challenging because of the morphological similarities with the other small round cell sarcomas, lymphomas, neuroendocrine carcinoma, and neuroblastoma. Detection of the t(X;18) translocation is an extremely useful tool.

Rhabdomyosarcoma

Rhabdomyosarcoma is the most common kind of sarcoma in children, and it occurs mainly in the deep muscles of the extremities. Among rhabdomyosarcoma, three different subtypes can be distinguished: embrional rhabdomyosarcoma (60% of all cases), alveolar rhabdomyosarcoma (20%), and pleomorphic rhabdomyosarcoma (20%) [37]. Alveolar rhabdomyosarcoma is known to be characterized by a specific translocation which leads to the fusion of 5¢ end of a member of the PAX gene family, encoding for a transcription factor that promotes the myogenesis in muscle stem cells, with the 3¢ end of the FOXO1A gene that provides the transactivation domain. The fusion's product causes an enhancement of cellular proliferation and invasion [38]. The t(2;13)(q35:q14), involving the PAX3 gene, is detected in almost 70% of alveolar rhabdomyosarcomas while the t(1;13)(p36;q14) translocation, involving PAX7, in 10%; the remaining 20% usually do not show any specific genetic aberration. Today, in the initial assessment of rhabdomyosarcoma, genetic analysis is mandatory because it allows a distinction between the embryonal and the alveolar form, which often share a similar immunohistochemical phenotype but have different biological behavior with the alveolar subtype being much more aggressive, and because the alveolar rhabdomyosarcoma harboring the PAX7–FKHR translocation has been proven to be associated with longer event-free survival and longer overall survival in both locoregional [39] and metastatic disease [40].

Alveolar Soft Part Sarcoma

Alveolar soft part sarcoma is a rare subtype of sarcoma that mainly affects children and young adult's deep muscles and is associated with an extremely poor prognosis. To date, the pathogenesis of alveolar soft part sarcoma is still poorly understood; however, cytogenetic studies have identified a recurrent translocation, t(X;17)(p11;q25), detected in majority of alveolar soft part sarcomas. This translocation fuses alveolar soft part sarcoma chromosome region candidate 1 (ASPSCR1; ASPL) gene on the long arm of chromosome 17 to the transcription factor for immunoglobulin heavy-chain enhancer 3 (TFE3) located at Xp11. Even if the function of the fusion product is still unknown, the consistent detection of the t(X;17) in alveolar soft part sarcoma is now considered a useful molecular diagnostic marker; moreover, recent studies have identified an MET activation mediated by ASPL–TFE3 fusion protein [41]. On the basis of these data, the efficacy of MET kinase inhibitors (such as sunitinib malate) [42] in the treatment of this aggressive tumor is currently under examination. In support of this hypothesis, a response has been reported in clear cell sarcoma in a phase II study of the MET inhibitor ARQ197 [43].

Dermatofibrosarcoma Protuberans

Dermatofibrosarcoma protuberans is a relatively common sarcoma characterized by a low-grade malignant behavior, which tends to recur locally (especially if incompletely resected) instead of giving distant metastasis. From the genetic viewpoint, it harbors a specific translocation, t(17;22)(q11;q13.1), or a supernumerary ring chromosome always derived from t(17;22), which determine the fusion of the COL1A1 gene on chromosome 17 with PDGFB1 gene (encoding the beta chain of PDGF, a homodimer) at 22q13 [44, 45]. The breakpoint is specific for the PDGFB1 gene while it is extremely changeable for the COL1A1 gene. This finding suggests that COL1A1 acts as an upregulator of PDGFR expression which exerts an auto or paracrine growth factor activity. Even if genetic analysis is usually not necessary for the diagnosis of dermatofibrosarcoma protuberans, the search for the translocation can be useful in order to identify the best therapeutic option. In fact, in cases harboring the translocation, the inhibition of PDGFR tyrosine kinase through imatinib has yielded extremely good results both in the locally advanced and in the metastatic setting [46]. In this disease, imatinib is thought to work through the inhibition of PDGFRB and can be useful in controlling locally advanced tumors with a response rate of almost 50%. Currently, sunitinib and sorafenib are under evaluation [47]. Conversely, in rare cases of fibrosarcomatous or pleomorphic sarcomatous transformation, in which the translocation is absent, the same therapy would be unsuccessful [48].

Low-Grade Fibromyxoid Sarcoma

Low-grade fibromyxoid sarcomas, also known as Evans tumors, arise in the deeper soft tissues of extremities (especially the thigh), characterized by an extremely heterogeneous morphologic appearance. In fact, many different histological subtypes once considered as different entities, such as hyalinizing spindle cell tumor with giant rosettes, have been included in this group after the identification of a common translocation, t(7;16)(q34; p11), with the 5′ part of FUS from chromosome 16 with 3′ part of CREB3L2 on chromosome 7. In a few cases, the translocation t(11,16)(p11;p11), with an FUS–CREB3L1 fusion gene, has been identified [49]. Of interest, in contrast to all the other translocation sarcomas, the breakpoint of the t(7;16) is mainly localized within the exons.

To date, no correlation has been found between the different morphological aspect of low-grade fibromyxoid sarcomas and its genetic signatures and more studies are needed to clarify the pathogenetic role and the potential therapeutic value of these findings.

Inflammatory Myofibroplastic Tumor

Inflammatory myofibroblastic tumor originates principally in the retroperitoneum mainly during the first decade of life, although some cases have been described in adult ages. This subtype of sarcoma has been proven to be associated with a translocation involving the C-terminal kinase domain of anaplastic lymphoma kinase (ALK) gene, located on 2p23. The ALK gene encodes for a tyrosine kinase oncogene which becomes constitutively activated in inflammatory myofibroblastic tumors, thanks to the fusion with many different partners, such as TPM3 or CLTC. Apart from the diagnostic utility of these genetic findings, recent studies have underlined the difference in the morphology and in the biological behavior of ALK-positive and ALK-negative inflammatory myofibroblastic tumors. In fact, ALK-negative tumors occur more frequently in adults and display more nuclear pleomorphism and atypical mitoses [50]. Moreover, given the pathogenetic relevance of ALK-mediated signaling in ALK-rearranged tumors, the efficacy of crizotinib in this subgroup of sarcomas is currently under evaluation [51].

Summary

The identification of the association between soft tissue sarcomas and translocations or specific genetic signatures makes this rare disease an extremely attractive field of research, both for the possibility of better understanding the pathogenetic processes (their onset and progression), both for diagnostic and therapeutical implications. To date, the use of modern molecular techniques in the diagnosis of soft tissue sarcomas appears mandatory, mainly in order to distinguish subtypes that share similar morphological phenotypes and immunohistochemical patterns. Furthermore, the identification of specific translocations leading to the upregulation of pathways probably exerting a key role in sarcomatogenesis makes soft tissue sarcomas particularly susceptible to molecularly targeted therapies, which represent today a promising hope for those patients affected by this poor-prognosis disease. Further studies are needed in order to better understand the pathogenetic mechanisms and the possible treatment options of these sarcomas, which are still defined as associated with complex genomic profiles.

References

1. Fletcher CD. The evolving classification of soft tissue tumours: an update based on the new WHO classification. Histopathology. 2006;48(1):3–12.
2. Wong FL, Boice Jr JD, Abramson DH, Tarone RE, Kleinerman RA, Stovall M, et al. Cancer incidence after retinoblastoma. Radiation dose and sarcoma risk. JAMA. 1997;278(15):1262–7.
3. Kansara M, Thomas DM. Molecular pathogenesis of osteosarcoma. DNA Cell Biol. 2007;26(1):1–18.
4. Williams VC, Lucas J, Babcock MA, Gutmann DH, Korf B, Maria BL. Neurofibromatosis type 1 revisited. Pediatrics. 2009;123(1):124–33.
5. Bajor J. Gastrointestinal stromal tumors in neurofibromatosis type 1. Orv Hetil. 2009;150(4):149–53.
6. Upton B, Chu Q, Li BD. Li-Fraumeni syndrome: the genetics and treatment considerations for the sarcoma and associated neoplasms. Surg Oncol Clin N Am. 2009;18(1):145–56.

7. Stuart NJ, Clark SK. Current ideas in desmoid tumours. Fam Cancer. 2006;5(3):275–85.
8. Cohen Jr MM. Beckwith-Wiedemann syndrome: historical, clinicopathological, and etiopathogenetic perspectives. Pediatr Dev Pathol. 2005;8(3):287–304.
9. Guillou L, Aurias A. Soft tissue sarcomas with complex genomic profiles. Virchows Arch. 2010;456(2):201–17.
10. Fletcher CDM, Unni KK, Mertens F. World Health Organization classification of tumours. Pathology and genetics of tumors of soft tissue and bone. Lyon: IARC; 2002.
11. Yang J, Du X, Chen K, et al. Genetic aberrations in soft tissue leiomyosarcoma. Cancer Lett. 2009;275:1–8.
12. Kawaguchi K, Oda Y, Saito T, et al. Mechanisms of inactivation of the p16INK4a gene in leiomyosarcoma of soft tissue: decrease. J Pathol. 2003;201:487–95.
13. Seidel C, Bartel F, Rastetter M, et al. Alterations of cancer related genes in soft tissue sarcomas: hypermethylation of RASSF1A is frequently detected in leiomyosarcoma and associated with poor prognosis in sarcoma. Int J Cancer. 2005;114:442–7.
14. Adamowicz M, Radlwimmer B, Rieker RJ, et al. Frequent amplifications and abundant expression of TRIO, NKD2, and IRX2 in soft tissue sarcomas. Genes Chromosomes Cancer. 2006;45:829–38.
15. Taylor BS, Barretina J, Socci ND, et al. Functional copynumber alterations in cancer. PLoS ONE. 2008;3:3179.
16. Singer S, Socci ND, Ambrosini G, et al. Gene expression profiling of liposarcoma identifies distinct biological types/subtypes and potential therapeutic targets in well-differentiated and dedifferentiated liposarcoma. Cancer Res. 2007;67:6626–36.
17. Bridge RS, Bridge JA, Neff JR, et al. Recurrent chromosomal imbalances and structurally abnormal breakpoints within complex karyotypes of malignant peripheral nerve sheath tumour and malignant Triton tumor: a cytogenetic and molecular cytogenetic study. J Clin Pathol. 2004;57:1172–8.
18. De Alava E, Gerald WL. Molecular biology of the Ewing's sarcoma/primitive neuroectodermal tumor family. J Clin Oncol. 2000;18:204–13.
19. Lin PP, Brody RI, Hamelin AC, Bradner JE, Healey JH, Ladanyi M. Differential transactivation by alternative EWS-FLI1 fusion proteins correlates with clinical heterogeneity in Ewing's sarcoma. Cancer Res. 1999;59:1428–32.
20. Antonescu CR, Nafa K, Segal NH, et al. EWS-CREB1: a recurrent variant fusion in clear cell sarcoma association with gastrointestinal location and absence of melanocytic differentiation. Clin Cancer Res. 2006;12:5356–62.
21. Zambrano E, Reyes-Mugica M, Franchi A, et al. An osteoclast-rich tumor of the gastrointestinal tract with features resembling clear cell sarcoma of soft parts: reports of 6 cases of a GIST simulator. Int J Surg Pathol. 2003;11:75–81.
22. Antonescu CR, Tschernyavsky SJ, Woodruff JM, et al. Molecular diagnosis of clear cell sarcoma: detection of EWSATF1 and MITF-M transcripts and histopathological and ultrastructural analysis of 12 cases. J Mol Diagn. 2002;4:44–52.
23. Rossi S, Szuhai K, Ijszenga M, et al. EWSR1-CREB1 and EWSR1-ATF1 fusion genes in angiomatoid fibrous histiocytoma. Clin Cancer Res. 2007;13:7322–8.
24. Ladanyi M, Gerald W. Fusion of the EWS and WT1 genes in the desmoplastic small round cell tumor. Cancer Res. 1994;54:2837–40.
25. Imatinib mesylate in patients with refractory Desmoplastic small round cell tumors. http://www.clinicaltrial.gov/ct2/results?term=desmoplastic+small+round+cell+tumour+PDGFR.
26. Stenman G, Andersson H, Mandahl N, Meis-Kindblom JM, Kindblom LG. Translocation t (9;22)(q22;q12) is a primary cytogenetic abnormality in extraskeletal myxoid chondrosarcoma. Int J Cancer. 1995;62:398–402.
27. Sjogren H, Meis-Kindblom JM, Orndal C, Bergh P, Ptaszynski K, Aman P, et al. Studies on the molecular pathogenesis of extraskeletal myxoid chondrosarcomacytogenetic, molecular genetic, and cDNA microarray analyses. Am J Pathol. 2003;162:781–92.
28. Hisaoka M, Ishida T, Imamura T, Hashimoto H. TFG is a novel fusion partner of NOR1 in extraskeletal myxoid chondrosarcoma. Genes Chromosomes Cancer. 2004;40:325–8.
29. Antonescu CR, Argani P, Erlandson RA, Healey JH, Ladanyi M, Huvos AG. Skeletal and extraskeletal myxoid chondrosarcoma: a comparative clinicopathologic, ultrastructural, and molecular study. Cancer. 1998;83:1504–21.
30. Filion C, Motoi T, Olshen AB, et al. The EWSR1/NR4A3 fusion protein of extraskeletal myxoid chondrosarcoma activates the PPARG nuclear receptor gene. J Pathol. 2009;217:83–93.
31. Antonescu CR, Ladanyi M. Myxoid Liposarcoma. In: Fletcher CDM, Unni KK, Mertens F, editors. World Health Organisation classification of tumours. Pathology and genetics of tumours of soft tissue and bone. Lyon: IARC; 2002. p. 40–3.
32. Knight JC, Renwick PJ, Dal Cin P, et al. Translocation t(12;16)(q13;p11) in myxoid liposarcoma and round cell liposarcma: molecular and cytogenetic analysis. Cancer Res. 1995;55:24–7.
33. Panagopoulos I, Hoglund M, Mertens F, et al. Fusion of the EWS and CHOP genes in myxoid liposarcoma. Oncogene. 1996;12:489–94.

34. Turc-Carel C, Dal Cin P, Limon J, Li F, Sandberg AA. Translocation X;18 in synovial sarcoma. Cancer Genet Cytogenet. 1986;23:93.
35. Crew AJ, Clark J, Fisher C, et al. Fusion of SYT to two genes, SSX1 and SSX2, encoding proteins with homology to the Kruppel-associated box in human synovial sarcoma. EMBO J. 1995;14:2333–40.
36. Ladanyi M, Antonescu CR, Leung DH, Woodruff JM, Kawai A, Healey JH, et al. Impact of SYT-SSX fusion type on the clinical behavior of synovial sarcoma: a multi-institutional retrospective study of 243 patients. Cancer Res. 2002;62:135–40.
37. Ge NWA, Webber BL, et al. Classification of Rhabdomyosarcoma and related sarcomas. Pathologic aspects and proposal for a new classification. An intragroup rhabdomyosarcoma study. Cancer. 1995;76:1073–85.
38. Anderson J, Ramsay A, Gould S, et al. PAX3-FKHR induces morphological change and enhances cellular proliferation and invasion in rhabdomyosarcoma. Am J Pathol. 2001;159:1089–96.
39. Kazanowska B, Reich A, Stegmaier S, et al. Pax3-fkhr and pax7-fkhr fusion genes impact outcome of alveolar rhabdomyosarcoma in children. Fetal Pediatr Pathol. 2007;26:17–31.
40. Sorensen PH, Lynch JC, Qualman SJ, et al. PAX3-FKHR and PAX7-FKHR gene fusions are prognostic indicators in alveolar rhabdomyosarcoma: a report from the children's oncology group. J Clin Oncol. 2002;20:2672–9.
41. Jun HJ, Lee J, Lim do H, Park JO, Ahn G, Seo SW, et al. Expression of MET in alveolar soft part sarcoma. Med Oncol. 2010;27(2):459–65.
42. Stacchiotti S, Tamborini E, Marrari A, Brich S, Rota SA, Orsenigo M, et al. Response to sunitinib malate in advanced alveolar soft part sarcoma. Clin Cancer Res. 2009;15(3):1096–104.
43. Goldberg JDG, Choy E, Rosen L, et al. Preliminary results from a phase II study of ARQ197 in patients with microphthalmia transcription family (MiT)-associated tumors. J Clin Oncol. 2009;27:Abstr 10502.
44. Sirvent N, Maire G, Pedeutour F. Genetics of dermatofibrosarcoma protuberans family of tumors: from ring chromosomes to tyrosine kinase inhibitor treatment. Genes Chromosomes Cancer. 2003;37:1–19.
45. Simon MP, Pedeutour F, Sirvent N, Grosgeorge J, Minoletti F, Coindre JM, et al. Deregulation of the platelet-derived growth factor B-chain gene via fusion with collagen gene COL1A1 in dermatofibrosarcoma protuberans and giant-cell fibroblastoma. Nat Genet. 1997;15:95–8.
46. Rutkowski P, Van Glabbeke M, Rankin CJ, Ruka W, Rubin BP, Debiec-Rychter M, et al. Imatinib mesylate in advanced dermatofibrosarcoma protuberans: pooled analysis of two phase II clinical trials. J Clin Oncol. 2010;28(10):1772–9.
47. Ostman A, Heldin CH. PDGF receptors as targets in tumor treatment. Adv Cancer Res. 2007;97:247–74.
48. McArthur GA. Molecular targeting of dermatofibrosarcoma protuberans: a new approach to a surgical disease. J Natl Compr Canc Netw. 2007;5:557–62.
49. Mertens F, Fletcher CD, Antonescu CR, et al. Clinicopathologic and molecular genetic characterization of low-grade fibromyxoid sarcoma, and cloning of a novel FUS/CREB3L1 fusion gene. Lab Invest. 2005;85:408–15. 149.
50. Coffin CM, Hornick JL, Fletcher CD. Inflammatory myofibroblastic tumor: comparison of clinicopathologic, histologic, and immunohistochemical features including ALK expression in atypical and aggressive cases. Am J Surg Pathol. 2007;31:509–20.
51. Butrynski JE, D'Adamo DR, Hornick JL, Dal Cin P, Antonescu CR, Jhanwar SC, et al. Crizotinib in ALK-rearranged inflammatory myofibroblastic tumor. N Engl J Med. 2010;363(18):1727–33.

Chapter 2
Heterogeneity of Breast Cancer: Gene Signatures and Beyond

Gaia Schiavon, Marcel Smid, Gaorav P. Gupta, Stefania Redana, Daniele Santini, and John W.M. Martens

Introduction

The initial steps into a better understanding of the heterogeneity and biology of breast cancer were made at the onset of 2000, with the first identification of distinct molecular subtypes of human breast tumors possessing different outcome [1–3]. Gene expression profiling and microarray analysis opened a road leading to the new molecular classification of breast cancer, recognizing at least five reproducible subtypes: luminal A, luminal B, ERBB2, basal, and normal-like [3–6]. This revolutionary concept was triggered from an intense research driven by the evidence that 60–70% of all breast cancers are classified as "not otherwise specified" infiltrating ductal carcinomas (IDC NOS) [4].

The methodology of microarray has been soon supported by other tools like array comparative genomic hybridization (array-CGH), single-nucleotide polymorphism (SNP), high-throughput screening (HTS) techniques and the increasing availability of multiple tools for pathways analysis. The combination of these advanced technologies is constantly applied to in vitro and in vivo research in order to improve our knowledge of breast cancer biology and our understanding of the complex process of metastasis. The ultimate goal is to create strategies and algorithms guiding a tailored management of patients with both early and advanced breast cancer.

The focus of this chapter is to provide a general summary of the genomic signatures available for breast cancer and to function as a tool for clinicians for the interpretation of gene signatures. Moreover, we also give a brief overview of the emerging tools designed to capture and study the heterogeneity of breast cancer (next-generation sequencing).

G. Schiavon (✉) • M. Smid • J.W.M. Martens
Department of Medical Oncology, Daniel den Hoed Cancer Center, Erasmus Medical Center, Rotterdam, The Netherlands
e-mail: gaiaschiavon@gmail.com

G.P. Gupta
Department of Radiation Oncology, Memorial Sloan-Kettering Cancer Center, New York, NY, USA

S. Redana
Department of Medical Oncology I, Institute for Cancer Research and Treatment, Fondazione Piemontese per L'Oncologia, Candiolo, Italy

D. Santini
Department of Medical Oncology, University Campus Bio-Medico, Rome, Italy

A. Russo et al. (eds.), *Diagnostic, Prognostic and Therapeutic Value of Gene Signatures*, Current Clinical Pathology, DOI 10.1007/978-1-61779-358-5_2,

Genomic Signatures and Microarray Analysis

From Binary to Bedside

In the last decade or so, there has been a slow switch from analyzing a gene-at-a-time, to higher throughput procedures. To establish which genes were active in a biological context, traditional methods like Northern and Southern Blotting were gradually replaced by the so-called microarrays. These started out as large membrane sheets spotted with cDNA, but these have evolved substantially to much smaller chips which can contain millions of oligonucleotides. mRNA of a biological sample can be hybridized to these chips, which yields the expression levels of thousands of genes in one single experiment.

This in turn necessitates a specialized field to measure, collect, transform, analyze, and evaluate these data using statistically sound methods. Enter bioinformatics. Although part of the bioinformatics field is concerned with the (pre)processing steps to extract reliable data from the microarrays, a big part focuses on applicable analyses. Demanding as it can be, dealing with millions of data-points, it does provide a very rich platform to build solid conclusions.

If one considers that the expression of a few markers like estrogen receptor (ER) or HER2/neu (together with tumor size, grading, nodal involvement, and other few other prognostic marker) will determine the best treatment regimen, it is easy to envision that by measuring thousands of genes in hundreds of samples, valuable markers or combinations of markers can be identified, linked with many, if not all, clinical aspects of the patients involved. However, teasing out statistically sound differentially expressed genes or reliable signatures hiding in the ranks of thousands of genes does entail a considerable challenge, one that has admittedly met with its pitfalls, but more so with considerable successes.

To start on a cautionary note, it is still a very good idea to uphold the scientific principle to validate one's results thoroughly. The most notorious example to date is the paper describing expression signatures guiding the choice of chemotherapy in cancer patients which was heralded in 2006 [7], but ultimately proved unreliable and was retracted in 2011. Indeed, bioinformatics is a powerful tool, which obliges researchers to wield it in a correct manner.

One of the most abundantly used bioinformatics analysis is "hierarchical clustering," in which order is created out of the chaos of gene expression patterns of tumor cells. By grouping tumors according to the similarity of their gene expression levels, Perou et al. were able to reflect the notorious heterogeneity which is found in the clinical outcome of breast cancer patients, in five distinct molecular subtypes [1]. This landmark paper was among the first which put the power of bioinformatics analysis on the map. The five molecular subtypes have since then been studied extensively, and many clinical relevant observations are ascribed to the subtypes, among which prognosis [1, 3] response to therapy [8], and site-specific relapse [9]. Thus, the overall gene expression patterns can be quite distinct in patients suffering from breast cancer. By linking the expression of specific genes to the clinical parameters of patients, any range of clinically relevant questions can be addressed. Milestone examples are gene expression signatures able to predict prognosis [10, 11], of which Food and Drug Administration (FDA) approval has been granted for the first multigene model, signatures for response to therapy [8, 12], and signatures for breast cancers relapsing to bone [13–15], lung [16, 17], and brain [18].

One dimension higher is the analysis of interacting genes. Similar gene functions or signaling cascades – pathways – can be identified from the expression data. Although more suited to increase understanding in the biological processes about breast cancer, clinical associations with specific pathways have been described [19]. Finally, similar bioinformatics analyses can be applied to miRNA, DNA copy-number, SNPs, sequence, and methylation data. Gathering all these data from the same tumor and integrating all this knowledge is the upcoming challenge. When available, a detailed tumor blueprint for each individual breast cancer patient could be constructed, which will guide the physician to treat future patients with the most tailored strategy.

Multiple Gene Signatures: Do Not Get Lost

As already mentioned, the identification of breast cancer molecular subtypes was the first insight into the biologic heterogeneity of breast cancer [20]. Gene-expression profiling resulted to be – and still is – a very appealing approach, but the clinical utility of classifying breast cancer into molecular subtypes using unsupervised cluster analysis has limitations. For example, with the addition of a new case to the data, the dendrograms of hierarchical cluster analysis are reorganized, and therefore it is not possible to prospectively classify new cases using this methodology [21]. If larger sample sets are used, more clusters and molecular subtypes of breast cancer could become evident [22]. In this perspective, several large studies gave the major clinical contribution of gene-expression profiling in predicting prognosis and response to therapies. Different approaches have been used and different questions have been addressed. For example, some signatures were derived from the gene profiling of human breast cancer cell lines with particular propensity to metastasize to one or another organ and/or mouse models of breast cancer and then applied to breast cancer patients. In other cases, the signature was directly derived from the profiling of fresh-frozen tissues from breast cancer patients (test set) and then validated in one or more independent datasets. Samples are selected on the basis of a specific aim: identification of molecular subtype, prediction of prognosis, resistance to chemo/hormonal-therapy, risk of relapse, etc. Table 2.1 provides a large overview of the most relevant signatures produced by breast cancer research. Of note, a source of concern has been the little or absent overlap between the different gene sets, when compared to each other. For example, the 70-gene signature and the 21-gene RS have only the SCUBE2 gene in common [10, 23, 32]. The reasons for this lower-than-expected overlap are not completely known, but they probably include differences in the patient cohorts (e.g., 70-genes and 76-genes prognostic signatures), microarray platforms, the large number of genes associated with prognosis, and bioinformatics–mathematical methods used for analysis [10, 11, 23]. To answer the question whether these predictors are concordant with respect to their predictions for individual patients, Fan et al. analyzed a single dataset on which five prognostic or predictive gene-expression-based models were simultaneously compared (the 70-gene signature model, the wound-response model, the 21-gene RS model, the intrinsic-subtype model, and the two-gene-ratio model) [1–3, 5, 10, 11, 23–26, 32, 33]. With this analysis, all gene-expression-based models with the exception of the two-gene ratio model, significantly predicted relapse-free survival and overall survival. A limitation of this study was that the 21-gene RS and two-gene ratio models were developed to be used in different clinical scenarios than was represented in this patient cohort [20]. In fact, the two-gene ratio and the 21-gene RS models were designed to predict outcomes in patients with ER-positive disease receiving tamoxifen as an adjuvant treatment. However, the Fan dataset included only 40 such patients and a substantial portion of it was used as the training set for the development of the intrinsic-type, 70-gene signature, and wound-response models.

Despite the lack of gene overlap, four of the five models showed significant agreement in predicting outcome for individual patients leading to the hypothesis that different gene sets may predict a biologically similar breast cancer phenotype. The concordance among the models in the identification of patients with a genomic high-risk for recurrence was excellent, but it remains unknown how much clinical utility these predictive models provide over standard clinicopathologic features. While in multivariate analysis, three of the gene-based models proved to be more predictive for outcome than standard clinicopathologic criteria, features routinely used by physicians, especially in clinically intermediate-risk cancers (e.g., progesterone receptor (PR) status, HER2 status, lymphovascular invasion, and mitotic rate) were not included in the analysis. On-going clinical trials will help to clarify the potential benefits of these genomic tools over standard clinicopathologic assessment. In fact, what we have learned from this field is the importance of clinical validation of gene signatures in independent datasets with long follow-up available and a subsequent prospective validation (see section "Clinical Application of Gene Signatures: Ongoing Trials").

Table 2.1 Gene-expression profiling models in breast cancer

Models	Aim of the signature	Tissue	Validation sets in the same paper	Assay	Number of genes in panel	Used in prospective studies
Intrinsic subtype [1–5]	Subtypes/prognosis	Fresh frozen	–	Microarray	534	None
Netherlands signature (MammaPrint™) [10, 22]	Prognosis	Fresh frozen	–	Microarray	70	MINDACT (on-going)
Recurrence score (OncotypeDX™) [23]	Relapse	Paraffin-embedded	–	RT-PCR	16 (+ 5 reference genes)	TAILORx (on-going)
Rotterdam Signature [11]	Prognosis	Fresh frozen	–	Microarray	76	None
Wound response [24, 25]	Progression/prognosis	Fresh frozen	–	Microarray	512	None
Two-gene ratio [26]	Prognosis	Paraffin-embedded	–	RT-PCR	2	None
Gene-expression grade index [27, 28]	Grading/prognosis	Fresh frozen	–	Microarray	97	None
p53 Signature [29]	p53 Status/prognosis	Fresh frozen	GSE:4382,1379 [10]	Microarray	32	None
p53 Signature [30]	Subtypes/prognosis/ chemo-response	Developed in [29]	GSE:3494,2034,7390,9195,16716	Microarray	39 (ER+) 30 (ER-)	None
Smid et al. [9]	Subtype/site of relapse	Developed in EMC344 (GSE:2034,5327)	–	Microarray	529 (bone), 67 (lung), 149 (brain), 18 (liver), 39 (pleura)	None
Jansen et al. [12]	Tamoxifen resistance	Fresh frozen (test+validation sets)	–	Microarray	44	None
Kang (bone) [13]	Bone relapse	human breast cancer cell line (highly metastatic to bone)	–	Microarray	102	None
Src responsive signature (SRS) [14]	Bone relapse	Developed in [31]	Fresh frozen EMC344, EMC189, MSK82 (GEO2603, GSE:5327,2034,12276)	Microarray	159	None
Predictor of Tumor Relapse to Bone [15]	Bone relapse	Fresh frozen	–	Microarray	31	None
Lung metastasis signature (LMS) [16]	Lung relapse	human breast cancer cell line (metastatic to lung) in a xenograft model	MSK-82 (fresh frozen)	Microarray	95 (54 for functional validation)	None
Lung metastasis signature (LMS) [17]	Lung relapse	Developed in [16]	MSK99, NKI295, EMC344	Microarray	18	None
Brain metastasis signature [18]	Brain relapse	brain-metastatic-derived (BrM) cells + MSK82/ EMC286 (training set)	EMC192, NKI295 (independent sets)	Microarray	17	None

An interesting point of view is that gene expression signatures can reflect the activation status of several oncogenic pathways. Yu et al. suggested that it might be more appropriate to interrogate the gene lists for biological themes, rather than individual genes [19]. Moreover, identification of the distinct biological processes between subtypes of cancer patients is more relevant to understand the mechanism of the tumorigenesis and metastatic capability and for targeted drug development. They resampled their dataset numerous times to get multiple gene lists whose expression correlated with patients' outcome. For example, based on these gene lists, they identified overrepresented pathways defined in gene ontology biological process (GOBP) for ER-positive or ER-negative breast cancer patients, separately. Then, they compared the pathways represented by different published prognostic gene signatures with the overrepresented pathways associated with metastatic capability. This study also demonstrated that it is feasible to construct a gene signature from the key pathways to predict clinical outcomes. Clustering tumors based on pathway signatures defines prognosis in different patient subsets, demonstrating that patterns of oncogenic pathway deregulation underlie the development of the oncogenic phenotype and reflect the biology and outcome of specific cancers. According to Bild et al., prediction of pathway deregulation in cancer cell lines is also able to predict the sensitivity to therapeutic agents that target components of the pathway. Linking pathway deregulation with sensitivity to therapeutics that target members of the pathway likely allows using these oncogenic pathway signatures to guide the choice of targeted therapeutics [31].

Triple Negative Breast Cancer: Any Signature Available?

Among the 4–5 molecular subgroups, basal-like tumors present the worst outcome. According to current estimates, triple negative breast cancers (TNBCs) account for 10–17% of all breast carcinomas, depending on thresholds used to define ER and PR positivity and HER2 overexpression [34]. In different series and patient populations TNBC may range 6–28% of breast cancers, but higher incidence rates are reported for some ethnical groups, such as African–Americans and for younger patients, as well as for BRCA-mutation carriers [35]. Despite its relatively small proportion among all breast cancers, TNBC is responsible for a large fraction of breast cancer deaths, because of the aggressive tumor phenotype(s), only partial response to chemotherapy and present lack of clinically established targeted therapies. It should be emphasized that TNBC currently includes a heterogeneous group of tumors. By simple morphology, a group of TNBC patients with a more favorable outcome can be identified, for example patients with invasive adenoid cystic, apocrine, and typical medullary tumors. Even within the relatively homogeneous group of patients with triple-negative invasive ductal carcinoma (IDC), patients with higher or lower risk may be identified, based on specific molecular markers. For example, Viale et al. found that epidermal growth factor receptor (EGFR) immunoreactivity significantly correlates with worse prognosis in 284 patients with triple-negative IDC [35]. This underscores the importance of defining underlying risk factors for TNBC as a crucial step toward its prevention. While several potential therapeutic targets have recently surfaced from the gene expression profiling of the triple-negative tumors, the search is still onto unravel the modifiable and nonmodifiable risk factors associated with this aggressive disease. Also, additional tumor markers might allow identification of patients at higher risk of relapse [36].

A TNBC metastasis-associated signature, currently nonavailable, would be extremely useful in clinical setting, to select patients that probably have benefit from treatment and patients that can avoid toxicity of not necessary treatments. Recent discoveries in this field have been presented from three groups at San Antonio Breast Cancer Symposium 2010.

Lehmann and colleagues analyzed 386 TNBC gene-expression profile training sets from 21 independent breast cancer studies and identified six stable clusters that display unique gene-expression patterns and gene ontologies [37]. These clusters are: two basal-like subtypes characterized by cell cycle and DNA damage response genes; two mesenchymal-like subtypes enriched in cell differentiation, epithelial–mesenchymal transition and growth factor pathways; an immunomodulatory subgroup defined by immune cell surface antigens, receptors, and signal transduction genes; and a luminal subgroup driven by androgen-receptor signaling. Lehmann et al. after the identification of representative cell lines to model each of the subgroups, treated xenografts of these TNBC subtypes and found that basal-like triple-negative disease is sensitive to cisplatin, mesenchymal-like TNBC may preferentially respond to Src and PI3K/mTOR inhibitors, and the luminal subtype is sensitive to the androgen-receptor antagonist bicalutamide and to HSP90 inhibitors. These data for the target selection in drug discovery, clinical trial design, and selection of biomarkers represent a potential approach to assign a personalized treatment to patients with TNBC.

Interestingly, another group of investigators evaluated 28 breast cancer datasets with gene-expression data and identified 12 different molecular phenotypes among 579 TNBC samples [38]. The analysis showed that 73% of TNBC are basal-like tumors, with the rest classified into phenotypes according to gene function (e.g., immune activity, angiogenesis, proliferation, apocrine activity, inflammation). Notably, there were no outcome differences between basal-like tumors and nonbasal-like tumors. However, high B-cell (immune system) and low IL-8 (inflammation) metagene expression were able to identify a subset of patients (32% of all tumors) with a favorable prognosis and a 5-year metastasis-free survival of 84%. The inhibition of the related pathways might provide new therapeutic approaches. Only the metagene ratio and lymph node status significantly predicted of prognosis in the multivariate analysis.

Goga A, on the basis of gene-expression arrays of 149 patients from the I-SPY trial, demonstrated that TNBC with high expression of MYC have worse clinical outcomes [39]. The oncogene MYC, whose genomic locus is amplified in 20–50% of all breast tumors, is associated with poorly differentiated tumors and abundant in TNBC. In the trial, patients with tumors expressing high MYC signatures had worse outcome. Disease-free survival at 5 years was approximately 95, 80, and 55% for patients with low, intermediate, and high MYC expression, respectively. In a multivariate analysis considering receptor status and MYC pathway activation as a continuous variable, triple-negative status, and MYC pathway activation had a hazard ratio of 1.5 and 16.7, respectively. The CDK1 inhibitor SCH-727965, currently in phase II trials, is a promising agent targeting MYC pathway via upregulation of the proapoptotic Bcl2 family member BIM. The cooperation between MYC overexpression and CDK1 inhibition induces cell death in triple-negative breast cancer cells and regression of tumor xenografts.

In conclusion, on the basis of transcriptomic analyses, TNBCs seem to represent a molecularly and clinically heterogeneous disease and not all of them have an unfavorable prognosis [37].

The results of these mentioned and other studies are enthusiastically awaited to discover driving pathways involved in TNBC progression and to better individualize therapy for these patients.

Clinical Application of Gene Signatures: Ongoing Trials

Adjuvant treatment for early-stage breast cancer is offered to the majority of patients after definitive surgery, assuming the presence of residual microscopic disease. Clinical–pathological prognostic factors (tumor stage, hormone receptors and HER2 expression, tumor grade, proliferative rate) are useful tools to estimate the risk of disease recurrence and thus, to decide whether to use adjuvant chemotherapy (CT), hormone therapy (HT), or biologic agents. However, during the last decade, a deeper insight into breast cancer biology led to the classification of breast cancer into different subtypes, characterized by different biological features and prognosis [1]. Several gene signatures proved to

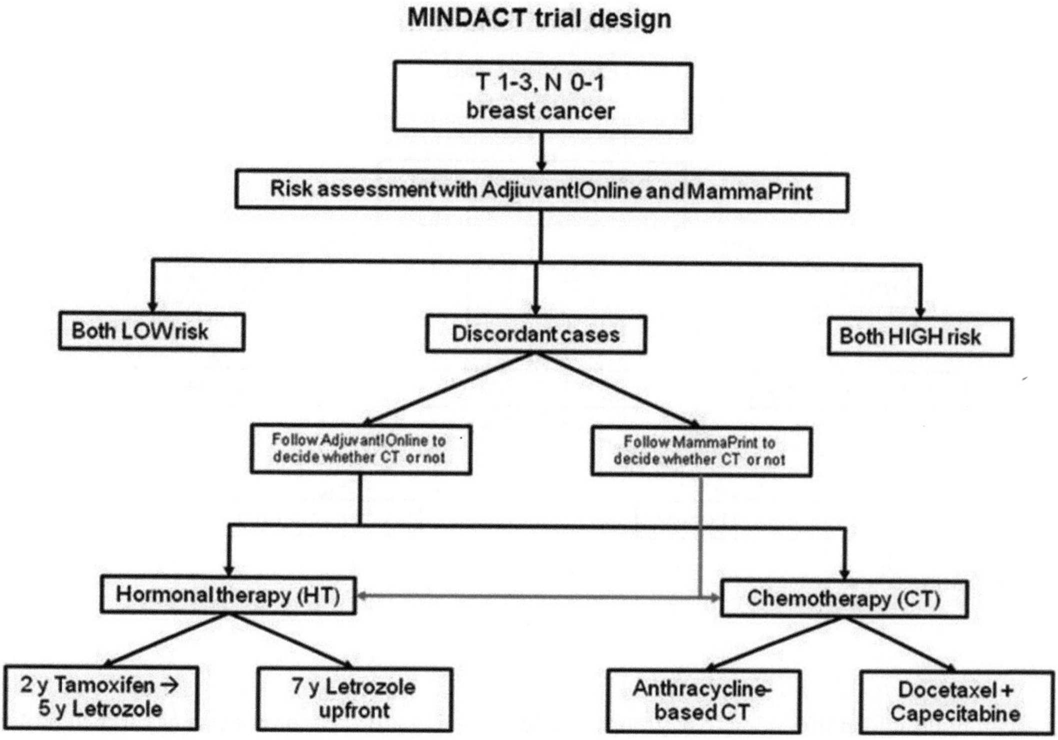

Fig. 2.1 MINDACT trial design

better estimate the risk of disease recurrence (Table 2.1), when compared to classic prognostic factors, and some of them are recommended from the American Society of Clinical Oncology (ASCO) and the NCCN Guidelines both as a prognostic and predictive tool in patients with node negative, endocrine positive disease. One of the priorities in breast cancer management is to identify patients with good-prognosis early-stage disease who could be spared adjuvant CT. Hence, the application of gene signatures in predicting the benefit of CT and identifying patients who will mostly benefit from a specific cytotoxic agent is an extremely active research field. Promising data suggesting a role in predicting benefit from adjuvant CT over endocrine therapy alone prompted the development of two large randomized phase III trials [MINDACT (Fig. 2.1) and TAILORx (Fig. 2.2)], actually ongoing.

MammaPrint™

The 70-gene expression profile was developed in the Netherlands, applying DNA-microarray technology to 78 frozen tumor samples from untreated node-negative breast cancer patients. Patients relapsing within 5 years from definitive surgery were defined as "poor prognosis," those who remained disease-free after 5 years were considered as "good prognosis." Researchers selected 70 genes that demonstrated to accurately classify tumors in either the poor or good-prognosis group [10]. The prognostic value of the 70-genes signature was validated in retrospective case series, showing that the use of MammaPrint could reduce misclassification of patients' risk, hence the overtreatment of the low-risk group [40].

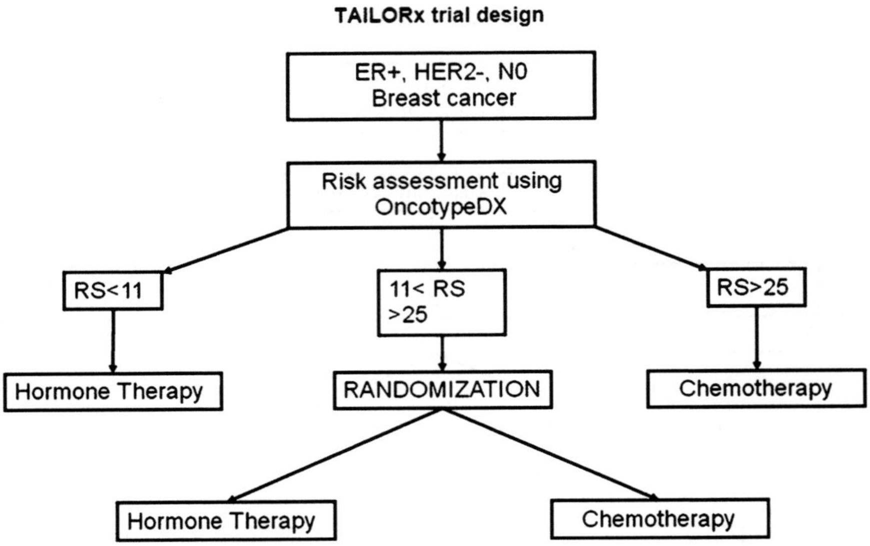

Fig. 2.2 TAILORx trial design

MammaPrint identifies a low-risk group of patients having a 10-year breast cancer survival probability ≥88% if their tumor had ER expression >1%, and of at least 92% if ER negative [41]. Recently, the role of MammaPrint to predict CT benefit in addition to HT was assessed: patient in the high-risk group derived a greater benefit from CT in terms of distant disease-free survival (DDFS) and breast cancer specific survival (BCSS), compared to high-risk patients treated with HT alone [42]. Moreover, the application of the 70-gene profile in the neoadjuvant setting further provided evidences of the predictive value of the signature [43].

A prospective validation of MammaPrint is ongoing in a large, multicentric, randomized, controlled, phase III trial: the microarray in node negative disease may avoid chemotherapy (MINDACT) trial. Primary objective of the trial is to confirm that patients with molecular low risk can safely be spared adjuvant CT even if they have clinical high-risk tumor.

The risk of relapse of 6,000 breast cancer patients (0–3 lymph nodes involved) will be assessed using both traditional clinical–pathological criteria (Adjuvant!Online) and the MammaPrint signature. Patients estimated low risk with both methods will be spared CT; patients estimated high risk with both methods will be proposed CT; if the methods are discordant, patients are randomized to be treated according to the clinical–pathological or the genomic result. Estimating a 35% rate of discordance between Adjuvant!Online and MammaPrint, it is expected that a third of them will not be treated with CT, while it would be recommended using the conventional risk assessment criteria. Patients defined as high clinical and low genomic risk will not receive CT, and will be closely followed. The role of MammaPrint in predicting benefit from a specific chemotherapeutic agent will be assessed in a second randomization: patients will be randomly treated with an anthracycline-based CT or the combination of docetaxel and capecitabine. A third randomization, offered to all ER positive patients, will compare 2 years of T followed by 5 years of letrozole to 7 years of upfront letrozole (see Fig. 2.1 for trial design). The whole genome will be analyzed for all the 6,000 patients, aiming at discovering new signatures with prognostic and predictive value.

Accrual started in February 2007 and is still ongoing. Results of this ambitious trial are eagerly awaited.

OncotypeDx

OncotypeDx is the most widely used gene signature in everyday practice. It is an RT-PCR assay performed on formalin-fixed, paraffin embedded tissues that evaluate expression of 21 genes (16 cancer related and 5 reference genes). Levels of gene expression are combined to provide a continuous variable: the recurrence score (RS). The assay was developed in a population of patients with node-negative ER positive disease, treated with tamoxifen (T). Thus, the RS quantifies the likelihood of 10-year distant recurrence (10 yDR) in patients with the aforementioned characteristics [32]. Patents with an RS <18 are considered low risk with a 10 yDR of 6.8%; an RS≥18 and <31 defines the intermediate risk group, with 10 yDR of 14.3%; finally, an RS≥31 is associated with a 10 yDR rate of 30.5%. The prognostic value of the assay was validated in several retrospective trials [44]. OncotypeDx to predict the benefit of CT over T in ER positive breast cancer patients has been assessed using samples from the NSABP B20 and the SWOG S8814 trials. In both cases, patients with a high RS derived the greater benefit from the addiction of CT to T, while it did not affect outcome of low-risk patients, suggesting that the later ones could safely be spared C [45, 46]. The later study was run in node-positive patients and proved an increase of both disease-free survival (DFS) and BCSS when CT was added to T only in the high-risk group (10-years BCSS in the high-risk group 73% vs. 54% for CAF+T vs. T alone, respectively) [46]. The predictive value of OncotypeDx has been confirmed in the neoadjuvant setting, proving that low RS tumors are less likely to obtain pathologic complete response than high RS [47].

Trial assigning individualized options for treatment (Rx) (TAILORx) trial is a prospective, randomized, controlled, phase III trial that aims at identifying the best individual treatment for patients with ER positive, HER2 negative, node negative breast cancer (see Fig. 2.2 for trial design). Primary objective of the trial is to compare DFS and distant recurrence-free interval (DRFI) of patients in the intermediate risk group treated with CH + HT vs. HT alone, other than create a tissue bank repository. Other objectives are to determine DFS, DRFI, and overall survival of patients with low RS and determine whether HT alone can be considered the optimal treatment for these patients.

The definitions of low, intermediate, and high risk utilized in this trial are slightly different than those previously defined:

- Low-risk patients (RS <11) will receive HT alone. A RS <11 is associated with a 10 yDR rate <5%, on average, if treated with T alone, and no benefit from CT has been demonstrated in this group.
- Intermediate-risk patients (RS 11–25) will be randomly assigned to receive CT and HT vs. HT alone. The risk of 10 yDR is approximately 10% if treated with T alone, and is considered sufficiently high to recommend CT even if CT benefit has not been clearly established for this group. The treating physician can chose the CT regimen among all those recommended from ASCO and NCCN guidelines.
- High-risk patients (RS > 26) will be assigned to receive CT. An RS >26 was selected because it is associated with a 10 yDR >20%, on average, and because CT has been shown to be beneficial in this group.

Accrual of the 11,248 was started in early 2007 and was very rapid. Actually, the study is ongoing but not accruing (clinicalTrials.gov – last assessed 4/1/2011).

Next-Generation Sequencing: An Emerging Tool to Characterize the Molecular Heterogeneity of Breast Cancer

Mutation Signatures: Mountains and Hills

Advances in whole genome sequencing technology have led us into an era of cancer genome discovery [48]. Systematic Sanger sequencing of over 13,000 genes in 11 human breast cancers has revealed preliminary insights into the breast cancer genome [49]. According to this analysis, the average tumor harbors approximately 90 point mutations in gene coding regions. Only small minorities of these mutations were recurrent, and were in genes already known to be important in breast cancer, including *p53* and *PIK3CA*. These relatively common mutations figuratively reflect "mountains" in the mutation landscape of breast cancer [50]. In contrast, much more numerous in this landscape are the "hills," which represent infrequent mutations in a diversity of genes that are much more numerous in any given tumor, and may or may not be contributing to the oncogenesis. Indeed, clarifying the drivers from the passengers in this complex landscape will likely be necessary to understand the heterogeneity of breast cancer, and remains one of the biggest challenges of cancer genomics.

Contrasting Paths of Genomic Evolution

Luminal and basal-like breast cancers exhibit distinct gene expression signatures, biology, and clinical behavior. Next-generation resequencing now suggests that genomic evolution during tumor progression might also follow divergent patterns between these biologic subtypes.

Comprehensive paired-end deep sequencing and transcriptome sequencing was performed of a metastatic lobular ER-positive breast cancer specimen [51]. Thirty-two nonsynonymous coding sequence mutations were identified. Significantly, 19 of these mutations were not present in the patient's primary tumor, which had been surgically excised 9 years prior to emergence of the metastasis. Notably, no significant genome rearrangements were identified in the tumor samples, when compared to normal tissue DNA. These findings suggested that a relatively small number of point mutations might be responsible for tumorigenesis in this breast cancer, and that additional mutations arose after intervening therapy and several years of dormancy. The contribution of these mutations to metastatic recurrence remains unclear, although it seems plausible that some of these might have biologic importance.

Contrasting insights have been gleaned from deep resequencing of a basal-like primary breast cancer and its associated brain metastasis that arose in an African–American woman who succumbed to a particularly rapid and aggressive clinical course of disease [52]. The primary breast cancer in this case was ER-negative, lacked ERBB2 amplification, and was classified as an inflammatory breast cancer. Clinically, this tumor exhibited resistance to chemotherapy, and gave rise to brain metastases within 8 months, resulting in fatality shortly thereafter. In this pair of primary and metastatic specimens, only two de novo mutations distinguished the metastasis from the primary tumor. Nonetheless, the relative abundance of shared mutations varied between the metastasis and primary tumor, suggesting that selective processes during metastasis emergence were operative on preexisting dominant clones within the primary tumor. Somatic genomic rearrangements and copy number alterations were also abundant in this primary tumor, and most of these were also shared in the metastasis. However, there was evidence of increased copy number alterations and structural variants during the course of tumor progression, which might contribute to metastasis emergence, or alternatively might be a byproduct of chemotherapy or ionizing radiation exposure.

Indeed, a comparison of genomic profiles from primary tumor to metastasis in luminal and basal-like breast cancer reveals more differences than similarities. Further analysis of primary tumor–metastasis–normal tissue triads will determine the generality of these observations. An improved understanding of the differing rates of genome mutation and evolution among breast cancers would likely have implications for prognosis, and may possibly also reveal novel approaches to therapeutic intervention.

Genome Rearrangement Profiles: Diversity and Complexity

Paired-end sequencing and structural genomic rearrangement analysis has been performed on a cohort of 24 primary breast tumors and immortalized breast cancer cell lines [53]. This high-resolution analysis revealed, for the first time, the landscape of chromosomal rearrangements in breast cancer. A striking discovery was the abundance of intrachromosomal rearrangements, which could not be readily characterized by previous techniques. Also apparent were distinct patterns of rearrangements in the various biologic subtypes of breast cancer. Hormone receptor-positive breast cancers exhibited either minimal genomic instability or rearrangements that almost exclusively involved amplified segments of the genome. In contrast, TNBCs exhibited rampant instability, frequently predominated by intrachromosomal tandem duplications. Sequence analysis of these breakpoints generally revealed short segments of microhomology at the junction of these rearranged segments, implicating DNA repair processes in their etiology. Of note, *BRCA1* and *BRCA2* null tumors did not exhibit this preponderance of tandem duplications, suggesting that a distinct pathway may be deranged in this subgroup of tumors.

Do these distinct patterns of genomic instability have implications for patient prognosis? An analysis by the Borreson-Dale group using multiple cohorts of array-CGH data suggests that it does. In this study, the authors characterized two quantitative indexes of genomic instability – a whole arm aberration index (WAAI) and a complex arm aberration index (CAAI) – that can be applied across array CGH platforms [54]. These two instability indices were used to segregate 595 breast tumors into distinct subgroups. Significantly, both indices correlated with poorer patient prognosis in univariate and multivariate analyses. These findings suggest that patterns of genomic instability may reflect an underlying biology of breast cancer that has implications for the likelihood of cancer recurrence, as well as the clinical responsiveness to DNA-damaging therapies.

Summary

Interestingly, particular gene classifiers able to predict the aggressiveness of a tumor and the tumor's ability to home and proliferate in an organ-specific manner have been identified. Gene-expression profiling has already provided important contribution into the biologic heterogeneity of breast cancer. The combination of several advanced genomic tools with new approaches described in this chapter has to be optimally incorporated. Whether this is feasible and effective will depend on trials, such as TAILORx and MINDACT. Moreover, technical and economic issues need to be examined, and optimization of the eventual use of these tools in decision-making process is necessary.

References

1. Perou CM, Sørlie T, Eisen MB, et al. Molecular portraits of human breast tumours. Nature. 2000;406:747–52.
2. Perou CM, Jeffrey SS, van de Rijn M, et al. Distinctive gene expression patterns in human mammary epithelial cells and breast cancers. Proc Natl Acad Sci USA. 1999;96:9212–7.
3. Sorlie T, Perou CM, Tibshirani R, et al. Gene expression patterns of breast carcinomas distinguish tumor subclasses with clinical implications. Proc Natl Acad Sci USA. 2001;98:10869–74.
4. Stingl J, Caldas C. Molecular heterogeneity of breast carcinomas and the cancer stem cell hypothesis. Nat Rev Cancer. 2007;7(10):791–9.
5. Sorlie T, Tibshirani R, Parker J, et al. Repeated observation of breast tumor subtypes in independent gene expression data sets. Proc Natl Acad Sci USA. 2003;100:8418–23.

6. Sotiriou C, Neo SY, McShane LM, et al. Breast cancer classification and prognosis based on gene expression profiles from a population-based study. Proc Natl Acad Sci USA. 2003;100:10393–8.
7. Potti A, Dressman HK, Bild A, et al. Genomic signatures to guide the use of chemotherapeutics. Nat Med. 2006;12:1294–300.
8. Rouzier R, Perou CM, Symmans WF, et al. Breast cancer molecular subtypes respond differently to preoperative chemotherapy. Clin Cancer Res. 2005;11:5678–85.
9. Smid M, Wang Y, Zhang Y, et al. Subtypes of breast cancer show preferential site of relapse. Cancer Res. 2008;68(9):3108–14.
10. Van't Veer LJ, Dai H, van de Vijver MJ, et al. Gene expression profiling predicts clinical outcome of breast cancer. Nature. 2002;415:530–6.
11. Wang Y, Klijn JG, Zhang Y, et al. Gene-expression profiles to predict distant metastasis of lymph-node-negative primary breast cancer. Lancet. 2005;365:671–9.
12. Jansen MP, Foekens JA, van Staveren IL, et al. Molecular classification of tamoxifen-resistant breast carcinomas by gene expression profiling. J Clin Oncol. 2005;23(4):732–40.
13. Kang Y, Siegel PM, Shu W, et al. A multigenic program mediating breast cancer metastasis to bone. Cancer Cell. 2003;3(6):537–49.
14. Zhang XH, Wang Q, Gerald W, et al. Latent bone metastasis in breast cancer tied to Src-dependent survival signals. Cancer Cell. 2009;16(1):67–78.
15. Smid M, Wang Y, Klijn JG, et al. Genes associated with breast cancer metastatic to bone. J Clin Oncol. 2006;24(15):2261–7.
16. Minn AJ, Gupta GP, Siegel PM, et al. Genes that mediate breast cancer metastasis to lung. Nature. 2005;436(7050):518–24.
17. Minn AJ, Gupta GP, Padua D, et al. Lung metastasis genes couple breast tumor size and metastatic spread. Proc Natl Acad Sci U S A. 2007;104(16):6740–5.
18. Bos PD, Zhang XH, Nadal C, et al. Genes that mediate breast cancer metastasis to the brain. Nature. 2009;459(7249):1005–9.
19. Yu JX, Sieuwerts AM, Zhang Y, et al. Pathway analysis of gene signatures predicting metastasis of node-negative primary breast cancer. BMC Cancer. 2007;7:182.
20. Stadler ZK, Come SE. Review of gene-expression profiling and its clinical use in breast cancer. Crit Rev Oncol Hematol. 2009;69(1):1–11.
21. Chang JC, Hilsenbeck SG, Fuqua SA. Genomic approaches in the management and treatment of breast cancer. Br J Cancer. 2005;92:618–24.
22. McShane LM, Radmacher MD, Freidlin B, et al. Methods for assessing reproducibility of clustering patterns observed in analyses of microarray data. Bioinformatics. 2002;18:1462–9.
23. van de Vijver MJ, He YD, van't Veer LJ, et al. A gene-expression signature as a predictor of survival in breast cancer. N Engl J Med. 2002;347:1999–2009.
24. Chang HY, Sneddon JB, Alizadeh AA, et al. Gene expression signature of fibroblast serum response predicts human cancer progression: similarities between tumors and wounds. PLoS Biol. 2004;2:E7.
25. Chang HY, Nuyten DS, Sneddon JB, et al. Robustness, scalability, and integration of a wound-response gene expression signature in predicting breast cancer survival. Proc Natl Acad Sci USA. 2005;102:3738–43.
26. Ma XJ, Wang Z, Ryan PD, et al. A two-gene expression ratio predicts clinical outcome in breast cancer patients treated with tamoxifen. Cancer Cell. 2004;5:607–16.
27. Sotiriou C, Wirapati P, Loi S, et al. Gene expression profiling in breast cancer: understanding the molecular basis of histologic grade to improve prognosis. J Natl Cancer Inst. 2006;98:262–72.
28. Loi S, Haibe-Kains B, Desmedt C, et al. Definition of clinically distinct molecular subtypes in estrogen receptor-positive breast carcinomas through genomic grade. J Clin Oncol. 2007;25:1239–46.
29. Miller LD, Smeds J, George J, et al. An expression signature for p53 status in human breast cancer predicts mutation status, transcriptional effects, and patient survival. Proc Natl Acad Sci USA. 2005;102:13550–5.
30. Coutant C, Rouzier R, Qi Y, et al. Distinct p53 gene signatures are needed to predict prognosis and response to chemotherapy in ER-positive and ER-negative breast cancers. Clin Cancer Res. 2011. doi:10.1158/1078-0432.CCR-10-1045.
31. Bild AH, Yao G, Chang JT, et al. Oncogenic pathway signatures in human cancers as a guide to targeted therapies. Nature. 2006;439(7074):353–7.
32. Paik S, Shak S, Tang G, et al. A multigene assay to predict recurrence of tamoxifen-treated, node-negative breast cancer. N Engl J Med. 2004;351:2817–26.
33. Fan C, Oh DS, Wessels L, et al. Concordance among gene expression-based predictors for breast cancer. N Engl J Med. 2006;355:560–9.
34. Podo F, Buydens LM, Degani H, et al. Triple-negative breast cancer: present challenges and new perspectives. Mol Oncol. 2010;4(3):209–29.

35. Viale G, Rotmensz N, Maisonneuve P, et al. Invasive ductal carcinoma of the breast with the "triple-negative" phenotype: prognostic implications of EGFR immunoreactivity. Breast Cancer Res Treat. 2009;116(2):317–28.
36. Maiti B, Kundranda MN, Spiro TP, et al. The association of metabolic syndrome with triple-negative breast cancer. Breast Cancer Res Treat. 2010;121(2):479–83.
37. Lehmann BD, Bauer JA, Chen X, Sanders ME, Chakravarthy AB, Shyr Y, et al. Identification of human triple-negative breast cancer subtypes and preclinical models for selection of targeted therapies. J Clin Invest. 2011;121(7):45014. doi:10.1172/JCI45014.
38. Rody A, Karn T, Liedtke C, et al. Identification of a clinically relevant gene signature in triple negative and basal-like breast cancer. 33rd Annual San Antonio Breast Cancer Symposium. Abstract S5-5. Presented December 11, 2010.
39. Goga A, Horiuchi D, Kusdra L, et al. Synthetic-lethality of triple-negative breast cancers via the MYC oncogene pathway. 33rd Annual San Antonio Breast Cancer Symposium. Abstract S5-4. Presented December 11, 2010.
40. Mook S, Van't Veer LJ, Rutgers EJ, et al. Individualization of therapy using Mammaprint: from development to the MINDACT Trial. Cancer Genomics Proteomics. 2007;4(3):147–55.
41. Bogaerts J, Cardoso F, Buyse M, et al. Gene signature evaluation as a prognostic tool: challenges in the design of the MINDACT trial. Nat Clin Pract Oncol. 2006;3(10):540–51.
42. Knauer M, Mook S, Rutgers EJ, et al. The predictive value of the 70-gene signature for adjuvant chemotherapy in early breast cancer. Breast Cancer Res Treat. 2010;120(3):655–61.
43. Straver ME, Glas AM, Hannemann J, et al. The 70-gene signature as a response predictor for neoadjuvant chemotherapy in breast cancer. Breast Cancer Res Treat. 2010;119(3):551–8.
44. Habel LA, Shak S, Jacobs MK, et al. A population-based study of tumor gene expression and risk of breast cancer death among lymph node-negative patients. Breast Cancer Res. 2006;8(3):R25.
45. Paik S, Tang G, Shak S, et al. Gene expression and benefit of chemotherapy in women with node-negative, estrogen receptor-positive breast cancer. J Clin Oncol. 2006;24(23):3726–34.
46. Albain KS, Barlow WE, Shak S, et al. Prognostic and predictive value of the 21-gene recurrence score assay in postmenopausal women with node-positive, oestrogen-receptor-positive breast cancer on chemotherapy: a retrospective analysis of a randomised trial. Lancet Oncol. 2010;11(1):55–65.
47. Gianni L, Zambetti M, Clark K, et al. Gene expression profiles in paraffin-embedded core biopsy tissue predict response to chemotherapy in women with locally advanced breast cancer. J Clin Oncol. 2005;23(29):7265–77.
48. Stratton MR. Exploring the genomes of cancer cells: progress and promise. Science. 2011;331:1553–8.
49. Sjöblom T, Jones S, Wood LD, et al. The consensus coding sequences of human breast and colorectal cancers. Science. 2006;314:268–74.
50. Wood LD, Parsons DW, Jones S, et al. The genomic landscapes of human breast and colorectal cancers. Science. 2007;318:1108–13.
51. Shah SP, Morin RD, Khattra J, et al. Mutational evolution in a lobular breast tumor profiled at single nucleotide resolution. Nature. 2009;461:809–13.
52. Ding L, Ellis MJ, Li S, et al. Genome remodeling in a basal-like breast cancer metastasis and xenograft. Nature. 2010;464:999–1005.
53. Stephens PJ, McBride DJ, Lin ML, et al. Complex landscapes of somatic rearrangement in human breast cancer genomes. Nature. 2009;462:1005–10.
54. Russnes HG, Vollan HK, Lingjaerde OC, et al. Genomic architecture characterizes tumor progression paths and fate in breast cancer patients. Sci Transl Med. 2010;2:38–47.

Chapter 3
Gene Signatures in CRC and Liver Metastasis

Daniele Fanale, Lidia Corsini, Sergio Rizzo, and Antonio Russo

Introduction

Colorectal cancer (CRC) is one of the most common causes of cancer-related death with a worldwide incidence of almost a million cases annually in both males and females [1].

The accelerated decrease in CRC incidence rates from 1998 to 2006 largely reflects the advances in diagnosis and treatment that have enabled to detect and remove precancerous polyps [2]. However, the screening technology has not resulted in major improvements in the prognosis of patients with advanced cancer, and liver metastasis remains the major cause of death in CRC [3]. Approximately 25% of patients have detectable liver metastasis at diagnosis, that are classified as "synchronous" lesions and approximately 70% of patients develop a liver recurrence during the course of their disease – identified as "metachronous" lesions [4]. Despite the development of different treatment modalities, the outcome for patients with unresectable metastatic lesions is still unfavorable and the metastatic spread to the liver is the major contributor to mortality in CRC [5]. Therefore, elucidation of the molecular mechanism involved in the development of metastases, by the identification of a specific gene signature for liver metastasis in CRC, could allow prediction of the onset of metastatic disease in patients with localized tumors and lead to designing new strategies for diagnosis and treatment of CRC.

Molecular Nature of Liver Metastasis in CRC

Most cancer, including CRC, might have a single clonal origin at the initial stage of the disease; however, a malignant tumor contains multiple cell populations with different properties and acquires the ability to invade and develop metastases. This heterogeneity determines a variation in clinical patterns and treatment efficacies because these cells have acquired the ability to regulate their adhesion or motility. Clones with metastatic potential show genetic properties that are different than clones without metastatic potential. Therefore, in recent years many researchers aimed to identify genetic markers of metastatic clones.

D. Fanale (✉) • L. Corsini • S. Rizzo • A. Russo
Section of Medical Oncology, Department of Surgical and Oncological Sciences, University of Palermo, Palermo, Italy
e-mail: fandan@libero.it

A. Russo et al. (eds.), *Diagnostic, Prognostic and Therapeutic Value of Gene Signatures*, Current Clinical Pathology, DOI 10.1007/978-1-61779-358-5_3, © Springer Science+Business Media, LLC 2012

Although little is known, two models have been proposed to explain the biological mechanisms of metastases. The genetic selection model proposes that only a subset of tumor cells acquire metastatic potential during the late stages of multistep tumorigenesis [6, 7]. Alternately, another model proposes that tumor cells acquire metastatic capacity during a relatively early stage of tumorigenesis depending on the genetic background. Recent gene expression analyses of clinical tumor samples support this hypothesis [8, 9]. Several biomarkers for CRC have been identified, including KRAS2, p53, p21/WAF/CEP1, cyclin D1, PCNA, COX-2, MMP-9, CD44, CK-19, VEGF-C, and E-cadherin [10–12].

Liver metastasis in CRC occurs in multiple steps, including the ability of cancer cells to release the primary site, to achieve a specific type of tissue through blood flow, and to establish a distant secondary tumor [13]. This event has been characterized as a complex process in which each step is characterized by several changes in gene expression profiling [14, 15] and in which multiple genes play a significant role. Carcinogenesis and progression of CRC involves multiple genetic and epigenetic changes in many genes, including common alterations in *TP53*, *K-RAS*, b-*catenin*, *APC*, and *AXIN2* [16, 17].

Microarray studies, through the analysis of gene expression profiling, have successfully shed light on various aspects of the molecular mechanisms involved in the development of different human tumors [18]. In CRC, the DNA array technology has allowed the identification of differences in gene expression profiles between normal mucosa, benign adenoma, and malignant carcinoma, and the creation of a molecular model of multistep carcinogenesis [19]. However, although much is known about genes in which mutation is responsible for the onset of CRC [16, 20], less is known about the specific molecular events that are crucial for the development of CRC liver metastasis. Nevertheless, although the mechanisms responsible for metastasis of CRC to the liver are yet to be elucidated, different recent microarray studies have identified several genes that are crucial for metastasis formation, leading to the identification of a possible gene expression signature that could differentiate between tumors that do or do not metastasize [21].

Gene Expression Profile Studies Related to CRC Liver Metastasis

The acquisition of a metastatic signature is often related to the loss of various activities, and genes associated with metastasis may already be expressed in early tumors. In order to identify specific metastatic gene signatures that can be used as a prognostic marker of metastatic spread using microarray technologies, several studies have been carried. The first studies were conducted comparing the gene expression profile of primary CRC from metastasis-free patients to those of patients affected by metastatic disease during a 5-year follow-up period [5].

Bertucci et al. have identified a total of 219 genes down- and 25 genes up-regulated, respectively, in metastatic samples as compared to nonmetastatic samples and additionally have identified a gene signature of 46 genes that discriminate CRC with and without lymph node metastases [22].

D'Arrigo et al. have found a different gene expression signature between ten primary CRCs from patients who did not develop metastases within a 5-year follow-up period and ten primary CRCs from patients with synchronous liver metastasis. This study identified 37 genes differentially expressed between the two groups of primary tumors, of which 29 genes distinguished nonmetastasizing tumors from metastases. In support of the existence of specific gene expression profiling distinguishing primary tumors from CRCs with a metastatic potential, the genes encoding for mannosyl (a-1,3-)-glycoprotein b-1, 4-*N*-acetyl-glucosamynyl-transferase were significantly up-regulated in metastatic tumors [23].

To identify specific metastatic gene signatures, another approach entails comparing the gene expression profile in primary tumors and in corresponding metastases.

These studies show the identification of a similar gene expression profile between primary tumors and their matched metastases, supporting the notion that the metastatic potential is genetically encoded in primary tumors [24].

To investigate whether the expression profile in primary CRC reflected the metastatic status, Yamasaki et al. have analyzed the chronological gene expression profiles of 104 colorectal samples corresponding to oncogenic development, including normal mucosa, localized and metastatic tumors, and liver metastasis, using microarray technologies. The hierarchical clustering obtained has shown that the metastatic potential of CRC is not acquired in correlation to cancer progression, but rather the gene expression profile in the primary tumors reflect the ability to trigger the metastatic process [21].

In their work, Lin et al. evaluated the gene expression changes in primary CRC and CRC after metastasis to the liver, in order to determine how the cancer cells adapt to the liver microenvironment [25]. Using genome-wide microarray analysis, the authors have compared the gene expression profile of 48 primary tumors and 28 liver metastases and have identified 778 genes differentially expressed. The gene ontology analysis has revealed that the genes involved in immune response (innate and adaptive) and tissue remodeling are statistically up-regulated in liver metastasis relative to primary tumors. The genes involved in immune response, including proteoglycan 3 (*PRG3*), natural killer cell receptor (*NCR3*), B-cell membrane protein CD22, chemokine receptor (*CXCR4*) and its ligand *CXCL12*, and the genes associated with tissue remodeling mainly encode extracellular matrix proteins such as fibronectin, osteopontin, and ADAM17. Additionally, this study has shown that the genes associated with proliferation are down-regulated in liver metastasis suggesting that the tumor cells are not proliferating in the liver metastasis as rapidly as in the primary tumors [25].

In order to identify genes that may be involved in tumor progression and liver metastasis of CRC patients, Ki et al. have analyzed gene expression profiles of 27 colorectal primary tumors, corresponding to 27 liver metastasis tumors, 25 normal colon mucosa, and 13 liver tissues from CRC patients with liver metastasis [26]. In this study, the investigators identified 46 liver metastasis-specific genes with an accuracy of 83.3% by comparing the gene expression levels of paired primary colorectal tumors and liver metastases using a cDNA microarray containing 17,104 known genes. These selected genes could help to more effectively pinpoint therapeutic targets for CRC. Several known oncogenes were included among the 46 selected genes. Of the 46 identified genes, 36 seem to be down-regulated and 8 up-regulated in liver metastasis tumors. *WNT5A*, a gene involved in carcinogenesis and several developmental processes, showed significantly lower expression in liver metastasis tumors and higher expression levels in primary tumors than in the normal colon.

Conversely, the most significantly up-regulated gene in liver metastasis was tissue inhibitor of metalloproteinase 1 (*TIMP1*), an inhibitor of *MMP-1* that plays an important role in cancer metastasis. Other identified cancer-related genes that showed variation in gene expression levels were *MMP-1*, *MMP-2*, *COX-2* and *HIF-1a, MMP-1* and *MMP-2*, known to be up-regulated in carcinogenesis, were up-regulated in primary tumors but not in normal colon tissue. However, they showed decreased expression levels in liver metastasis tumors. Furthermore, 21 of the 46 genes were differentially expressed in primary tumors with synchronous liver metastasis compared with primary tumors without liver metastasis. These 21 genes were mainly involved in the regulation of cellular process (47.1%), cell development (35.3%), and cellular morphogenesis (17.7%).

Moreover, 4,583 selected organ-specific genes, differentially expressed between 13 paired normal colon and normal liver tissue, were identified; 2,236 of these genes were up-regulated and 2,347 were down-regulated in normal liver compared with the normal colon tissues. The genes showing differences in gene expression were those related to metabolism, cellular physiological process, cell communication, response to stimulus, and organism physiological processes [26].

In their study, Pantaleo et al. have evaluated the gene expression profiling of ten synchronous and eight metachronous liver metastatic lesions, by using the Affymetrix platform [27]. The gene expression analysis identified 49 up-regulated genes in metachronous and 55 up-regulated genes in synchronous metastases under which the main alterations were found in two pathways – EGFR signaling and eicosanoid metabolism. The key genes involved in this pathway are EGFR, COX-2, and COX-1; specifically, EGFR was overexpressed in metachronous lesions and the COX-2 gene was overexpressed in synchronous metastases. These results showed that the molecular background of liver metastases may be differentially related to differential expression of these genes and this observation may have clinical implications, allowing a more targeted treatment of patients with synchronous or metachronous lesions. Therefore, the identification of a specific metastatic gene signature could result in the choice of selective and differential chemotherapy treatment [27].

In a recent study, Koh et al. have analyzed 12 matched primary and metastatic colorectal carcinomas, and have identified 80 genes differentially expressed by means of a supervised hierarchical clustering [28]. The metastasis was distinguished from the corresponding primary tumor in half of the metastases analyzed using a specific set of genes identified. Among 80 genes, *MMP1*, *MMP-2*, *MMP3*, *MMP-13*, *COL1A2*, and *CXCL3* were up-regulated in primary colon carcinoma, and *SOX15*, *LIMS1*, *SERPINA3*, *CYP1B1*, *NAT5*, *SPP1*, and *SERPINA1* were up-regulated in metastatic colon carcinoma. SERPINA1 and SERPINA3 were associated with a poor prognosis in colon cancer and are overexpressed in metastatic breast carcinoma. Furthermore, the up-regulated genes in metastatic colon cancer include genes involved in embryonic development (*GA17*), cell adhesion (*ADRM1*), RNA binding (*SNRPB2*), transcriptional activity (*TWIST1* and *ETV4*), cell cycle and proliferation (*CKS2*), DNA repair (*RPA3*), signal transduction (*PRDX4*), and prefolin complex (*VBP1*). Down-regulated genes in metastatic colon cancer included genes involved in the cell–cell adhesion (*ICAM4*), extracellular region (*GUCA2A*) and carbonate dehydratase activity (*CA4*). Genes defined with high metastatic potential encode proteins influencing cell growth and proliferation (GAS1, ITLN1, IL1B, and IL24), extracellular proteases (MMP family), cell adhesion (COL12A1 and PAP), and cell motility (VIP and WNT5A). This molecular profiling may be useful for the differentiation of primary and metastatic carcinoma and the proteins encoded by these genes could be effective biomarkers for early metastasis detection.

Recently, Fritzmann et al. conducted a gene expression-profiling experiment to identify genetic markers of risk and to elucidate the molecular mechanisms of CRC metastasis. In particular, they have compared the gene expression patterns between metastatic and nonmetastatic matched CRC by microarray analysis. This study allowed the identification of 115 gene signatures that differentiated metastatic from nonmetastatic primary tumors. Among these, the transforming growth factor (TGF) b inhibitor BAMBI was noted to be highly expressed in half of the metastatic primary tumors and metastases but not in nonmetastatic tumors and it was observed an inverse correlation between the level of BAMBI expression and metastasis-free survival time of patients. BAMBI is a target of Wnt signaling and its inhibition of TGF-b pathway could directly drive the metastasis formation in CRC. Therefore, the metastatic gene signatures obtained in this work made it possible to identify a specific gene that could be used as a powerful prognostic indicator in CRC [29]. The most important genes involved in liver metastasis formation in CRC and reported in several studies are shown in Table 3.1.

Gene Signature: Potential and Limits

Studies of gene signatures have the potential to identify specific genes that may play a key role in the onset of liver metastases and, therefore, be used both to clarify the molecular mechanisms responsible for the development of metastases, and as prognostic indicators in CRC. However, the determination of metastatic expression signatures by microarray technologies must take into consideration different factors that can lead to differences among the results obtained by various groups.

Table 3.1 The most important genes involved in liver metastasis formation whose expression is altered in colorectal cancer

Gene	Gene name	Function
PRG3	Proteoglycan 3	Immune response
NCR3	Natural killer cell receptor	Immune response
CD22	T-cell surface antigen Leu-14	Immune response
CXCR4	Chemokine receptor 4	Response to stimulus
CXCL12	CXCR4 ligand	Response to stimulus
CXCL3	Chemokine (C–X–C motif) ligand 3	Response to stimulus
FN1	Fibronectin 1	Tissue remodeling
OPN	Osteopontin	Tissue remodeling
ADAM17	A disintegrin and metalloproteinase domain 17	Tissue remodeling
TIMP1	Tissue inhibitor of metalloproteinase 1	Tissue remodeling
MMP-1	Matrix metallopeptidase 1	Proteolysis
MMP-2	Matrix metallopeptidase 2	Proteolysis
MMP-3	Matrix metallopeptidase 3	Proteolysis
MMP-7	Matrix metallopeptidase 7	Proteolysis and invasion
MMP-13	Matrix metallopeptidase 13	Proteolysis
COX-1	Cyclooxygenase 1	Cellular metabolism
COX-2	Cyclooxygenase 2	Cellular metabolism
CYP1B1	Cytochrome P450, family 1, subfamily B, polypeptide 1	Cellular metabolism
HIF-1	Hypoxia-inducible factor 1	Signal transduction
BAMBI	BMP and activin membrane-bound inhibitor homolog	Signal transduction
WNT5A	Wingless-type MMTV integration site family, member 5A	Signal transduction and cell motility
PRDX4	Peroxiredoxin 4	Signal transduction
GUCA2A	Guanylate cyclase activator 2A	Signal transduction
COL1A2	Collagen, type I, alpha 2	Cell adhesion
LIMS1	LIM and senescent cell antigen-like domains 1	Cell adhesion
ADRM1	Adhesion regulating molecule 1	Cell adhesion
ICAM4	Intercellular adhesion molecule 4 (Landsteiner–Wiener blood group)	Cell adhesion
SERPINA3	Serpin peptidase inhibitor, clade A (alpha-1 antiproteinase, antitrypsin), member 3	Blood coagulation
SERPINA1	Serpin peptidase inhibitor, clade A (alpha-1 antiproteinase, antitrypsin), member 1	Blood coagulation
SNRPB2	Small nuclear ribonucleoprotein polypeptide B	RNA binding
TWIST1	Twist homolog 1 (Drosophila)	Transcriptional activity
ETV4	ETS variant 4	Transcriptional activity
RPA3	Replication protein A3	DNA repair
VBP1	Von Hippel-Lindau binding protein 1	Transport
CA4	Carbonic anhydrase IV	Carbonate dehydratase activity
VIP	Vasoactive intestinal peptide	Cell motility
CKS2	CDC28 protein kinase regulatory subunit 2	Cell cycle and proliferation
GAS1	Growth arrest-specific 1	Cell growth and proliferation
ITLN1	Intelectin 1 (galactofuranose binding)	Cell growth and proliferation
IL1B	Interleukin 1, beta	Cell growth and proliferation
IL24	Interleukin 24	Cell growth and proliferation
EGFR	Epidermal Growth Factor Receptor	Cell growth
VEGF	Vascular Endotelial Growth Factor	Angiogenesis
SOX15	Sex determining region Y (SRY)-box 15	Embryonic development

The main differences are due to the use of different array platforms (Affymetrix, cDNA nylon membranes) or experimental conditions. The ability to have tissue samples suitable for the analysis is a major problem. Availability of frozen tissues is not the norm in many institutions. Formalin-fixed or paraffin-embedded tissues usually yield low quality RNA and/or DNA. This stressed the need for creating frozen-tissue tumor banks. In addition, different methodologies for RNA isolation can lead to varying results, and the number of samples used varies enormously among the different studies. Selection of homogeneous samples among heterogeneous tumors can often also be a problem. Macrodissection techniques include tumor tissue with both tumor cells and tumor stroma, which may justify the variability of CRC gene expression profiles obtained. Microdissection techniques help to avoid this problem. Laser capture microdissection (LCM) allows isolation of only tumor cells and is considered the gold standard in microdissection procedures [30].

The analysis of data obtained by microarray is another point that may lead to impairment of the results of gene signature. Analysis of differentially expressed genes can be altered by the use of different normalization procedures, different baseline references for ratio calculations, and arbitrary criteria for cut-off values applied to fold-change and significance level. During this selection, information about the quantitative levels of gene expression can be lost. For this reason, different selection algorithms should be tested in order to improve the accuracy of the gene analyzed [19].

In conclusion, to obtain a genetic signature for liver metastases in CRC need to be implemented different measures in order to improve reproducibility, to increase consistence of data and to validate the results.

References

1. Jemal A, Siegel R, Xu J, et al. Cancer statistic 2010. Cancer J Clin. 2010;60:277–300.
2. Edwards BK, Ward E, Kohler BA, et al. Annual report to the nation on the status of cancer 1975–2006, featuring colorectal cancer trends and impact of interventions (risk factors, screening, and treatment) to reduce future rates. Cancer. 2010;116:544–73.
3. Hanahan D, Weinberg RA. The hallmarks of cancer. Cell. 2000;100:57–70.
4. Millikan KW, Staren ED, Doolas A. Invasive therapy of metastatic colorectal cancer to the liver. Surg Clin N Am. 1997;77:27–48.
5. Nadal C, Maurel J, Gascon P. Is there a genetic signature for liver metastasis in colorectal cancer? World J Gastroenterol. 2007;13(44):5832–44.
6. Fidler IJ. Critical factors in the biology of human cancer metastasis: twenty-eighth G.H.A. Clowes memorial award lecture. Cancer Res. 1990;50:6130–8.
7. Fidler IJ, Kripke ML. Genomic analysis of primary tumors does not address the prevalence of metastatic cells in the population. Nat Genet. 2003;34:23.
8. Ramaswamy S, Ross KN, Lander ES, et al. A molecular signature of metastasis in primary solid tumors. Nat Genet. 2003;33:49–54.
9. van't Veer LJ, Dai H, van de Vijver MJ, et al. Gene expression profiling predicts clinical outcome of breast cancer. Nature 2002;415:530–6.
10. Andre T, Kotelevets L, Vaillant JC, et al. Vegf, Vegf-B, Vegf-C and their receptors KDR, FLT-1 and FLT-4 during the neoplastic progression of human colonic mucosa. Int J Cancer. 2000;86:174–81.
11. Eccles SA, Modjtahedi H, Box G, et al. Significance of the c-erbB family of receptor tyrosine kinases in metastatic cancer and their potential as targets for immunotherapy. Invasion Metastasis. 1994;14:337–48.
12. Karube H, Masuda H, Ishii Y, et al. E-Cadherin expression is inversely proportional to tumor size in experimental liver metastases. J Surg Res. 2002;106:173–8.
13. Yanagawa R, Furukawa Y, Tsunoda T, et al. Genome-wide screening of genes showing altered expression in liver metastases of human colorectal cancers by cDNA microarray. Neoplasia. 2001;3:395–401.
14. Gutman M, Fidler IJ. Biology of human cancer metastasis. World J Surg. 1995;19:226–34.
15. Ec W, Chuaqui RF, Liotta LA. General mechanisms of metastasis. Cancer. 1997;80:1529–37.
16. Kinzler KW, Vogelstein B. Lessons from hereditary colorectal cancer. Cell. 1996;87:159–70.
17. Liu W, Dong X, Mai M, et al. Mutations in AXIN2 cause colorectal cancer with defective mismatch repair by activating beta-catenin/TCF signalling. Nat Genet. 2000;26:146–7.

18. Mohr S, Leikauf GD, Keith G, et al. Microarray as cancer keys: an array of possibilities. J Clin Oncol. 2002;20:3165–75.
19. Cardoso J, Boer J, Morreau H, et al. Expression and genomic profiling of colorectal cancer. Biochim Biophys Acta. 2007;1775:103–37.
20. Bienz M, Clevers H, et al. Linking colorectal cancer to Wnt signaling. Cell. 2000;103:311–20.
21. Yamasaki M, Takemasa I, Komori T, et al. The gene expression profile represents the molecular nature of liver metastasis in colorectal cancer. Int J Oncol. 2007;30:129–38.
22. Bertucci F, Salas S, Eysteries S, et al. Gene expression profiling of colon cancer by DNA microarrays and correlation with histoclinical parameters. Oncogene. 2004;23:1377–91.
23. D'Arrigo A, Belluco C, Ambrosi A, et al. Metastatic transcriptional pattern revealed by gene expression profiling in primary colorectal carcinoma. Int J Cancer. 2005;115:256–62.
24. Koehler A, Bataille F, Schimid C, et al. Gene expression profiling of colorectal cancer and metastases divides tumors according to their clinicopathological stage. J Pathol. 2004;204:65–74.
25. Lin HM, Chatterjee A, Lin YH, et al. Genome wide expression profiling identifies genes associated with colorectal liver metastasis. Oncol Rep. 2007;17:1541–9.
26. Ki DH, Jeung HC, Park CH, Kang SH, Lee GY, Lee WS, et al. Whole genome analysis for liver metastasis gene signatures in colorectal cancer. Int J Cancer. 2007;121:2005–12.
27. Pantaleo MA, Astolfi A, Nannini M, et al. Gene expression profiling of liver metastases from colorectal cancer as potential basis for treatment choice. Br J Cancer. 2008;99:1729–34.
28. Koh KH, Rhee H, Kang HJ, Yang E, You KT, Lee H, et al. Differential gene expression profiles of metastases in paired primary and metastatic colorectal carcinomas. Oncology. 2008;75:92–101.
29. Fritzmann J, Morkel M, Besser D, et al. A colorectal cancer expression profile that includes transforming growth factor b inhibitor BAMBI predicts metastatic potential. Gastroenterology. 2009;137(1):165–75.
30. Ma XJ, Salunga R, Tuggle JT, et al. Gene expression profiles of human breast cancer progression. Proc Natl Acad Sci USA. 2003;100:5974–9.

Chapter 4
Gene Signatures in Gastrointestinal Stromal Tumors

Piotr Rutkowski, Giuseppe Badalamenti, Laura La Paglia, Joanna Przybył, and Maria Debiec-Rychter

Introduction

Gastrointestinal stromal tumors (GISTs) constitute a rare heterogeneous group of the most common mesenchymal neoplasm of gastrointestinal tract (GI). GISTs have emerged during the recent years as a distinct sarcoma entity due to advances in the understanding of molecular mechanism of their pathogenesis [1–4]. They are believed to originate from precursors shared with interstitial cells of Cajal (ICC) – the pacemaker cells of the gut (for which CD117 antigen is the immunohistochemical marker), and they may arise along all GI (most commonly in the stomach or the small bowel) or rarely elsewhere. Their biological behavior is difficult to predict, ranging from clinically benign to malignant. The treatment of choice in primary resectable GISTs is radical surgery, but majority of GISTs are associated with a risk of recurrences [1, 5]. The main criteria of aggressive behavior of GIST are based on the presence of invasion of surrounding structures and/or metastases (overtly malignant cases), as well as on primary tumor site, size, and mitotic index [6]. A Consensus Conference held at the National Institutes of Health (NIH) in 2001 provided the first evidence-based definition and a practical scheme for the assessment of the risk in the clinical course of this disease. The risk categorization was based on evaluation of the size and mitotic rate (evaluated per 50 high-powered fields; HPF) of the tumors as the most reliable prognostic factors [7]. Its utility has been supported by analyses of some large series of patients [1, 8]. Additional analyses of patients with primary tumor after complete macroscopic resection confirmed the significance of tumor anatomic location as independent prognostic factor. Miettinen and Lasota from the Armed Forces Institute of Pathology (AFIP) created their own classifications for risk assessment in gastric, duodenal, and intestinal GISTs [9–12], adopted then for use by National Comprehensive Cancer Network (NCCN)

P. Rutkowski (✉)
Department of Soft Tissue/Bone Sarcoma and Melanoma, Maria Sklodowska-Curie Memorial Cancer Center and Institute of Oncology, Warsaw, Poland
e-mail: rutkowskip@coi.waw.pl

G. Badalamenti • L. La Paglia
Section of Medical Oncology, Department of Surgical and Oncological Sciences, University of Palermo, Palermo, Italy

J. Przybył
Department of Molecular Biology, Maria Sklodowska-Curie Memorial Cancer Center and Institute of Oncology, Warsaw, Poland

M. Debiec-Rychter
Department of Human Genetics, Catholic University of Leuven, Leuven, Belgium

A. Russo et al. (eds.), *Diagnostic, Prognostic and Therapeutic Value of Gene Signatures*, Current Clinical Pathology, DOI 10.1007/978-1-61779-358-5_4, © Springer Science+Business Media, LLC 2012

and constituting the basis for new staging system of American Joint Committee on Cancer (AJCC) [13]. In particular, this system reflects the fact that gastric GISTs show a much lower rate of aggressive behavior than jejunal and ileal GISTs of comparable size and/or mitotic rate [9, 12]. In the series of 1,055 gastric, 629 small intestinal, 144 duodenal, and 111 rectal GISTs from the pre-imatinib era, large (>10 cm) gastric GISTs with a low mitotic rate (<5 per 50 HPF) had only a 12% risk for metastasis, whereas similar tumors originating from the small bowel had a high risk for aggressive behavior (>50%). Recurrent and/or metastatic and/or unresectable cases had a very poor prognosis until the beginning of twenty-first century, when advances in the understanding of the molecular mechanisms of GIST pathogenesis have resulted in the development of a treatment approach, which has become a model of targeted therapy in oncology. The introduction of imatinib mesylate [inhibiting KIT/platelet-derived growth factor receptor-a (PDGFRA) and their downstream signaling cascade] has revolutionized the therapy of advanced (inoperable and/or metastatic) GISTs [13–18]. Imatinib at initial dose of 400 mg daily has now become the standard of care in the treatment of patients with advanced GIST and as compared to historical clinical data with median survival of advanced patients being 10–19 months, current survival dramatically improved with median overall survival (OS) reaching approximately 5 years and median progression-free survival (PFS) in the range of 2–3 years. Recently, imatinib has been also registered for use in adjuvant therapy in patients after resection of primary GIST at significant risk of relapse, based on the published results of one clinical trial demonstrating significant reduction of the risk of recurrence without impact on OS [19]. The spectacular response to imatinib therapy is time-limited and secondary resistance to imatinib therapy (after initial stabilization or response) develops in majority of patients. There are several therapeutic strategies in patients showing progression during imatinib treatment, such as escalation of the dose of imatinib to 800 mg daily, surgical removal of focus progression-lesions, and therapy with registered second-line drug sunitinib malate (multitargeted tyrosine kinase inhibitor with antiangiogenic properties) [20, 21].

Molecular Aspects

The majority of GISTs are associated with activating, somatic, mutually exclusive mutations of two genes, *KIT* and *PDGFRA*, which are the early oncogenic events during GIST development [22–30]. Cytogenetically, GISTs are characterized by losses of chromosomes and this feature is accumulating with tumor progression, usually starting from chromosome 14, then chromosome 22 and short arm of chromosome 1. In highly aggressive/overtly malignant cases, are present losses of chromosomes 13, 15, and 18, simultaneously with partial deletions of chromosome regions 9p [where suppressor genes *CDKN2A* (*p16INK4A* and *p14ARF*) and *CDKN2B* (*p15INK4B*) are located] and 11p, altogether with gain of 5p, 8q, and 17q [22, 31–38].

The permanent, ligand-independent activation of transmembrane receptors with tyrosine kinase activity – KIT (receptor for stem-cell factor; SCF) (Fig. 4.1) or PDGFRA as a consequence of gene mutation is a characteristic molecular feature of GISTs [22, 24, 26, 30, 38]. This process leads to changes in receptor conformation, cross-phosphorylation and subsequent triggering of intracellular signaling transduction pathways, such as PI3K/AKT, MEK-mitogen-activated protein kinase (MAPK), and signal transducer and activator of transcription (STAT). The reported frequency varies widely, depending on fresh/embedded tissue used for screening, methods of detection, and tumor selection. In 70–80% of cases, tumor cells have detectable mutations in *KIT* and usually these mutations are heterozygous. The most frequent mutations in juxtamembrane domain coded by exon 11 of *KIT* gene were reported in the range of 20–92%. Exon 11 *KIT* mutations include deletions, deletions associated with insertions, point mutations, and duplications (most often in gastric GISTs).

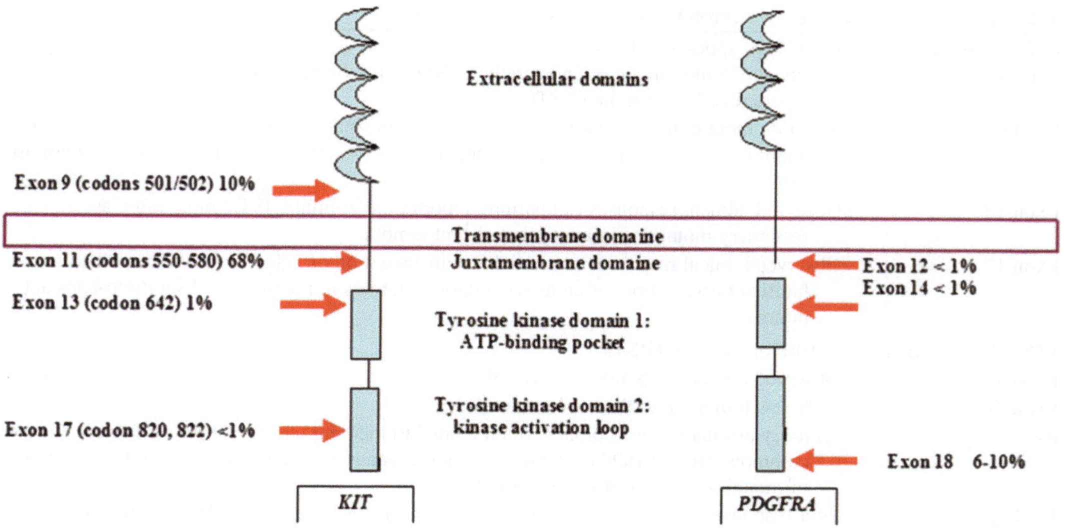

Fig. 4.1 Activating mutations of KIT receptor and PDGFRA described in human primary GISTs

Less frequent are mutations in extracellular domain coded by exon 9 of *KIT* (5–12%). They are mainly present as insertions that result in duplication of Ala^{501} and Tyr^{502}. Exon 9 *KIT* mutations occur preferentially in GISTs originating from small bowel and are related to more aggressive behavior [39, 40], as well as poorer response to imatinib (see below). In rare cases (1–2%), mutations in kinase I domain (exon 13 *KIT*) or the activation loop of kinase domain (exon 17 *KIT*) are detected. The alternative mutations in *PDGFRA* gene are identified in 5–7% GIST cases. The most common *PDGFRA* mutations occur in activation loop kinase domain coded by exon 18, followed by mutations in juxtamembrane domain coded by exon 12 or in first kinase domain coded by exon 14. GIST with *PDGFRA* mutations are often CD117-immunonegative and originate from the stomach, presenting low-aggressive clinical behavior and epithelioid morphology [41, 42]. In approximately 10–15% of GISTs, *KIT* and *PDGFRA* mutations are not detected (the so-called wild-type (WT) GISTs). In this subgroup of patients, excessive activation of KIT also occurs, but the mechanism has not been elucidated yet. Wild-type GISTs constitute the majority of pediatric and adolescent GISTs, GIST in neurofibromatosis type 1 (von Recklinghausen disease), Carney's triad and GISTs in Striatiakis-Carney syndrome [43–46]. In some of these wild-type cases (especially pediatric), overexpression of insulin-like growth factor 1 receptor (IGF1R) has been observed [47]. Few cases of adult wild-type GISTs were reported to carry mutations in *BRAF* exon 15.

The determination of mutation types in GISTs may have prognostic role in primary GISTs, although at present the insufficient data exist to incorporate the kinase mutation status into stratification of the risk of primary tumors. Several authors reported an association between the presence of some types of *KIT* mutations (especially deletions in exon 11 involving codons 557–558 and homozygotic KIT exon 11 mutants) and more aggressive clinical course of disease [48–51]. Similarly, exon 9 *KIT* mutants seem to be characterized by more aggressive behavior. However, the true prognostic significance of these mutations is not well elucidated because KIT mutations are common even in small, incidentally discovered, clinically benign GISTs [28, 29].

In advanced GISTs, the associations between tumor mutational status and PFS during tyrosine kinase inhibitors therapy are well established as well as the molecular characteristic of primary gain-of-function genes encoding KIT or PDGFRA is the most important predictive factor for the response to imatinib [18, 52]. Patients with tumor harboring exon 11 *KIT* mutants have the best

Table 4.1 Molecular features of gastrointestinal stromal tumors (GISTs)

KIT mutations	80–85% of sporadic GISTs
Exon 11	The most common mutation in sporadic GISTs, with the best response to imatinib; reported also in familial GISTs
Exon 9	Mutation more common in GISTs originating from small bowel; intermediate response to imatinib, patients may benefit from higher dose (800 mg) of imatinib; good response to sunitinib
Exon 13	Observed clinical responses to imatinib; reported in familial GISTs; more often as secondary mutations in imatinib-resistant tumors
Exon 17	Observed clinical responses to imatinib (with exception of D816V); it was described in familial GISTs; more often as secondary mutations in imatinib- and sunitinib-resistant tumors
PDGFRA mutations	5–10% of sporadic GISTs
Exon 12	Observed clinical responses to imatinib
Exon 14	Only few tumors described in literature
Exon 18	Majority originate from stomach, often related to indolent clinical behavior; D842V is most common (80% of *PDGFRA* mutants) and resistant to imatinib and sunitinib; other types of mutations are sensitive to imatinib
Wild-type	Poor response to imatinib, better to sunitinib; frequent in pediatric GISTs, typical for GISTs related to neurofibromatosis type 1 or Carney's triad (gastric GIST+pulmonary chondromas +/– paraganglioma); in part of the cases associated with amplification of *IGF1R* or *BRAF* mutation

response to imatinib, with the highest rate of objective responses (70–85% of patients) and the longest overall and PFSs [53–55]. On the contrary, approximately 15–30% of cases with tumors with exon 9 *KIT* mutations and 25–50% of patients without detectable *KIT* or *PDGFRA* mutations show primary resistance to imatinib therapy. Moreover, the response of patients with GIST with exon 9 *KIT* is dependent on the dose of the drug and these patients been identified as needing higher dosages of imatinib (800 mg/day) for achieving longer PFS [56, 57]. Clinical and laboratory studies demonstrated that tumors with exon 18 *PDGFRA* D842V mutation are insensitive to imatinib and sunitinib, whereas other PDGFRA-mutant GISTs show variable response [23].

There are two forms of resistance to the first-line therapy with imatinib, with different pathophysiological mechanisms: primary and secondary resistance. Early tumor progression within the first 6 months of therapy is primarily caused by an intrinsic mechanism related to the presence of specific *KIT* or *PDFRA* mutations (which favor the kinase in the active conformation which is not conductive for imatinib binding) or to lack of mutations in both kinase genes [52, 53]. Tumors demonstrating primary resistance to imatinib include WT GISTs, tumors with exon 9 *KIT* mutation and GISTs with point mutations of the codon 842 of *PDGFRA* gene (D842V). Secondary resistance to imatinib occurs mainly due to selection or acquisition of secondary *KIT* or *PDGFRA* mutations, which hamper imatinib affinity [58–61]. The most common secondary *KIT* mutations are reported in the ATP-binding pocket of the kinase domain (coded by exon 13 and 14) or in the kinase activation loop (exon 17 and 18). In at least some cases, multiple secondary *KIT* mutations were detected with a distinct new mutation in each separate anatomical site of progressing disease. Secondary mutations develop more frequently in tumors harboring primary exon 11, rather than exon 9 mutated *KIT* (60% and 20% of cases, respectively), probably because patients with initially imatinib-sensitive tumors have been treated for longer periods, providing both the selection pressure and time for the emergence of imatinib resistant clones. Table 4.1 shows a summary of the molecular features of GISTs and their behavior with imatinib related to the mutational status. There are also other possible mechanisms for the development of resistance to imatinib, such as (1) bypass of inhibitory effect of imatinib through activation of other, alternative genes (AXL, IGFR-1, BRAF, IGFBP-3,

FAK) and signaling pathways, and usually associated with loss of KIT expression in previously KIT-positive tumors, (2) genomic amplification or overexpression of KIT, outweighing inhibitory capacity of imatinib, (3) overexpression of drug-efflux pump, leading to decreased intratumoral imatinib levels, (4) high blood level of a_1-acid glycoprotein, which binds and inactivates imatinib, (5) increased clearance of imatinib over time, causing decreased systemic imatinib concentrations. As it was proved for imatinib, *KIT* mutation status appears to serve as a predictor of tumor response also to sunitinib [62]. Contrary to imatinib, however, there is evidence that tumors initially (pre-imatinib) bearing an exon 9 *KIT* mutation or with wild-type genotype have a higher chance to respond to sunitinib. Furthermore, patients with tumors harboring a secondary mutation in *KIT* exon 13 or exon 14 have a longer PFS than patients with resistant to sunitinib therapy *KIT* exon 17 or 18 mutations. Ex vivo assays have demonstrated that sunitinib potently inhibits KIT kinase activity of V654A and T670I mutants and suppresses proliferation of the cells expressing these mutations, which was also confirmed by further clinical experience. In contrast, sunitinib did not potently inhibit the D842V *PDGFRA* mutant or secondary mutations in the KIT activation loop (such as D820Y, D820E, N822K, or D823A) [63].

Treatment of children and adolescents with GISTs is an additional clinical challenge. Primary resistance or lesser responsiveness (stable disease) to imatinib is reported in this group of patients, which is predictive given the fact that most cases are wild-type. Sunitinib has been reported to show activity in a small series of imatinib-refractory pediatric patients [64].

Gene Signatures

General Overview

In the past decade, gene expression profiling has been applied to GIST to better understand their pathogenesis and to identify clinically useful diagnostic, prognostic, and predictive markers.

One of the earliest studies compared GIST gene expression profiles with gene signatures of other high-grade sarcomas of different histopathological subtypes [65, 66]. Of interest was the up-regulation of KIT and genes in the KIT signaling pathway, such as the protein kinase C Theta (PRKCQ) and G-protein-coupled receptor 20 (GPR20), both being suggested as mediators of KIT kinase function. Several subsequent sarcoma gene expression studies included significant numbers of GISTs [67–72]. In all these studies, GISTs showed a uniform and rather noncomplex gene expression profile, with homogeneous unsupervised hierarchical clustering of a set of defined genes, which is consistent with the notion of GIST separate biological entity. The list of top ranking genes dissected from these studies is shown in Table 4.2.

Most of these studies have been set up to enhance delineation of diagnosis or to identify expression differences according to *KIT* or *PDGFRA* mutation status (Table 4.3). The later identified not only gene expression profiles characteristic of different mutation types [70, 77], but also associated with distinct tumor anatomic site [74]. Accordingly, Subramanian et al. [77] performed gene expression profiling on 26 GISTs with known KIT/PDGFRA status and identified gene expression profiles characteristic of different mutation types. To investigate the consequences of the distinct KIT genotype and anatomic site of tumors on gene expression profiles, Antonescu et al. [74] characterized transcriptional levels in a cohort of 24 GISTs, including tumors with different KIT genotypes and origin. Notably, GISTs signature was highly dependent on the anatomic site of the tumor. A number of genes involved in muscle contraction and development, genes involved in modulating digestive enzymes and secretion, as well as cell cycle regulators, growths factors, and mediators of growth factor signaling were found to be on the list of discriminatory genes. Interestingly, the set of genes

Table 4.2 Comparison of GIST signature with signatures of other sarcoma types

First author [Reference]	Reference tissue	E	Gene symbol	Gene function/Biological process
Nielsen [66]	Synovial sarcoma, liposarcoma, leiomyosarcoma, malignant fibrous histiocytoma, schwannoma	↑	KIT, PRKCQ	Signal transduction
		↑	PIK3CG	Negative regulation of apoptosis, Inhibition of KIT signaling
		↑	ABCB1, ABCC4	Multidrug resistance
Segal [67]	Fibrosarcoma, leiomyosarcoma, round-cell liposarcoma, pleomorphic liposarcoma, dedifferentiated liposarcoma, clear-cell sarcoma, synovial sarcoma	↑	KIT, ACVR2A, GPR20, PRKCQ, ANXA3, PRKAR2B	Signal transduction
		↑	SLC4A4, ATP1B1	Ion transport
		↑	PENK, Preproenkephalin	Neuropeptide precursor
		↑	PIK3CG	Negative regulation of apoptosis, Inhibition of KIT signaling
Skubitz [73]	Synovial sarcoma, leiomyosarcoma, liposarcoma, schwannoma, meningioma, DFSP, sclerosing epithelioid sarcoma, mesothelioma, fibromatosis, MFH, NOS (high-grade sarcoma, not otherwise specified)	↑	ABCA3	Transporter activity
		↑	ATP1B1, KCNK3, LOC155066	Ion transport
		↑	KIT	Signal transduction
		↑	ATP2A3	Calcium signaling pathway
		↑	ENTPD1	Cell adhesion, cell–cell signaling
		↑	FACL3	Fatty acid metabolic process, lipid metabolic process
		↑	IGF2	Development and growth
Antonescu [68]	Leiomyosarcoma	↑	KIT, PRKCQ, GPR20, ANXA3, P70S6KB, PIK3R1	Signal transduction
		↑	ATP1B1, TASK, CACNB2	Ion transport
		↑	PENK	Neuropeptide precursor
		↑	AKT/PKB	Seronin threonin kinase
		↑	SRC	Protein kinase cascade, signal complex formation, intracellular signaling
		↑	RAC1	Cell adhesion, cell motility
		↑	KRAS	Regulation of progression through cell cycle
		↑	ERK (p38)	Growth factor signaling mediator
		↑	FKHRL1	Transcription factor
		↑	CA2 (Carbonic anhydrase II)	Carbonate dehydratase activity, zinc ion binding, lyase activity
Li [71]	Familial vs. sporadic GISTs, leiomyosarcomas, synovial sarcomas, and DFSP	↑	KIT, PRKCQ	Signal transduction
		↑	DOG1	Ion channel activity

Baird [69]	ASPS, clear-cell sarcoma, chondrosarcoma, DFSP, Ewing sarcoma, fibrosarcoma, malignant HPC, leiomyosarcoma, liposarcoma, MFH, mixed Mullerian, osteosarcoma, MPNST, benign schwannoma, rhabdomyosarcoma, synovial sarcoma, previously unclassified sarcomas	↑	KIT	Signal transduction
		↑	LYN	Receptor signaling protein tyrosine kinase activity
		↑	ZNF41, FHL2	Regulation of transcription
		↑	HRASLS3	Negative regulation of progression through cell cycle
		↑	PLAT	Plasminogen activator activity
		↑	IGF2, GHR	Development and growth
		↑	FAT	Gene regulation by peroxisome proliferators via PPARa(alpha)
		↑	PBX3	Homeobox and early developmental gene
Francis [72]	Malignant fibrous histiocytoma (MFH), leiomyosarcoma, synovial sarcoma, liposarcoma, malignant peripheral nerve sheath tumor, extraskeletal osteosarcoma, fibrosarcoma, epithelioid sarcoma, malignant mesenchymoma	↑	KIT, PRKCQ	Signal transduction
		↑	BMP4, FGF2, IGF2, SFRP1, TLE4	Developmental pathways
		↑	SMPD1, HOXA4, CIT, HOXA9, SIM2, NPTX1, NEDD5, DCTN1	Neurogenesis and neural differentiation
		↑	PROM1 (prominin 1)	Stem cell marker; organization of plasma membrane
Antonescu [74]	Leiomyosarcoma	↑	PRKAR2B	Intracellular signaling cascade, signal transduction.
		↑	PEX11B	Peroxisome fission

E level of gene expression in reference to control group of tumors ↑ = gene up-regulation, ↓ = gene down-regulation identified was distinct from that described by Subramanian et al., which most likely reflects the impact of used methodology and biological differences between diverse subset of analyzed tumors. Similarly, the differences in expression profile of 22 gastric GISTs were investigated by Kang et al. [70] and 70 genes were differentially expressed in GISTs with KIT mutations compared to GISTs with PDGFRA mutations. In parallel, Li et al. [71] provided the first evidence that gene expression profile is indistinguishable in familial and sporadic GISTs.

More recently, based on the analysis of 31 GISTs, Ostrowski and coworkers [75] compared the expression signature of tumors with low *versus* high KIT expression level. The expression of genes annotated to synaptic transmission, blood level development, and G-protein signaling were at least twofold higher in the former compared to later.

One of the few studies that attempted the integrative analysis of the genome and transcription profiling in GISTs was based on 25 mutants or KIT/PDGFRA wild-type tumors [76]. This study aimed to identify target genes located within commonly altered genomic regions that have been described in GISTs within the last 15 years. The majority of discriminative genes were down-regulated

Table 4.3 GIST signatures according to different genotype and/or tumor anatomic site

First author [Reference]	GIST type	E	Gene symbol	Gene function/Biological process
Allander [65]	KIT mutants	↑	KIT, GPR20, PRKCQ, ANXA3	Signal transduction
		↑	BCHE	Cocaine metabolic process
		↑	DMN	Structural support in muscle
		↑	PFKM	Regulation of glycolysis, glycogen metabolic process
Subramanian [72]	KIT exon 11 mutants	↑	TM4SF12	Unknown
		↑	DIO2	Selenocysteine incorporation, thyroid hormone generation, hormone biosynthetic process
		↑	CAPNS2	Integrin signaling pathway
		↑	CCKBR	Calcium signaling pathway, neuroactive ligand–receptor interaction
		↑	LCN2	Transporter activity
	KIT exon 9 mutants	↑	IGSF4, DSG2	Cell adhesion
		↑	ALDH1A2	Retinol metabolism
		↑	MAP1B	Protein folding
		↑	EPHA4	Signal transduction, transmembrane receptor protein tyrosine kinase signaling pathway
	PDGFRA mutants	↑	SPON1	Cell adhesion, multicellular organismal development
		↑	IGFBP5	Signal transduction, regulation of cell growth
		↑	ANGPTL1	Signal transduction, transmembrane receptor protein tyrosine kinase signaling pathway
		↑	LUM	Collagen binding, extracellular matrix structural constituent, transferase activity
		↑	KIAA0534	Receptor activity, sugar binding

(continued)

Table 4.3 (continued)

First author [Reference]	GIST type	E	Gene symbol	Gene function/Biological process
Antonescu [68]	Mutant GIST	↑	RAC2, Shp1	Growth factor signaling genes
	Wild-type GIST	↑	BCL2, GLUT1/VEGF, MCSF, IL2, MAGE1	Apoptosis/angiogenesis and proliferation
	KIT Exon 9 mutants	↑	GGT1/MPF/FZD2, FZD3	mRNA processing/cell adhesion/Wnt signaling
	KIT Exon 11 mutants	↑	STAT3, CTNNB1, NTAK (NRG2), PDGF1	Signal transduction
	Familial GIST	↑	CALCYON, NOS1	Synaptic transmission
		↑	PIB5PA	Phosphatidylinositol signaling system
		↑	CACNA1H	Muscle contraction and development
		↑	GLIPR1	GLI pathogenesis-related 1 (glioma)
	Gastric GIST	↑	TNNI2, TPM1, SMTN, LAMM, ESG	Muscle contraction and development
		↑	CCKBR, PLAG2G4B	Digestive enzymes modulation and secretion
		↑	PDGFRA, TGFRBR3, LTBP-4, TSC22	Growth factor receptors
		↑	CD34	Cell-cell adhesion, leukocyte migration
	Small bowel GIST	↑	MyHC-eo (Myosin heavy-chain polypeptide)	Muscle contraction and development
		↑	PIK3C2B, VAV2, Shp1, RAC1, RAC2, RAC3	Mediators of growth factor signaling
	Epithelioid GIST	↑	SOX11	Nervous system development, regulation of transcription
		↑	Cancer Testis Antigen 2 (CAMEL)	–
		↑	VEGF/PDGF1/BCL2, BCL-G, CASP10	Proliferation/proliferation and signal transduction/apoptosis
Kang [70]	PDGFRA mutants (gastric)	↑	ASB2, PDGFRA, MX1/ISG15/SPON1, TGFBI, THY1	Signal transduction/protein metabolism/cell adhesion
	KIT mutants (gastric)	↑	S100A1	Signal transduction
Ostrowski [75]	KIT mutants	↑	PRKCQ (PKC-theta)	Regulation of cell growth, signal transduction
		↓	PKC-alpha	Induction of apoptosis, regulation of progression through cell cycle
Astolfi [76]	Mutant GISTs(both KIT and PDGFRA)	↓	RTN1	Neuron differentiation, apoptosis
		↓	DACT1	Inhibition of Wnt/beta-catenin signaling
		↓	DAAM1	Cell motility, action cytoskeleton

E level of gene expression in reference to control group of tumors, ↑ = gene up-regulation, ↓ = gene down-regulation in mutated samples. Not surprisingly, many of these genes have a role in cancer progression. RTN1, DAAM1, and DACT1, located in 14q23.1 cyto-band, showed the strongest down-regulation in mutant samples.

It has been proved that GISTs from children and young adults express different expression profile compared to adult GISTs. Thus, Agaram et al. [78] performed recently microarray analysis using a U133A Affymetrix chip platform on 13 pediatric GIST tumor nodules derived from 8 patients. The expression values were compared to a control group of adult GISTs, with available array data. In an unsupervised hierarchical clustering analysis, pediatric GIST tumors formed a tight cluster distinct from the adult GIST tumors. A number of the overexpressed genes in the pediatric GISTs showed a high fold change difference compared to adult tumors, including *BAALC, IGF1R, CRLF1, PLAG1, FGF4*, and *FGF3*. Also IGF1 expression was fivefold higher in pediatric GIST compared with adult WT GIST. The authors speculated that this might be one of the mechanisms that induce the expression of IGF1R.

GIST Signature for Diagnosis

Expression profiling was used as a diagnostic tool in soft tissue sarcoma/GIST classification. Along these lines, West et al. [79] identified DOG1 (named also TMEM16A) as a validated marker for CD-117-immunonegative GIST. Recently, Price et al. [80] performed gene expression profiling on a large collection of GISTs and leioyosarcomas (LMS) and created a novel paired gene analysis that was highly discriminative for the two tumor types. Two genes were identified (Obscurin and Prune 2) that could accurately distinguish GIST and LMS from one another.

GIST Signatures for Prognosis

The development of a reliable method of GIST prognostication is essential for the proper clinical management of GISTs patients. This is especially imperative given the necessity of appropriate selection of patients for imatinib mesylate therapy to avoid possible development of tumor resistance to treatment, the existence of possible alternative treatment and resource waste. There is a need to more deeply understand the biology underlining the aggressiveness of GISTs in order to identify objective biomarkers that enhance the specificity and the reproducibility of outcome prediction. To achieve this purpose, genomic and expression profiling has been used, but heterogeneous results have been obtained (Table 4.4).

One of the earliest expression profiling study [81] detected changes related to GISTs progression and revealed 27 overexpressed genes in malignant GISTs compared with benign tumors. These genes included proliferation markers, cell cycle regulators, and several kinases (CCNB1, CENP-F, FAK, HMG2, TSG101, DTRK2, and ezrin).

At the genomic level, it has been shown that the genome complexity level increases with tumor stage, but no threshold has ever been defined and no specific alteration has been proposed except for $p16^{INK4A}$ alterations [87, 88]. Inactivation of the later was confirmed on the transcript and protein level to be associated with increased risk for GIST progression [87].

To identify prognosis associated gene-expression signature, Yamagushi and colleagues [17] performed expression profiling in a series of 32 GISTs and identified CD26 as a prognosis marker in GISTs of gastric origin. Nevertheless, this signature is limited as it has not been compared to histopathologic staging methods considered as the "gold standard." Arne et al. [86] performed global

Table 4.4 Genomic and expression profiling of different genes in GIST related to different biological function in which they are involved to

First author [Reference]	Reference tissue	E	Gene symbol	Gene function/Biological process
Koon [81]	Malignant vs. benign GISTs	↑	CCNB1, CENPF	Cell cycle and mitosis control
		↑	FAK (PTK2)	Angiogenesis, apoptosis, microtubule cytoskeleton organization
		↑	HGM2 (high-mobility group box 2)	DNA ligation, DNA repair, DNA replication
		↑	TSG101	Cell growth and differentiation
		↑	EZR (VIL2)	Regulation of actin cytoskeleton, cell adhesion molecule binding
Yamaguchi [82]	Gastric GISTs validation set	↑	DPP4 (dipeptidyl peptidase IV) gene encoding CD26 protein	Positive regulation of cell proliferation, cell adhesion
Hur [83]	Low vs. intermediate vs. high-risk GIST groups	↑	DNASE1L2/NCBP1, HSF2BP/FGF17, GCGR, RIPK2/MUC11/HYAL4/PPP3CA/A4GNT	Cell proliferation/regulation of transcription/signal transduction/regulation of cell growth/carbohydrate metabolism/cell cycle control, calcium ion transport/transferase activity
		↓	CPE/HTRA1/ANXA9/PTH/CYB5R3/TLR6, GABRA5, CD48/HSPA5/HIVEP2	Biosynthesis of hormones and neurotransmitters/regulation of cell growth/cell adhesion/signal transduction, cell–cell signaling/metabolism of vitamins and cofactors/inflammatory response/regulation of transcription
Chibon [84]	Low vs. intermediate vs. high-risk GIST groups	↑	ANLN, ASPM, AURKA, AURKB, BIRC5, BUB1, BUB1B, C13orf34, CCNA2, CCNB1, CCNB2, CDC2, CDC20, CDC45L, CDC6, CDC7, CDCA2, CDCA3, CDCA8, CENPA, CENPE, CENPL, CEP55, CHEK1, CKS2, ECT2, ESPL1, FBXO5, FOXM1, H2AFX, HP1BP3, KIAA1794, KIF11, KIF14, KIF15, KIF18A, KIF20A, KIF23, KIF2C, KIF4A, KIFC1, MADL2L1, MCM2, MCM7, MELK, NCAPH, NDE1, NEK2, NUF2, OIP5, PAK3, PBK, PLK4, PRC1, PTTG1, RAD51AP1, RNASEH2A, RRM2, SGOL2, SMC2, SPAG5, SPBC25, TOP2A, TPX2, TRIP13, TTK, ZWINT	Mitosis and control of chromosome integrity
Ylipaa [85]	Leiomyosarcoma	↓	AKAP13	Intracellular signaling, regulation of Rho signal transduction
		↓	C15orf5	Unknown
		↓	OXA1L	Protein export
		↑	SMARCA3	Chromatin modification, regulation of transcription
Arne [86]	KITex11 del vs. KITex9 and PDGFRA and wild-type GISTs	↑	CD133 (prominin – 1)	Stem cell marker; organization of plasma membrane

E level of gene expression in reference to control group of tumors, ↑ = gene up-regulation, ↓ = gene down-regulation gene expression profiling on GISTs arranged into two separate groups, i.e., tumors with *KIT* exon 11 deletions and without *KIT* exon 11 mutations. Among the differentially regulated genes were previously known markers for GIST, including KIT, DOG1, and CD34, which were up-regulated in GISTs carrying KIT exon 11 deletions. Notably, one highly up-regulated gene in the group of tumors carrying KIT exon 11 deletions was CD133 (prominin-1), which was also associated with gastric location and poor prognosis. Hur et al. [83] reported the gene expression profiling according to the GISTs malignant potential through direct comparison between various risks of GISTs (based on NIH consensus criteria) and human universal RNA as a reference. A total of 181 genes were identified to be expressed differentially according to GIST risk category.

Importantly, Chibon and coworkers [84] recently reported a 67 genes-expression prognostic signature related to genome complexity (CINSARC for Complexity INdex in SARComas). Gene ontology analysis of CINSARC genes showed that all annotated genes are involved in the same biological process, i.e., cell cycle/mitosis and control of chromosome integrity. As a continuation of this work, using a training/validation set strategy gathering 99 GISTs, Lagerde et al. [89] demonstrated that CINSARC signature and a new one-gene-expression signature (AURKA expression level) predict metastatic outcome in GIST, and that their combination with genome imbalances outperform current histopathological grading method in determining patient prognosis. More specifically, these molecular signatures identify "at risk patients" within cases stratified as of intermediate risk according to the AFIP classification. Application of the signature will permit more selective imatinib mesylate adjuvant therapy leading to decreased iatrogenic morbidity and improved outcomes for individual patients.

GIST Signatures Response to IMATINIB

GISTs signatures may be used not only to better understand the molecular basis of GISTs tumorigenesis and for the identification of prognostic biomarkers, but also to identify a novel target molecule for effective curative treatments.

Evaluation of surrogate markers of imatinib response in clinical GIST samples, using core biopsies from patients both before and after initiation of therapy, was performed by Frolov and coworkers [90]. Down-regulation of Sprouty homolog 4 (SPRY4A) and up-regulation muscle atrophy F-box protein (MAFbx) transcripts were found to be highly reliable predictors of immediate response to the drug.

Based on expression profiling analysis of treated with imatinib GIST cell lines, Trent et al. [91] identified insulin-like growth factor binding protein 3 (IGFBP3) as an important early marker of antitumor activity of imatinib in GIST.

Mahadevan et al. [92] performed global expression profiling of GIST882 cell line subclones, sensitive (expressing KIT) and resistant (that lost KIT expression) to imatinib mesylate. Differential expression of a number of cytoskeletal, cell adhesion, and extracellular matrix proteins genes were hypothesized to contribute to the imatinib-resistant phenotype. Importantly, AXL receptor tyrosine kinase overexpression was found in the subline with the KIT expression loss, pointing to kinase switch as an alternative mechanism of imatinib resistance and AXL as a new possible therapeutic target.

Affymetrix microarray analysis was performed also on ten nodules from three responsive GIST patients and compared with a group of 34 nontreated GISTs [93]. Genomic signature of imatinib response identified alterations of genes involved in cell cycle control and overexpression of genes involved in muscle differentiation and function. Galanin receptor 2 (GALR2) and glypican 3 (GPC3) were the two top genes overexpressed in responsive GIST.

Recently, Rink et al. [94] were able to elucidate a gene expression profile that is unique to patients whose tumors are less responsive to imatinib mesylate in comparison with those that rapidly respond. This profile consists of 32 annotated genes, 18 of which encoded Krüppel-associated box (KRAB) domain containing zinc finger (ZNF) transcriptional repressors. Importantly, ten KRAB-ZNF genes mapped to a single locus on chromosome 19p, and a subset predicted response to imatinib mesylate-based therapy in a naïve panel of GIST.

References

1. Miettinen M, Lasota J. Gastrointestinal stromal tumors – definitions, clinical, histological, immunohistochemical and molecular genetic features and differential diagnosis. Virchows Arch. 2001;438:1–12.
2. Kindblom LG, Remotti HE, Aldenborg F, Meis-Kindblom JM. Gastrointestinal pacemaker cell tumor (GIPACT): gastrointestinal stromal tumors show phenotypic characteristics of the interstitial cells of Cajal. Am J Pathol. 1998;152(5):1259–69.
3. Joensuu H. Gastrointestinal stromal tumor (GIST). Ann Oncol. 2006;17(10):280–6.
4. Nilsson B, Bumming P, Meis-Kindblom JM, et al. Gastrointestinal stromal tumors: the incidence, prevalence, clinical course, and prognostication in pre-imatinib mesylate era. Cancer. 2005;103:821–9.
5. Rutkowski P, Nowecki ZI, Michej W, et al. Risk criteria and prognostic factors for predicting recurrences after resection of primary gastrointestinal stromal tumors (GISTs). Ann Surg Oncol. 2007;14(7):2018–27.
6. Miettinen M, Lasota J. Gastrointestinal stromal tumors: pathology and prognosis at different sites. Semin Diagn Pathol. 2006;23(2):70–83.
7. Fletcher C, Berman JJ, Corless C, et al. Diagnosis of gastrointestinal stromal tumors: a consensus approach. Hum Pathol. 2002;33:459–65.
8. Joensuu H. Risk stratification of patients diagnosed with gastrointestinal stromal tumor. Hum Pathol. 2008;39:1411–9.
9. Miettinen M, Sobin LH, Lasota J. Gastrointestinal stromal tumors of the stomach: a clinicopathologic, immunohistochemical, and molecular genetic study of 1765 cases with long-term follow-up. Am J Surg Pathol. 2005;29:52–68.
10. Miettinen M, Furlong M, Sarlomo-Rikala M, et al. Gastrointestinal stromal tumors, intramural leiomyomas, and leiomyosarcomas in the rectum and anus: a clinicopathologic, immunohistochemical, and molecular genetic study of 144 cases. Am J Surg Pathol. 2001;25:1121–33.
11. Miettinen M, Kopczynski J, Makhlouf HR, et al. Gastrointestinal stromal tumors, intramural leiomyomas, and leiomyosarcomas in the duodenum: a clinicopathologic, immunohistochemical, and molecular genetic study of 167 cases. Am J Surg Pathol. 2003;27:625–41.
12. Miettinen M, Makhlouf H, Sobin LH, et al. Gastrointestinal stromal tumors of the jejunum and ileum: a clinicopathologic, immunohistochemical, and molecular genetic study of 906 cases before imatinib with long-term follow-up. Am J Surg Pathol. 2006;30:477–89.
13. Edge SB, Byrd DR, Compton CC, Fritz AG, Greene FL, Trotti A, editors. American joint committee on cancer staging manual. 7th ed. New York: Springer; 2009.
14. van Oosterom AT, Judson I, et al. Safety and efficacy of imatinib (STI571) in metastatic gastrointestinal stromal tumours: a phase I study. Lancet. 2001;358:1421–3.
15. Demetri GD, von Mehren M, Blanke CD, et al. Efficacy and safety of imatinib mesylate in advanced gastrointestinal stromal tumors. NEJM. 2002;347:472–80.
16. Verweij J, Casali PG, Zalcberg J, et al. Progression-free survival in gastrointestinal stromal tumours with high-dose imatinib: randomised trial. Lancet. 2004;364:1127–34.
17. Blanke CD, Demetri GD, Von Mehren M, et al. Long-term follow-up of a phase II randomized trial in advanced gastrointestinal stromal tumor (GIST) patients (pts) treated with imatinib mesylate. J Clin Oncol (Meeting Abstracts). 2006;24(18 suppl):9528.
18. Rutkowski P, Nowecki ZI, Dębiec-Rychter M, et al. Predictive factors for long term effects of imatinib therapy in patients with inoperable/metastatic CD117(+) gastrointestinal stromal tumors (GISTs). J CA Res Clin Oncol. 2007;133(9):589–97.
19. DeMatteo R, Ballman KV, Antonescu CR, et al. Adjuvant imatinib mesylate after resection of localised, primary gastrointestinal stromal tumour: a randomised, double-blind, placebo-controlled trial. Lancet. 2009;373:1079–104.
20. Demetri GD, van Oosterom AT, Garrett CR, et al. Efficacy and safety of sunitinib in patients with advanced gastrointestinal stromal tumour after failure of imatinib: a randomised controlled trial. Lancet. 2006;368:1329–38.

21. Andtbacka RHI, Ng CS, Scaife CL, et al. Surgical resection of gastrointestinal stromal tumors after treatment with imatinib. Ann Surg Oncol. 2007;14(1):14–21.
22. Corless CL, Fletcher JA, Heinrich MC. Biology of gastrointestinal stromal tumors. J Clin Oncol. 2004;22(18):3813–25.
23. Corless CL, Schroeder A, Griffith D, et al. PDGFRA mutations in gastrointestinal stromal tumors: frequency, spectrum and in vitro sensitivity to imatinib. J Clin Oncol. 2005;23(23):5357–64.
24. Heinrich MC, Corless CL, Duensing A, et al. PDGFRA activating mutations in gastrointestinal stromal tumors. Science. 2003;299(5607):708–10.
25. Hirota S, Ohashi A, Nishida T, et al. Gain-of-function mutations of platelet-derived growth factor receptor alpha gene in gastro-intestinal stromal tumors. Gastroenterology. 2003;125(3):660–7.
26. Hirota S, Isozaki K, Moriyama Y, et al. Gain-of-function mutations of c-kit in human gastrointestinal stromal tumors. Science. 1998;279(5350):577–80.
27. Hirota S, Nishida T, Isozaki K, et al. Gain-of-function mutation at the extracellular domain of KIT in gastrointestinal stromal tumours. J Pathol. 2001;193(4):505–10.
28. Corless CL, McGreevey L, Haley A, Town A, Heinrich MC. KIT mutations are common in incidental gastrointestinal stromal tumors one centimeter or less in size. Am J Pathol. 2002;160(5):1567–72.
29. Agaimy A, Wunsch PH, Hofstaedter F, et al. Minute gastric sclerosing stromal tumors (GIST tumorlets) are common in adults and frequently show c-KIT mutations. Am J Surg Pathol. 2007;31:113–20.
30. Lasota J, Miettinen M. KIT and PDGFRA mutations in gastrointestinal stromal tumors (GISTs). Sem Diagn Pathol. 2006;23:91–102.
31. Wozniak A, Sciot R, Guillou L, et al. Array CGH analysis in primary gastrointestinal stromal tumors: cyto genetic profile correlates with anatomic site and tumor aggressiveness, irrespective of mutational status. Genes Chromosomes Cancer. 2007;46:261–76.
32. Sandberg AA, Bridge JA. Updates on the cytogenetics and molecular genetics of bone and soft tissue tumors. Gastrointestinal stromal tumors. Cancer Genet Cytogenet. 2002;135:1–22.
33. Debiec-Rychter M, Lasota J, Sarlomo-Rikala M, et al. Chromosomal aberrations in malignant gastrointestinal stromal tumors: Correlation with c-KIT gene mutation. Cancer Genet Cytogenet. 2001;128:24–30.
34. Bergmann F, Gunawan B, Hermanns B, et al. Cytogenetic and morphologic characteristics of gastrointestinal stromal tumors. Recurrent rearrangement of chromosome 1 and losses of chromosomes 14 and 22 as common anomalies. Verh Dtsch Ges Pathol. 1998;82:275–8.
35. Fukasawa T, Chong JM, Sakurai S, et al. Allelic loss of 14q and 22q, NF2 mutation, and genetic instability occur independently of c-kit mutation in gastrointestinal stromal tumor. Jpn J Cancer Res. 2000;91:1241–9.
36. El-Rifai W, Sarlomo-Rikala M, Andersson LC, Knuutila S, Miettinen M. DNA sequence copy number changes in gastrointestinal stromal tumors: tumor progression and prognostic significance. Cancer Res. 2000;60:3899–903.
37. Schneider-Stock R, Boltze C, Lasota J, Miettinen M, Peters B, Pross M, et al. High prognostic value of p16INK4 alterations in gastrointestinal stromal tumors. J Clin Oncol. 2003;21:1688–97.
38. Rutkowski P, Debiec-Rychter M, Ruka W. Gastrointestinal stromal tumors: key to diagnosis and choice of therapy. Mol Diagn Ther. 2008;12(3):131–43.
39. Antonescu CR, Sommer G, Sarran L, et al. Association of KIT exon 9 mutations with nongastric primary site and aggressive behavior: KIT mutation analysis and clinical correlates of 120 gastrointestinal stromal tumors. Clin Cancer Res. 2003;9:3329–37.
40. Lasota J, Wozniak A, Sarlomo-Rikala M, et al. Mutations in exons 9 and 13 of KIT gene are rare events in gastrointestinal stromal tumors: a study of 200 cases. Am J Pathol. 2000;157:1091–5.
41. Lasota J, Dansonka-Mieszkowska A, Sobin LH, Miettinen M. A great majority of GISTs with PDGFRA mutations represent gastric tumors of low or no malignant potential. Lab Invest. 2004;84:874–83.
42. Debiec-Rychter M, Wasag B, et al. Gastrointestinal Stromal tumours (GISTs) negative for KIT (CD117 antigen) immunoreactivity. J Pathol. 2004;202:430–8.
43. Prakash S, Sarran L, Socci N, et al. Gastrointestinal stromal tumors in children and young adults. J Pediatr Hematol Oncol. 2005;27:179–87.
44. Maertens O, Prenen H, Debiec-Rychter M, et al. Molecular pathogenesis of multiple gastrointestinal stromal tumors in NF1 patients. Hum Mol Genet. 2006;15:1015–23.
45. Carney JA, Stratakis CA. Familial paraganglioma and gastric stromal sarcoma: a new syndrome distinct from the Carney triad. Am J Med Genet. 2002;108(2):132–9.
46. Carney JA. Gastric sarcoma, pulmonary chondroma and extra-adrenal paraganglioma (Carney's triad): natural history, adrenocortical component and possible familial occurrence. Mayo Clin Proc. 1999;74:543–52.
47. Tarn C, Rink L, Merkel E, Flieder D, Pathak H, Koumbi D, et al. Insulin-like growth factor 1 receptor is a potential therapeutic target for gastrointestinal stromal tumors. PNAS. 2008;105:8387–92.
48. Lasota J, Jasinski M, Sarlomo-Rikala M, et al. Mutations in exon 11 of c-Kit occur preferentially in malignant versus benign gastrointestinal stromal tumors and do not occur in leiomyomas or leiomyosarcomas. Am J Pathol. 1999;154:53–60.

49. Andersson J, Bumming P, Meis-Kindblom JM, et al. Gastrointestinal stromal tumors with KIT exon 11 deletions are associated with poor prognosis. Gastroenterology. 2006;130:1573–81.
50. Martin J, Poveda A, Llombart-Bosch A, et al. Deletions affecting codons 557–558 of the c-KIT gene indicate a poor prognosis in patients with completely resected gastrointestinal stromal tumors: a study by the Spanish group for sarcoma research (GEIS). J Clin Oncol. 2005;23:6190–8.
51. Wardelmann E, Losen I, Hans V, et al. Deletion of Trp-557 and Lys-558 in the juxtamembranedomain of the c-kit protooncogene is associated with metastatic behavior of gastrointestinal stromal tumors. Int J Cancer. 2003;106:887–95.
52. Rutkowski P, Symonides M, Zdzienicki M, Siedlecki JA. Developments in targeted therapy of advanced gastrointestinal stromal tumors. Recent Pat Anticancer Drug Discov. 2008;3:88–99.
53. Heinrich MC, Corless CL, Demetri GD, et al. Kinase mutations and imatinib mesylate response in patients with metastatic gastrointestinal stromal tumor. J Clin Oncol. 2003;21:4342–9.
54. Heinrich MC, Corless CL, Blanke CD, et al. Molecular correlates of imatinib resistance in gastrointestinal stromal tumors. J Clin Oncol. 2006;24:4764–74.
55. Debiec-Rychter M, Dumez H, Judson I, et al. Use of c-KIT/PDGFRA mutational analysis to predict the clinical response to imatinib in patients with advanced gastrointestinal stromal tumours entered on phase I and II studies of the EORTC Soft Tissue and Bone Sarcoma Group. Eur J Cancer. 2004;40(5):689–95.
56. Debiec-Rychter M, Sciot R, et al. KIT mutations and dose selection for imatinib in patients with advanced gastrointestinal stromal tumours. Eur J Cancer. 2006;42:1093–103.
57. Gastrointestinal Stromal Tumor Meta-Analysis Group (MetaGIST). Comparison of two doses of imatinib for the treatment of unresectable or metastatic gastrointestinal stromal tumors: a meta-analysis of 1,640 patients. J Clin Oncol. 2010;28(7):1247–53.
58. Debiec-Rychter M, Cools J, Dumez H, et al. Mechanisms of resistance to imatinib mesylate in gastrointestinal stromal tumors and activity of the PKC412 inhibitor against imatinib-resistant mutants. Gastroenterology. 2005;128:270–9.
59. Antonescu CR, Besmer P, Guo T, et al. Acquired resistance to imatinib in gastrointestinal stromal tumor occurs through secondary gene mutation. Clin Cancer Res. 2005;11:4182–90.
60. Wardelmann E, Merkelbach-Bruse S, et al. Polyclonal evolution of multiple secondary KIT mutations in gastrointestinal stromal tumors under treatment with imatinib mesylate. Clin Cancer Res. 2006;12:1743–9.
61. Wakai T, Kanda T, Hirota S, et al. Late resistance to imatinib therapy in a metastatic gastrointestinal stromal tumour is associated with a second KIT mutation. Br J Cancer. 2004;90:2059–61.
62. Heinrich MC, Maki RG, Corless CL, Antonescu CR, Harlow A, Griffith D, et al. Primary and secondary kinase genotypes correlate with the biological and clinical activity of sunitinib in imatinib-resistant gastrointestinal stromal tumor. J Clin Oncol. 2008;26:5352–9.
63. Guo T, Hajdu M, Agaram NP, Shinoda H, Veach D, Clarkson BD, et al. Mechanisms of sunitinib resistance in gastrointestinal stromal tumors harboring KITAY502-3ins mutation: an in vitro mutagenesis screen for drug resistance. Clin Cancer Res. 2009;15(22):6862–70.
64. Janeway KA, Albritton KH, Van Den Abbeele AD, D'Amato GZ, Pedrazzoli P, Siena S, et al. Sunitinib treatment in pediatric patients with advanced GIST following failure of imatinib. Pediatr Blood Cancer. 2009;52(7):767–71.
65. Allander SV, Nupponen NN, Ringnér M, Hostetter G, Maher GW, Goldberger N, et al. Gastrointestinal stromal tumors with KIT mutations exhibit a remarkably homogeneous gene expression profile. Cancer Res. 2001;61:8624–8.
66. Nielsen TO, West RB, Linn SC, Alter O, Knowling MA, O'Connell JX, et al. Molecular characterisation of soft tissue tumours: a gene expression study. Lancet. 2002;359(9314):1301–7.
67. Segal NH, Pavlidis P, Antonescu CR, Maki RG, Noble WS, DeSantis D, et al. Classification and subtype prediction of adult soft tissue sarcoma by functional genomics. Am J Pathol. 2003;163(2):691–700.
68. Antonescu CR, Viale A, Sarran L, Tschernyavsky SJ, Gonen M, Segal NH, et al. Gene expression in gastrointestinal stromal tumors is distinguished by KIT genotype and anatomic site. Clin Cancer Res. 2004;10(10):3282–90.
69. Baird K, Davis S, Antonescu CR, Harper UL, Walker RL, Chen Y, et al. Gene expression profiling of human sarcomas: insights into sarcoma biology. Cancer Res. 2005;65(20):9226–35.
70. Kang HJ, Nam SW, Kim H, Rhee H, Kim NG, Kim H, et al. Correlation of KIT and platelet-derived growth factor receptor alpha mutations with gene activation and expression profiles in gastrointestinal stromal tumors. Oncogene. 2005;24(6):1066–74.
71. Li FP, Fletcher JA, Heinrich MC, Garber JE, Sallan SE, Curiel-Lewandrowski C, et al. Familial gastrointestinal stromal tumor syndrome: phenotypic and molecular features in a kindred. J Clin Oncol. 2005;23(12):2735–43.
72. Francis P, Namløs HM, Müller C, Edén P, Fernebro J, Berner JM, et al. Diagnostic and prognostic gene expression signatures in 177 soft tissue sarcomas: hypoxia-induced transcription profile signifies metastatic potential. BMC Genomics. 2007;14(8):73.
73. Skubitz KM, Skubitz AP. Role of gene expression arrays in sarcomas. Curr Oncol Rep. 2004;6(4):309–14.

74. Antonescu CR, Wu K, Xing GL, Cao M, Turpaz Y, Leversha MA, et al. DNA copy number analysis in gastrointestinal stromal tumors using gene expression microarrays. Cancer Inform. 2008;6:59–75.
75. Ostrowski J, Polkowski M, Paziewska A, Skrzypczak M, Goryca K, Rubel T, et al. Functional features of gene expression profiles differentiating gastrointestinal stromal tumours according to KIT mutations and expression. BMC Cancer. 2009;9:413.
76. Astolfi A, Nannini M, Pantaleo MA, Di Battista M, Heinrich MC, Santini D, et al. A molecular portrait of gastrointestinal stromal tumors: an integrative analysis of gene expression profiling and high-resolution genomic copy number. Lab Invest. 2010;90(9):1285–94.
77. Subramanian S et al. Gastrointestinal stromal tumors (GISTs) with KIT and PDGFRA mutations have distinct gene expression profiles. Oncogene. 2004;23:7780–90.
78. Agaram NP, Laquaglia MP, Ustun B, Guo T, Wong GC, Socci ND, et al. Molecular characterization of pediatric gastrointestinal stromal tumors. Clin Cancer Res. 2008;14(10):3204–15.
79. West RB, Corless CL, Chen X, Rubin BP, Subramanian S, Montgomery K, et al. The novel marker, DOG1, is expressed ubiquitously in gastrointestinal stromal tumors irrespective of KIT or PDGFRA mutation status. Am J Pathol. 2004;165(1):107–13.
80. Price ND, Trent J, El-Naggar AK, Cogdell D, Taylor E, Hunt KK, et al. Highly accurate two-gene classifier for differentiating gastrointestinal stromal tumors and leiomyosarcomas. Proc Natl Acad Sci USA. 2007;104:3414–9.
81. Koon N, Schneider-Stock R, Sarlomo-Rikala M, Lasota J, Smolkin M, Petroni G, et al. Molecular targets for tumour progression in gastrointestinal stromal tumours. Gut. 2004;53(2):235–40.
82. Yamaguchi U, Nakayama R, Honda K, Ichikawa H, Hasegawa T, Shitashige M, et al. Distinct gene expression-defined classes of gastrointestinal stromal tumor. J Clin Oncol. 2008;26(25):4100–8.
83. Hur K, Lee HJ, Woo JH, Kim JH, Yang HK. Gene expression profiling of human gastrointestinal stromal tumors according to its malignant potential. Dig Dis Sci. 2010;55(9):2561–7.
84. Chibon F, Lagarde P, Salas S, Pérot G, Brouste V, Tirode F, et al. Validated prediction of clinical outcome in sarcomas and multiple types of cancer on the basis of a gene expression signature related to genome complexity. Nat Med. 2010;16:781–7.
85. Yang D, Ylipaa A, Yang J, Hunt K, Pollock R, Trent J, et al. An integrated study of aberrant gene copy number and gene expression in GIST and LMS. Technol Cancer Res Treat. 2010;9(2):171–8.
86. Arne G, Kristiansson E, Nerman O, Kindblom LG, Ahlman H, Nilsson B, et al. Expression profiling of GIST: CD133 is associated with KIT exon 11 mutations, gastric location, and poor prognosis. Int J Cancer. 2010;129:1149–61.
87. Schneider-Stock R, Boltze C, Lasota J, Miettinen M, Peters B, Pross M, et al. High prognostic value of p16INK4 alterations in gastrointestinal stromal tumors. J Clin Oncol. 2003;21:1688–97.
88. Wozniak A, Sciot R, Guillou L, Pauwels P, Wasag B, Stul M, et al. Array CGH analysis in primary gastrointestinal stromal tumors: cytogenetic profile correlates with anatomic site and tumor aggressiveness, irrespective of mutational status. Genes Chromosomes Cancer. 2007;46(3):261–76.
89. Legarde P, Pérot G, Kauffmann A, Céline B, Dapremont V, Hostein I, et al. Validated prediction of clinical outcome in gastrointestinal stromal tumors based on gene expression signatures and on genome complexity. Paris: CTOS Meeting; 2010.
90. Frolov A, Chahwan S, Ochs M, Arnoletti JP, Pan ZZ, Favorova O, et al. Response markers and the molecular mechanisms of action of Gleevec in gastrointestinal stromal tumors. Mol Cancer Ther. 2003;2(8):699–709.
91. Trent JC, Ramdas L, Dupart J, Hunt K, Macapinlac H, Taylor E, et al. Early effects of imatinib mesylate on the expression of insulin-like growth factor binding protein-3 and positron emission tomography in patients with gastrointestinal stromal tumor. Cancer. 2006;107(8):1898–908.
92. Mahadevan D, Cooke L, Riley C, Swart R, Simons B, Della Croce K, et al. A novel tyrosine kinase switch is a mechanism of imatinib resistance in gastrointestinal stromal tumors. Oncogene. 2007;26(27):3909–19.
93. Agaram NP, Besmer P, Wong GC, Guo T, Socci ND, Maki RG, et al. Pathologic and molecular heterogeneity in imatinib-stable or imatinib-responsive gastrointestinal stromal tumors. Clin Cancer Res. 2007;13(1):170–81.
94. Rink L, Skorobogatko Y, Kossenkov AV, Belinsky MG, Pajak T, Heinrich MC, et al. Gene expression signatures and response to imatinib mesylate in gastrointestinal stromal tumor. Mol Cancer Ther. 2009;8(8):2172–82.

Chapter 5
Pancreatic Cancer Genetics

Juan Iovanna, Ezequiel Luis Calvo, Jean Charles Dagorn, and Nelson Dusetti

Introduction

Cancer of the pancreas is the fourth leading cause of death by cancer in Western countries. In France, 7,200 persons die from this disease every year and more than 70,000 in Europe; in the USA, the annual death rate is about 30,000 [1]. With an incidence/death ratio of 0.99, pancreatic cancer is a crucial public health issue [2]. Although perioperative radiation therapy and chemotherapy can be slightly advantageous in terms of survival, they have no proven effect on prognosis [3]. Surgery offers the only possible cure if resection margins remain healthy. But only 10–15% of patients have localized resectable tumors at diagnosis [4] and most operated patients rapidly develop locoregional or metastatic disease progression [5]. Screening for pancreatic cancer, which could result in early diagnosis and thus increase the chances of curative treatment, is not currently available. There is no known biological or clinical screening test with a proven efficacy and the risk factors associated with the sporadic forms of the disease are unknown. Thus, this tumor has a very poor prognosis. A better understanding of the mechanisms of development of pancreatic cancer could probably contribute to the identification of new molecular targets and the improvement of diagnosis and treatment. A large body of work has been devoted to the identification of genetic alterations and their role in the development of pancreatic cancer. In this chapter, we focus on genetic anomalies associated with two well-recognized precancerous pancreatic lesions, pancreatic intraepithelial neoplasia (PanIN) and intraductal papillary mucinous neoplasm (IPMN). Both anomalies mainly involve alterations of oncogenes or tumor suppressor genes. Oncogenes are genes which favor the expression of malignant transformation. These genes are either mutated genes or mutation-free genes which are abnormally overexpressed. Tumor suppressor genes are genes which normally protect cells from degeneration. Their mutation or their inhibition favors the development of cancer cells. We first focus on gene expression changes occurring in pancreatic cancer tissues and on microRNA (miRNA) expression changes in pancreatic cancer cells. Then, we describe genes involved in the rare familiar cases of pancreatic adenocarcinoma. Finally, we describe all known genetic anomalies present in germ line DNA that increase susceptibility to pancreatic adenocarcinoma development.

J. Iovanna (✉) • J.C. Dagorn • N. Dusetti
INSERM U624, Marseille, France
e-mail: juan.iovanna@inserm.fr

E.L. Calvo
Centre de Recherche du CHUL (CHUQ), Quebec, PQ, Canada

A. Russo et al. (eds.), *Diagnostic, Prognostic and Therapeutic Value of Gene Signatures*, Current Clinical Pathology, DOI 10.1007/978-1-61779-358-5_5, © Springer Science+Business Media, LLC 2012

Genetics of PanIN and IPMN Precancerous Lesions

Pancreatic Intraepithelial Neoplasia

Although precursor lesions of pancreatic cancer were already well documented more than a century ago, a rigorous classification of these lesions was not obtained until the last decade. Morphological analyses of resected pancreatic cancers suggest that pancreatic adenocarcinomas do not develop de novo but are a result of a stepwise progression leading to the generation of invasive lesions [6–8]. In the late 1990s, a plethora of often biologically imprecise terms were used to describe these lesions. In 2001, an international agreement [9] was reached on the nomenclature used for a vast spectrum of lesions ranging from low-grade lesions (PanIN-1) to in situ carcinomas (PanIN-3), considered as the stage preceding invasion of the neighboring stroma. A detailed description of the histological features of PanINs is beyond the scope of this chapter. However, it should be noted that the histological progression of PanINs is (with a few exceptions) associated with progressive accumulation of the same molecular anomalies than those observed in invasive cancer [10]. These molecular alterations have been helpful in demonstrating that PanINs are clonal precursors of pancreatic adenocarcinoma. Although the exact natural history of PanINs has been difficult to establish, these lesions are thought to exist well before the appearance of the adenocarcinoma. Because of the poor prognosis of pancreatic cancer, it is important to obtain a genetic characterization of the pancreatic tissue harboring PanIN lesions, and in particular PanIN-3 lesions, before they transform into malignant lesions. The identification of molecular anomalies specific to PanINs would enable early screening of the pancreatic transformation process and might provide important information for selecting new therapeutic targets. While animal models have not yet been developed for IPMNs, several genetically modified mouse models of PanINs are currently available. These models have shown that PanIN lesions develop into pancreatic adenocarcinomas. The process has always involved expression of the mutated *KRAS* gene in pancreas [11–13]. Expression of the mutated *KRAS* gene induces the formation of PanINs but not their malignant transformation, which is only obtained after loss of activity of a tumor suppressor gene (*CDKN2A/p16*, *SMAD4* or other molecules participating in the *TGF beta* and *TP53* signaling pathways) [14–20] or in the presence of chronic pancreatic inflammation [21].

Genetic Alterations in PanIN

The characterization of genetic alterations observed in PanINs has progressed further than that for IPMN.

Oncogenes

ERBB2 encodes a tyrosine kinase growth factor receptor, whose activation induces cell proliferation. The ERBB2 protein is thus considered as a powerful oncogene. Its overexpression is one of the earliest events in the oncogenic process in the pancreas. It is detected in 82% of PanIN-1A lesions and in 100% of PanIN-3 lesions [22]. The *KRAS* oncogene is activated in approximately 90% of all pancreatic adenocarcinomas. These mutations affect codons 12 (the most common mutation), 13, and 61 [23]. The mutated KRAS protein facilitates progression along the cell cycle via activation of

MAP kinases and AKT kinase [24]. These *KRAS* gene mutations also appear very early in pancreatic cancer. They are found in 36% of PanIN-1A lesions, 44% of PanIN-1B and PanIN-2 lesions, and 87% of PanIN-3 lesions [25]. It is important to note that PanIN lesions carrying different mutations can coexist in the same organ; these mutations belonging to different clones [26]. Nevertheless, only one of the clones progresses toward an invasive form, the clone with the mutations always found in pancreatic adenocarcinomas.

Tumor Suppressor Genes

A certain number of tumor suppressor genes are inactivated in PanINs. They are also inactivated in adenocarcinomas. The *CDKN2A/p16* gene encodes a protein which binds to cyclin-dependent kinases Cdk4 and Cdk6, thus inhibiting their linkage with cyclin D1 and stopping the cell cycle at G1/S [27]. This loss of activity, which is observed in more than 90% of pancreatic adenocarcinomas, can occur through at least three mechanisms (a) homozygous deletion, (b) mutation of one allele plus deletion on the second, or (c) hypermethylation of its promoter resulting in nearly total suppression of its expression [28–30]. Loss of CDKN2A/p16 expression is also observed in PanIN lesions with an incidence of 30% in PanIN-1A and PanIN-1B, 55% in PanIN-2 and more than 70% in PanIN-3 [31]. *TP53* tumor gene suppressor is inactivated in more than 50% of pancreatic adenocarcinomas. The mechanism of inactivation is often associated with the loss of one allele and an inactivating mutation of the other [32]. Inactivation of *TP53* is rarely observed in PanIN-1A or PanIN-1B lesions, indicating that the loss of TP53 function occurs late in the tumorigenesis process [33]. *SMAD4* tumor suppressor gene is frequently inactivated in pancreatic adenocarcinomas [34]. The SMAD4 protein is implicated in the signaling cascade of TGF-b; its inactivation blocks the inhibitor effect of TGF-b and other members of its family on cell growth, thus allowing uncontrolled growth of tumor cells [35]. SMAD4 expression is preserved in PanIN-1 and PanIN-2 lesions, but lost in 40% of PanIN-3 lesions [36].

Other Anomalies Observed

Loss of telomere integrity in the ductal epithelium is probably the cause of the genomic instability observed in PanIN [37]. Telomeres are TTAGGG repeats that cap the ends of chromosomes, inducing a certain degree of stability during cell division [38]. Telomere shortening is one of the first genetic aberrations detected in PanIN lesions, occurring in more than 90% [37]. Intact telomeres act like “guardians” of the pancreatic ductal genome; their shortening in PanIN leads to progressive accumulation of chromosomal anomalies and finally malignant transformation. In addition, other proteins are overexpressed during the stepwise progression toward PanIN; these proteins could thus be interesting markers of pancreatic carcinogenesis. Expression of protein Ki-67 is associated with cell proliferation. Ki-67 is expressed more frequently in the nuclei of high-grade PanIN cells (22% in PanIN-3) than in early forms (0.7% in PanIN-1A) [33]. The index is 35–40% in adenocarcinomas [39]. Expression of topoisomerase II, necessary for DNA relaxation before replication, follows that of protein Ki-67 [33]. Cyclin D1, which is a key element of the cell cycle, is overexpressed in 80% of pancreatic adenocarcinomas, 50% of PanIN-3, and 30% of PanIN-2 [40]. The expression of these markers is not due to a genomic or epigenetic event. It is simply the consequence of accelerated cell proliferation due to the malignant transformation at an intensity directly related to the transformation.

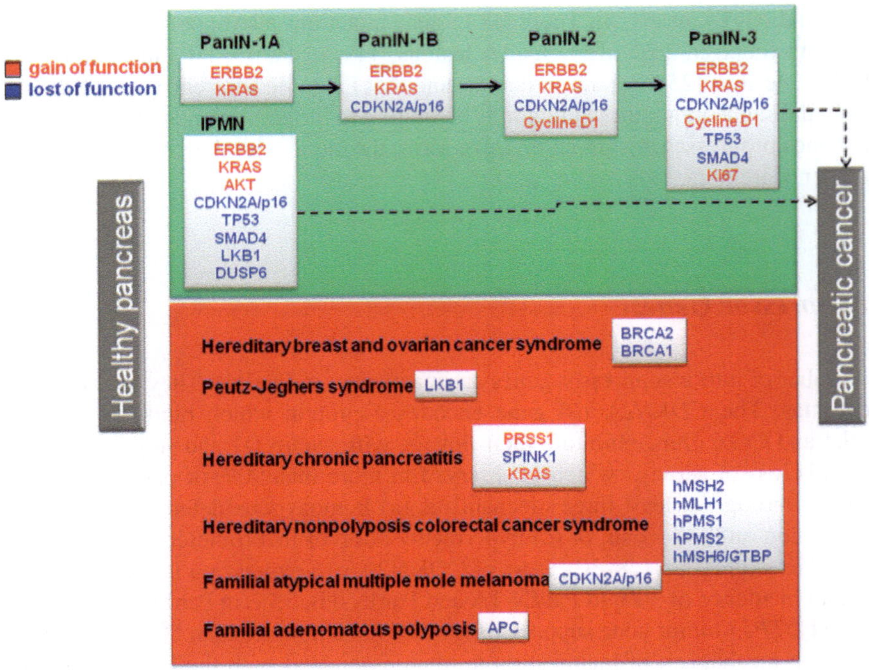

Fig. 5.1 Genetic alterations in pancreas cancer. At the *upper part* are shown the genetic alterations present in the precancerous lesions PanIN and IPMN, and at the *lower part* are showed the genes implicated in the familiar forms of pancreas cancer. Gain of function is represented in *red* and loss of function is represented in *blue*

PanIN and Accumulation of Genetic Anomalies

Progression from PanIN-1A to PanIN-3 then to pancreatic adenocarcinoma is associated with progressive accumulation of genomic alterations. This model of progression has been described by several laboratories [33, 41, 42] and appears to be well established, with new characterizations of new anomalies in precancerous lesions regularly published. Activation of the *KRAS* oncogene and amplification of ERBB2 are the earliest genetic alterations observed in PanIN-1A. Telomere shortening is also observed at an early stage. Loss of CDKN2A/p16 activity comes slightly later in stages PanIN-1B and PanIN-2. Cyclin D1 overexpression is seen in PanIN-2. The latest events, observed in PanIN-3, are inactivation of *TP53* and *SMAD4* and expression of Ki-67 (Fig. 5.1).

Intraductal Papillary Mucinous Neoplasm

IPMN is a relatively uncommon lesion. The macroscopic appearance is a dilated pancreatic duct filled with mucus; there is a strong potential for degeneration [43]. IPMN can also be seen as a small cyst in the lumen of a secondary pancreatic duct or as a set of large multicystic lesions involving the main pancreatic duct and several secondary ducts. Malignant transformation is mainly seen in this second type [44]. IPMNs occur more readily in older men and less frequently in women [45]. Discovery is generally fortuitous; sometimes made because of acute obstructive pancreatitis resulting from mucus-filled ducts [43].

Microscopically, IPMNs are composed of a mucin-producing ductal type epithelium. IPMNs are classed by architecture into four types: gastric, intestinal, pancreaticobiliary, and oncocytic [46]. Ductal cells show diverse degrees of atypia ranging from low-grade to high-grade lesions, the latter corresponding to in situ carcinoma. In 1996, the World Health Organization defined four groups of IPMN as a function of cell atypia: adenomas, borderline tumors, in situ carcinomas, and invasive tumors. Disease prognosis depends upon the presence of invasive carcinoma at diagnosis. Unfortunately, invasive carcinoma progresses much like classic adenocarcinoma.

Genetic Alterations in IPMN

Three oncogenes (*KRAS*, *ERBB2*, and *AKT*) and five tumor suppressor genes (*CDKN2A/p16*, *TP53*, *SMAD4*, *LKB1*, and *DUSP6*) are involved.

Oncogenes

The oncogene *KRAS* is activated by mutation in approximately 72% of IPMN lesions [47]. This mutation has been identified in low-grade and high-grade lesions. Although it is not particularly specific, this mutation appears to be necessary for the development of IPMN. It is interesting to note that different mutations which activate *KRAS* are sometimes observed in the same patient at different IPMN foci. There could be several explanations for this, including polyclonal development of IPMN and *KRAS* mutation secondary to the development of the lesions. Yoshizawa and colleagues have suggested that low-grade forms arise via a polyclonal mechanism while high-grade forms develop from low-grade lesions in a clonal manner [48]. Overexpression of ERBB2, which is a common and early event in pancreatic cancer as mentioned above, has been observed in 63% of IPMN lesions [49]. AKT is a protein kinase involved in KRAS signaling. It plays a key role in cell growth and survival. A recent study showed that the activated (phosphorylated) form of this kinase is found in 63% of lesions (versus 70% in adenocarcinomas). In addition, AKT activation is slightly more frequent in high-grade than in low-grade forms [50].

Tumor Suppressor Genes

As mentioned above, *CDKN2A/p16* plays a major role in cell cycle arrest. *CDKN2A/p16* expression is lost in about half of all IPMN lesions [51]. Such diminished expression results in most cases from hypermethylation of its promoter [52]. The tumor suppressor gene *TP53* is a transcription factor which induces the expression of several genes involved in cell cycle arrest, DNA repair, and apoptosis induced by DNA damage. TP53 activation is lost in half of human tumors, either by deletion or mutation of its gene, or by proteosomal hyperdegradation. Loss of TP53 activity is observed in half of IPMNs with a higher frequency in high-grade lesions [46, 47, 53]. The current hypothesis is that loss of TP53 activity in this type of lesion induces loss of genome integrity which in turn leads to malignant transformation. Chromosome region 18q21.1 frequently exhibits deletion of its two alleles in pancreatic adenocarcinomas [54–56]. The tumor suppressor *SMAD4* is located in this region. Nevertheless, homozygous mutations of *SMAD4* are rare in IPMN. Moreover, expression of the SMAD4 protein is generally preserved in these lesions, independent of the degree of atypicity [57]. When IPMN transformation involves the SMAD4 pathway, loss of SMAD4 function is a very late event. The *LKB1* gene is responsible for Peutz-Jeghers syndrome (transmitted by autosomal dominant inheritance, this syndrome associates periorificial lentiginosis and intestinal polyposis; the development of several types of cancer, including pancreatic cancer can be observed during the course of the disease). The role of *LKB1* involves preservation of cell polarity [58]. Mutation of *LKB1* has been identified in 25% of IPMN lesions in patients without Peutz-Jeghers syndrome and also in a few cases of pancreatic adenocarcinomas [59]. *LKB1* might thus play a role in the development of IPMNs in some patients. Finally, the expression of DUSP6, a phosphatase which interacts with the kinase MAPK1 regulating its activity, appears to be lost or greatly diminished in a few IPMNs [52].

Differences Between PanINs and IPMNS

PanINs and IPMNs are both intraductal lesions with a potential for malignant degeneration. They are nevertheless very different lesions. The most important difference is their size. PanINs are microscopic lesions while IPMNs are macroscopic. At the present time, PanINs cannot be detected by routine endoscopic or radiological methods. They are discovered fortuitously during histological analysis of surgical specimens. Immunoreactivity can also distinguish between the two types of lesions. MUC2 is found in most IPMNs but never in PanINs. SMAD4 expression, often lost in PanIN-3, is almost systematically preserved in both degenerated and nondegenerated IPMNs [57]. DUSP6 expression is preserved in PanINs but strongly reduced in most IPMNs [46]. These differences suggest that the developments of these two precancerous lesions occur through distinct molecular mechanisms (Fig. 5.1).

Hence, PanINs and IPMNs are different precancerous lesions. For the time being, the diagnosis of PanIN is based on pathological analysis of a surgical specimen. A set of genetic markers of PanINs is needed to obtain a simple, specific, and early diagnosis so that prophylactic measures can be taken, particularly in the context of familial pancreatitis. Although the diagnosis of IPMN is more and more early, the decision for surgery is still a major challenge. If genomic or proteomic markers of IPMNs could be detected in endoscopic ultrasound fine needle aspirations, patients at high risk of malignant degeneration could be identified soon enough that the most appropriate preventive intervention could be proposed. These observations illustrate the importance of obtaining a better knowledge of the genetic alterations in PanINs, which could be useful for familial screening. If the genetic alterations of IPMNs were better understood, patients could be selected for close monitoring or surgery. It is also important to note that most genetic anomalies observed in PanINs and IPMNs do not differ from those observed in established pancreatic adenocarcinomas, except for major chromosomal alterations (duplications, deletions, rearrangements) which are only present in adenocarcinomas. These anomalies may result from anomalous mitosis, with poor redistribution of the genetic material between the two daughter cells, which most probably occurs in advanced adenocarcinoma. This suggests that genetic alterations are necessary for the malignant transformation of precancerous lesions but are not sufficient for progression to cancer. Other factors are required. Characterizing these factors, expected to be similar to other predisposing factors for cancer (chronic inflammation, cigarette smoking, certain diets, etc.), could result in better prophylaxis for pancreatic cancer.

Gene Expression Changes Occurring in Pancreatic Cancer

Since the development of the first microarray systems in 1995 [60], their use was widespread in cancer research in general and especially in pancreatic cancer. Pan-transcription profiling of pancreatic cancer tissues were conducted with different types of platforms and approaches, such as microarrays,

SAGE, and more recently by exhaustive exome sequencing using second-generation sequencing analysis. Results have suggested that, unlike other types of tumors whose appearance is triggered by altering a single oncogene, as in certain types of leukemia, in pancreatic cancer development results from genetic alteration of a large number of genes although these genes are involved in a limited number of specific pathways and processes (see above). They also suggest that the number of paths or processes involved increases during progression of the lesion.

The numerous studies on pancreatic cancer involving a critical postgenomic vision can be classified into three main categories (1) studies using a single platform type; (2) studies that combined data from two or more types of platforms; and (3) studies that analyzed data already available in public databases, or studies in silico.

IGH-Throughput Genomic Studies

The capacity to analyze the entire transcriptome of tumors has given researchers a powerful tool to better understand the pathophysiology of pancreatic cancer. Soon, these tools have enabled the exploration of diagnostic and therapeutic targets. Thus, overall, genomic studies can be divided according to their objectives in three main categories: (a) identification of specific genes for diagnostic, prognostic, or therapeutic purposes; (b) decoding of complex patterns of gene expression to better understand the pathophysiology of the disease; and (c) use of the genetic profiling to identify new subtypes or classes of diseases that are not detectable by routine clinical procedures. These three groups are also referred to as candidate gene selection, class prediction, and class discovery, respectively [61]. Several postgenomic studies have been performed to search for new prognostic markers or diagnostic tools of pancreatic cancer (a comprehensive compilation of these studies is given in papers [62–64]). The most recent of these studies [64], analyzed the expression profiles of 44,000 genes in 34 primary tumors and their metastases, and selected a small set of six genes (FOSB, KLF6, NFKBIZ, ATP4A, GSG1, and SIGLEC11) whose expression seems to predict metastatic potential. The expression pattern of these genes used as a "training set," was subsequently validated on a "test set" of 67 new tumors. According to the results, this "gene signature" would select pancreas cancer patients with high or low survival potential. The first group had a survival at 1 year of 55%, compared to 91% for those classified as low risk. This detection system appears to be more sensitive than the current TNM classification and could be particularly useful for targeting more aggressive therapeutic measures.

Studies Using Several Types of Platforms Simultaneously

Because of the vast genetic heterogeneity of pancreatic cancer [65], genetic alterations may be very different among pancreatic tumors. In fact, a finding common in publications of global gene expression analysis is that gene lists reported vary greatly between studies and even between patients within a group. In this context, a remarkable study provides some interesting answers. This study analyzed 24 human pancreatic cancers by performing simultaneously exome sequencing, transcriptional analysis, and quantification of amplifications and deletions by microarray [66]. Coding regions of 20,661 genes were sequenced in the tumors and respective normal tissues of 24 individual presenting with advanced pancreatic adenocarcinoma. Analysis of exomes has detected 1,562 changes in tumor DNA compared to DNA from normal tissue ("germline sequence variations"). More than 60% of these mutations produce a change in amino acids (missense mutation). Of 924 mutations described, 55 affected the function of genes carrying the mutation, and 160 had already been associated with tumorigenesis in other studies.

Moreover, the authors also investigated the variations in the copy number of genes and their expression levels in the genomes of 24 individuals. This high-throughput method identified 144 amplifications associated with overexpression of the respective genes, and 198 homozygous deletions of DNA fragments with an average size of 335,000 base pairs. These structural changes are followed by changes in expression levels of 541 genes, grouped in 69 sets, and present in 90% of the 24 tumors. Among these genes, 54 encode secretory proteins or cell surface proteins, and their expression level is more than 10 times overexpressed compared to normal ductal cells. This could provide new targets for diagnosis and treatment of pancreatic cancer. More interestingly, the colossal amount of information obtained by combining these three types of technology, demonstrated that 12 signaling pathways and cellular processes were altered in 67% of pancreatic cancer (apoptosis, DNA damage, regulation of G1/S phase transition, hedgehog signaling, homophilic cell adhesion, integrin signaling, c-Jun N-terminal kinase signaling, KRAS signaling, regulation of invasion, small GTPase-dependent signaling other than KRAS, TGF-b signaling, and Wnt/Notch signaling). The key to identifying these 12 pathways was to assume that any alteration of genes involved in the same pathway or process results in similar tumor effects, rather than considering the function of each gene individually. In fact, it is important to say that the identified genes are not common to all tumors, but the experimental information currently available allows them to alter the pathway. However, the key to interpreting data from pancreatic cancer is integration and reduction. Integrating genomics, epigenetics, the transcriptome and proteomics studies, and reducing the mass of information on functional, metabolic and ontologic pathways.

In a second study from the same team, authors studied the germ line exome from an individual with a hereditary pancreatic cancer [67]. After identification of the candidate gene, they sequenced the same region in 96 individuals with hereditary pancreatic cancer. They could show that a 4 bp deletion within the gene "Partner and Localizer of BRCA2" or PALB2 was present in the germ line DNA of all these individuals. This discovery makes PALB2 the second most frequently mutated gene in hereditary pancreatic cancer after BRCA2.

In spite of the limitation that the exome represents only 1% of the genome, the two works described above clearly demonstrate the potential of this approach in studying pancreatic cancer by identifying genes responsible for triggering sporadic and hereditary pancreatic cancer.

In Silico Studies

Recently, efforts have been made to integrate the wealth of data produced by high-throughput technologies and undertake systematic validation of most frequently detected molecules, with the aim to discover their potential clinical utility [62, 68, 69]. In this line, results from high-throughput studies at the genomic, epigenomic, transcriptomic, and proteomic level in pancreatic cancer were reanalyzed by Ranganathan and colleagues. A subset of ten molecules frequently overexpressed in pancreatic tumors was identified (CEACAM5, ITGA2, MMP11, MSLN, MUC5AC, S100A11, S100P, SPARC, TFF2, and SFN). Almost all of these genes are involved in cell cycle, cell growth and proliferation, and represent potential markers for diagnosis and prognosis (see Table 5.1).

A similar study was specifically directed to the identification and validation of specific markers of pancreatic tumors [68]. It showed that more than 441 genes overexpressed at the mRNA and protein level, measured and validated by different methods are reported in at least four different publications. Antibodies against 60 of these genes, all coding for proteins associated with the cell membrane, are being tested as potential tumor markers, and possibly as therapeutic tools.

Databases to compile different types of "omics" for cancer of the pancreas are available on the Web owing to international efforts (http://www.pancreasexpression.org) [70], which allows integrated data mining. It is hoped that similar efforts are made on other pancreatic diseases and on other types of cancer, which would allow working on more complex queries.

Table 5.1 Ten molecules frequently overexpressed in pancreatic tumors

Symbol	Gene name
CEACAM5	Carcinoembryonic antigen-related cell adhesion molecule 5
ITGA2	Integrin, alpha 2 (CD49B, alpha 2 subunit of VLA-2 receptor)
MMP11	Matrix metallopeptidase 11 (stromelysin 3)
MSLN	Mesothelin
MUC5AC	Mucin 5AC, oligomeric mucus/gel-forming
S100A11	S100 calcium binding protein A11
S100P	S100 calcium binding protein P
SPARC	Secreted protein, acidic, cysteine-rich (osteonectin)
TFF2	Trefoil factor 2
SFN	Stratifin

miRNA Expression Changes Occurring in Pancreatic Cancer

Structure, Genomic Organization, and Biosynthesis of miRNAS

miRNAs are a particular class of noncoding RNA molecules implicated in various gene silencing pathways. Their size ranges from 19 to 24 nucleotides and they are expressed in a large variety of organisms. The first miRNA described in 1993 was lin-4. It was shown to be a key regulator in the developmental timing of *C. elegans* [71]. Lin-4 was shown to interact with the 3¢ untranslated region (3¢UTR) of the lin-14 mRNA and to repress its expression [72]. Today, we know that miRNAs regulate their targets by direct mRNA cleavage or translational inhibition. miRNAs are coded by genes and are transcribed by RNA polymerase II (pol II). They have their own regulatory elements and appear as transcriptional units containing either unique (e.g., miR-21) or multiple miRNAs (e.g., the miR-17-92-1 cluster) [73, 74]. The primary miRNA transcript (pre-miRNA) contains a cap structure and a poly(A) tail. It is initially transcribed as part of a much longer primary transcript (>1 kb). miRNAs represents 1–3% of the human genome [75–77], controlling about 20–30% of protein-coding genes in the human genome. Most of the miRNAs (70%) are located in introns and/or exons (miR-10b, miR-33, and miR-198), and the remaining 30% are situated in intergenic regions [78, 79]. miRNA primary transcripts are cleaved in the nucleus by the enzyme Drosha before being exported to the cytoplasm [80, 81]. They are further processed by the enzyme Dicer, resulting in a mature duplex containing 19–24 base pairs, one strand of which is incorporated into an effector complex called RNA-Induced Silencing Complex (RISC) [82, 83], whereas the opposite is eliminated [84, 85]. Animal miRNAs are often only partially complementary to the 3¢UTR of their target sites and the level of base pairing conditions the efficacy of sequence-specific cleavage by RISC. The miRNA–mRNA interaction is restricted to the miRNA 5¢ end sequence. Strong complementarity between nucleotides 2–8 of the miRNA ("seed" sequence) and the 3¢UTR of the targeted mRNA is essential for recognition [86]. Translational regulation by miRNAs of a given mRNA is also a function of the number of target sites of these miRNAs into the 3¢UTR of the mRNA [87].

When miRNA and mRNA interact, three scenarios are possible (1) the target can be cleaved; (2) the translation of the target can be inhibited by the miRNA at the initiation step or at a postinitiation step or (3) the miRNA can lead to the deadenylation or decapping of the mRNA [88]. In consequence, the level of target mRNA in the cell can remain constant, be increased or diminished, the protein level decreasing in all cases. On a structural standpoint, miRNAs are grouped into several families that share common seed sequences and have similar target specificity. miRNAs expressed in the same polycistron do not belong to the same structural family but frequently act in synergy to regulate the expression of functionally related genes.

miRNA Expression Profiling in Pancreatic Ductal Adenocarcinoma

Pancreatic cancer is characterized by multiple genetic alterations associated with the different steps in pancreatic cancer progression, from the noninvasive precursor lesions to pancreatic ductal adenocarcinoma (PDAC) [89]. Understanding decisive shifts in the behavior of early lesions and how they relate to underlying molecular alterations indicative of the disease stage would help designing new markers and therapeutic targets that could have a profound impact on clinical practice and patient outcome [90, 91]. The problem with pancreatic cancer is that specific markers allowing its detection are scarce. Analysis of miRNA expression in PDAC showed that it presented with a specific miRNA signature that can be monitored by profiling miRNAs at different stages of cancer. The expression pattern of miRNAs seems to be a better way to identify the cancer type than mRNA expression patterns [92]. Over the past years, a number of different approaches, including DNA microchips and RT-PCR have been described to quantify miRNAs. By these techniques, several studies demonstrated the tissue-specificity of miRNA expression and deregulation of miRNA expression in pancreatic cancer [93, 94]. Monitoring differential miRNA expression might therefore be useful in the differential diagnosis of pancreatic cancer from other tumors. Expression of the miR-376 precursor was highest in the human pancreatic cancer cell line Panc-1, compared to other cell lines studied [95, 96]. Large-scale miRNA profiling in 540 samples of solid tumors (breast, colon, lung, pancreas, prostate, and stomach) showed that the spectrum of miRNA expression varied in different solid tumors and was different from that of normal cells (43 of 137 miRNAs, 31%). miR-21, miR-191, and miR-17-5p were significantly overexpressed in all six tumor types, whereas miR-218-2 was consistently downregulated in colon, stomach, prostate, and pancreatic cancers, but not in lung and breast carcinomas. This observation indicated that colon, pancreas, prostate, and stomach have similar miRNA signatures, different from those of breast and lung cancer. Similarly, upregulation of miR-142-3p, miR-142-5p, miR-155, and miR-146a expressions was observed in human pancreatic neuroendocrine tumors (PNETs) as compared with normal human islets [97]. Lee and colleagues reported the aberrant expression of one hundred miRNA precursors in pancreatic cancer or desmoplasia, including the miRNAs previously reported in other human cancers (miR-155, miR-21, miR-221, and miR-222) as well as the first reported miR-376a and miR-301 for the differential expression of cancer [98]. A significant upregulation of miR-196a, miR-190, miR-186, miR-221, miR-222, miR-200b, miR-15b, and miR-95 in most pancreatic cancer tissues and cell lines was reported. miR-155 and miR-21 were significantly upregulated in 15 IPMNs vs. matched controls [99]. The expression pattern of miR-155 in 53 of 64 (83%) IPMNs compared to 4 of 54 (7%) normal ducts only, and that of miR-21 in 52 of 64 (81%) IPMNs compared to 1 of 54 (2%) normal ducts, respectively. The expression of miR-216 has also shown to be specific to the pancreas [100]. Let-7 miRNA was found expressed in pancreatic cancer cells but did not inhibit the epithelial–mesenchymal transition (EMT) [101]. Therefore, miRNAs play vital roles not only in different kinds and stages of pancreatic tumors, but also in many diseases, including cancers, cardiovascular diseases, and immune disorders. Epigenetic modifications, DNA copy number changes and genetic mutations may regulate the expression of miRNAs. The conclusion of these data is that PDAC miRNAomes associated with normal and tumor tissues are different. Differences are tumor-specific and, in some cases, indicators of prognosis. These findings suggest that miRNA expression patterns constitute a signature of the disease which could offer new clues about pancreatic cancer occurrence and also provide new molecular markers that would improve diagnosis and orient the treatment. A very promising diagnostic strategy could arise from miRNAs if they are found in serum and can be detected by RT-PCR. In a remarkable study Lu and colleagues [92] showed that expression data for 217 miRNAs only performed better at identifying cancer types than analysis of 16,000 mRNAs. They concluded that miRNAs might help detecting cancer better than other strategies presently available because miRNAs are only several hundred, compared to tens of thousands for mRNAs and proteins. One single miRNA can modulate the expression of many genes rendering miRNAs powerful

Table 5.2 MicroRNAs frequently misregulated in pancreatic cancers

Up	Down
miR-155	miR-375
miR-100	miR-345
miR-376a	miR-142-P
miR-125b-1	miR-139
miR-181a	miR-148a
miR-181c	miR-148b
miR-146a	miR-141
miR-196a	miR-96
miR-25	miR-29c
miR-214	miR-130b
miR-222	miR-216
miR-29b-2	miR-217
miR-128b	miR-107
miR-200	miR-34a
miR-95	Let-7
miR-15b	miR-218-2
miR-32	
miR-30c	
miR-21	
miR-17-92	
miR-191	
miR221	
miR190	
miR186	

molecules regulating several related pathways, each miRNA playing multiple but coherent roles in the cell. Also, miRNAs are shorter than mRNAs, and therefore more resistant to ribonuclease degradation. miRNAs can remain intact in formalin-fixed, paraffin-embedded (FFPE) tissues or in serum. Pilot studies using endoscopic ultrasound-guided fine needle aspiration and FFPE specimen help evaluating the diagnostic value of candidate biomarker miRNAs. Finally, miRNA analysis requires no expensive and time-consuming detection strategies using antibodies or mass spectrometry. Simple, extremely sensitive methods, such as PCR, in situ hybridization, or real-time PCR, are sufficient for their detection. A summary of up- and downregulated miRNAs in pancreas cancer tissue is presented in Table 5.2.

miRNAS in the Molecular Mechanisms Associated with PDAC

Recent studies have shown deregulation of many miRNAs during pancreatic cancer and their associated targets as well as underlying molecular mechanisms now begin to be elucidated.

miR-34a: miR-34a expression was assessed in pancreatic cancer cells which frequently exhibit p53 loss of function. Two nontransformed pancreatic ductal epithelial cell lines and 15 pancreatic cancer cell lines were analyzed by northern blotting. miR-34a was highly expressed in nontransformed pancreatic ductal epithelial cell lines, showing that this miRNA is normally expressed in this cell type. By comparison, all 15 pancreatic cancer cell lines presented at least a twofold reduction in miR-34a expression. Eleven of the 15 cell lines exhibited a tenfold reduction or complete absence of this miRNA. Although p53 loss is expected to reduce miR-34a expression, it could hardly account for the reduced miR-34a expression observed in all pancreatic cancer cell lines. In fact, there is no direct correlation between biallelic loss of p53 and the magnitude of miR-34a downregulation, and cell lines with wild-type p53 status also exhibit low levels of miR-34a. It is therefore likely that other mechanisms, in addition to p53 inactivation, contribute to the reduction in miR-34a abundance. In fact, miR-34a expression was absent from several types of tumors due to the aberrant CpG methylation of its promoter [102]. Nineteen out of 24 primary prostate carcinomas displayed CpG methylation of the promoter sequence of the miR-34a gene, which resulted in the loss of its expression. CpG methylation of the miR-34a promoter was also detected in breast, lung, colon, kidney, bladder, pancreatic carcinoma cell lines, melanoma cell lines, and primary melanoma samples. After DNA damage, silencing of miR-34a was dominant over its transactivation by p53. In summary, miR-34a plays an important role in modulation and fine-tuning of gene expression initiated by p53. Lack of transcriptional transactivation by p53, deletion or CpG methylation should contribute to the decreased expression of miR-34a that acts as a tumor suppressor gene in pancreatic cancer.

miR-155: Another interesting study concerns the association between miR-155 and TP53INP1 (Tumor Protein 53-Induced Nuclear Protein 1). miR-155 is known as oncogenic (oncomir) and it is overexpressed in pancreatic cancer as well as in other tumors. The stress-induced gene TP53INP1 is downregulated in pancreatic cancer [103]. This gene is transcriptionally induced by p53 and regulates p53 activity [104–106]. TP53INP1 is also implicated in colorectal cancer since TP53INP1-deficient mice show high susceptibility to colorectal cancer induction [107]. Moreover, TP53INP1 is downregulated in other tumors, such as breast and gastric cancer, and was proposed as a tumor suppressor gene [108, 109]. In our laboratory, we have demonstrated that the TP53INP1 protein disappears at a precancerous step during PDAC development, whereas its mRNA expression persists in tumors. In fact, we have shown that TP53INP1 loss is due to the activity of the oncogenic miR-155 which is overexpressed in pancreatic cancer cells. Finally, we identified the target site for miR-155 into the 3¢UTR of TP53INP1 and confirmed its functionality in vitro and in vivo. These results are very exciting because TP53INP1 is the first miR-155 target gene with antitumoral activity which could account for the oncogenic potential of miR-155.

Let-7: Let-7 is downregulated in PDAC samples, as compared to adjacent tissue and its complete loss in poorly differentiated cancer samples. Overexpression of let-7 in pancreatic cancer-derived cell lines strongly inhibits cell proliferation, K-ras expression, and mitogen-activated protein kinase activation [110].

miR-21: miR-21 was found significantly upregulated in pancreatic cancer [99, 111] and targets the phosphatase and tensin homologue 2 (PTEN), programmed cell death 4 (PDCD4), trophomyosin 1 (TPM1), and tissue inhibitor of metalloproteinases 3 (TIMP3) leading to inhibition of apoptosis and therefore to an increased tumorogenicity [112]. Overexpression of miR-21 precursor in pancreatic cancer cells showed increased proliferation, invasion, and chemoresistance to gemcitabine compared with control cells, and the reverse was observed when the mir-21 was knocked down in pancreatic cancer cells [113].

miR-17-92: The miR-17-92 cluster is upregulated in lymphomas and breast, lung, colon, stomach, and pancreatic cancers. This miRNA cluster targets E2F1, BIM, and PTEN leading to increased proliferation of cancer cells [100, 114, 115].

miR-107: Ectopic expression of miR-107 in pancreatic cell lines MiaPACA-2 and Panc-1 results in decreased growth rate due to downregulation of cyclin-dependent kinase 6 by miR-107 [116].

miRNA-200: Transcriptional suppression of miRNA-200 family members, such as miR-141 and miR-200c by zinc-finger E-box binding homeobox 1 (ZEB1) [117] strongly activate epithelial differentiation in pancreatic, colorectal, and breast cancer cells by targeting EMT activators, such as transforming growth factor beta 2 and ZEB1. These results indicate that ZEB1 triggers miRNA-mediated stabilization of EMT and promotes the invasion of cancer cells. Questions regarding the association of miRNAs and their role in cancer development remain unresolved but it is suggested that a set of miRNAs might be associated with pancreatic tumorigenesis by acting as tumor suppressors or oncogenes.

Targeting miRNA Expression in PDAC as a Potential Therapeutic Strategy

When miRNA overexpression is associated with PDAC progression, synthetic, chemically modified antisense oligonucleotides targeting mature miRNAs or their precursors might be considered for therapy. Several modifications of anti-miRNA oligonucleotide structure have been designed to stabilize the molecules and protect them from degradation [118]. They include 2′-*O*-methyl, 2′-*O*-methoxyethyl, cholesterol conjugated (antagomirs) or locked nucleic acid (LNA) oligonucleotides. Recent publications illustrate the application of these inhibitors. Cheng and colleagues [119] used a library of modified anti-miRNA oligonucleotides to successfully identify miRNAs involved in the control of cell growth and apoptosis in various cell lines. Chan and colleagues used 2′-*O*-methyl and DNA/LNA-mixed oligonucleotides to specifically knock down an miRNA and demonstrate that its aberrant expression contributes to the malignant phenotype of glioblastoma cells [120]. In a more recent publication, Krutzfeldt and colleagues have shown that antagomirs inhibit miRNA function in vivo [121]. The authors observed that silencing this single miRNA results in an increased expression of several hundred genes, many of them containing in their 3′ UTRs the corresponding miRNA recognition sequences. However, about 300 genes were downregulated in response to this antagomir. This led the authors to speculate that the mechanism related to such downregulation could reflect the suppression of a transcriptional repressor. This example illustrates the complexity of the miRNA regulation pathways. Furthermore, these works indicate that therapeutic adjustment of miRNA overexpression may become feasible but the precise mechanisms by which modified oligonucleotides cause the depletion of targeted miRNAs remains unknown. Data presently available are not conclusive but provide some information. For example, when high levels of miRNA and anti-miRNA duplexes are present in the cell, degradation products of these duplexes can be detected [121]. This observation is in agreement with previous works showing that gene-silencing mechanisms are triggered by short double-stranded RNA species (a process known as RNA interference). Thus, the simplest explanation for the effectiveness of the anti-miRNA is its hypothetical ability to bind and promote rapid degradation of the target miRNAs by nucleases normally present in the cells. Treatment of mice with anti-miRNA was shown to selectively remove the targeted miRNA but not other miRNA species. Furthermore, another interesting observation is that antagomir injection leads to broad and sustained distribution (up to 23 days), allowing effective and long-lasting silencing of the targeted miRNA in most tissues.

Conversely, if the function of underexpressed miRNAs, due to deletion or loss of function, has to be restored, a therapeutic approach could involve exogenous delivery of corrective synthetic miRNAs. They could be delivered in the form of (siRNA-like) double-strand oligoRNAs, directly or via viral systems [122]. One example was given by Takamizawa and colleagues, who showed that forced expression of let-7 inhibited in vitro the growth of the lung adenocarcinoma cell line A549 [123]. This holds the promise that constructs designed to promote the synthesis of mature miRNAs could be useful in cancer therapy.

Genetics of the Familial Pancreatic Adenocarcinoma

In addition to sporadic adenocarcinoma, an increased risk of pancreatic cancer has been demonstrated among persons with a family history of pancreatic cancer. Coughlin and colleagues [124] reported an increased risk of developing pancreatic cancer for individuals who reported a positive family history of pancreatic cancer at baseline, with an RR of 1.5 after adjusting for age. Furthermore, a population-based cohort study demonstrated that the risk of pancreatic cancer increased 1.72-fold for individuals with a parent who developed a pancreatic cancer. The risk was not elevated when a more distant relative had been diagnosed with pancreatic cancer. Thus, some studies strongly support the hypothesis that familial aggregation and genetic susceptibility play an important role in the development of pancreatic cancer. However, the relative contributions of genetic risk factors and environmental risk factors to pancreatic cancer risk that cluster within the families (i.e., smoking) remain unclear (see below).

Discovering the genetic basis of inherited pancreatic cancer is an active area of research. In 2001, a multicenter linkage consortium, PACGENE, was established to conduct linkage studies aiming at localization and identification of pancreatic cancer susceptibility genes [125]. Other groups have used linkage studies to suggest that the paladin gene (*PALD*) on chromosome 4q32 predisposes to pancreatic cancer [126]; however, this finding has not been validated in subsequent studies [127–131].

The complex nature of pedigree data makes it difficult to accurately assess risk based upon the simple counting of the number of affected family members, as it does not account for family size, current age or age of onset of pancreatic cancer, and the exact relationship between affected family members. Computer-based, risk-assessment tools have been developed to integrate this complex risk factor and pedigree data into risk assessment. These models can provide more precise risk assessment than guidelines or models that rely on counts of affected family members, such as the Bethesda Guidelines for hereditary nonpolyposis colorectal cancer (HNPCC) or myriad tables for hereditary breast and ovarian cancer. In April 2007, the first risk prediction tool for pancreatic cancer, PancPRO, was released [132].

Although the genetic basis for most instances of aggregation of pancreatic cancer in families is unknown, the genes responsible for a small portion of familial pancreatic cancer are known (see Fig. 5.1). Germ line mutations in the *BRCA2*, *CDKN2A/p16*, *STK11*, and *PRSS1* genes have all been shown to increase the risk of pancreatic cancer [133–136]. In addition, some studies have described pancreatic cancers developing among individuals with HNPCC; however, the association between HNPCC syndromes and pancreatic cancer is not as well defined as it is for some of the other syndromes [137, 138]. While most patients with sporadic and familial pancreatic cancer have classic infiltrating ductal (tubular) adenocarcinoma, some inherited syndromes are associated with a specific histologic type. Although rare, these cases provide a unique opportunity to correlate genetics with histology. For example, many pancreatic cancers that develop in patients with HNPCC syndrome have a medullary phenotype [139–141] and individuals with the Peutz-Jeghers syndrome appear to be predisposed to IPMNs [59, 142, 143]. These associations between phenotype and genotype are important because tumor phenotype can be used to identify at-risk families.

Hereditary Breast and Ovarian Cancer Syndrome

Hereditary breast and ovarian cancer syndrome is an autosomal, dominantly inherited disease characterized by early-onset breast and/or ovarian cancers. Germ line mutations in *BRCA1* and *BRCA2* are responsible for the breast and ovarian cancer syndrome in most families [144]. Point mutations account for most germ line *BRCA1* and *BRCA2* mutations in these families, but germ line deletions of these genes also occur [145–147]. Germ line *BRCA2* mutations have been clearly associated with an increased risk of pancreatic cancer. Analysis of a large series of *BRCA2* mutation-positive families, ascertained for young age at onset of breast and/or ovarian cancer, demonstrated a 3.5-fold increased risk of pancreatic cancer in mutation carriers. Furthermore, the probability that a patient with pancreatic cancer has a germ line mutation in *BRCA2* increases as the number of family members with pancreatic cancer increases. To date, mutations in the *BRCA2* genes are considered the most common known genetic mutations associated with pancreatic cancer. Germ line *BRCA2* mutations do not appear to be associated with a specific type of pancreatic cancer [148] as most pancreatic cancers that develop in *BRCA2* carriers are traditional ductal adenocarcinomas. There are, however, significant clinical differences between BRCA2-deficient and BRCA2-intact pancreatic cancers. The BRCA2 functions in the repair of DNA interstrand cross-links and double-strand breaks. Pancreatic cancer cells with mutations in the BRCA2 pathway are hypersensitive to DNA-interstrand cross-linking agents, such as mitomycin C, cisplatin, chlorambucil, and melphalan [149], as well as to inhibitors of poly (ADP-ribose) polymerase [150, 151]. Therefore, the *BRCA2* gene could be a potential target for a genotype-based anticancer therapy. Large studies of *BRCA1* mutation-positive families, ascertained for young age of onset of breast and/or ovarian cancers, suggest that *BRCA1* gene mutation carriers have a twofold increased risk of pancreatic cancer [152, 153]. *BRCA1* gene mutations, however, appear to be substantially less common in families with pancreatic cancer without a significant breast cancer history [154] so that the possibility that adenocarcinoma of the pancreas is an incidental finding in *BRCA1* mutation carriers cannot be ruled out.

Peutz-Jeghers Syndrome

Peutz-Jeghers syndrome is an autosomal, dominantly inherited disease characterized by hamartomatous polyps of the gastrointestinal tract and pigmented macules of the lips and buccal mucosa. A variety of cancers have been associated with Peutz-Jeghers syndrome, including gastrointestinal, gynecologic, lung, breast, and pancreatic cancer [155–158]. Inherited mutations in the *STK11/LKB1* gene are responsible for most cases of Peutz-Jeghers syndrome, and as many as 80% of patients with Peutz-Jeghers syndrome have a germ line *STK11/LKB1* mutation. The hamartomatous polyps found in patients with Peutz-Jeghers syndrome most commonly occur in the small intestine; however, they can also involve the stomach, colon, and rectum. Patients with Peutz-Jeghers syndrome have a greater than 132-fold increased risk of developing pancreatic cancer [133]. Interestingly, these cancers may progress through an IPMN precursor pathway. In addition, *STK11/LKB1* gene inactivation is more frequently seen in sporadic IPMNs than it is in conventional ductal adenocarcinoma [142, 159]. The association of Peutz-Jeghers syndrome with IPMN precursor lesions has significant ramifications for screening because most IPMNs are detectable with currently available imaging technologies.

Hereditary Pancreatitis

Hereditary pancreatitis is a rare inherited form of chronic pancreatitis characterized by repeated attacks of acute pancreatitis, usually starting early in childhood, and leading to long-term exocrine and endocrine failure [160]. It is now generally accepted that patients with chronic pancreatitis have a higher risk of developing adenocarcinoma of the pancreas [135], and more particularly if there is hereditary chronic pancreatitis [161]. Chronic inflammation might cause DNA damage which accumulates over time, as observed in PanINs. Several teams have described PanIN-like lesions in chronic pancreatitis tissue [162, 163]. This could be highly significant for the well-known problem of the differential diagnosis between “tumor-like” chronic pancreatitis and cancer: despite the contribution of endoscopic ultrasound, a definite diagnosis of adenocarcinoma is difficult to obtain. Furthermore, a negative specimen does not necessarily exclude a diagnosis of cancer in these patients [164–166]. The usefulness of searching for genetic anomalies in samples obtained from patients with chronic pancreatitis has already been explored: the search for the *KRAS* oncogene mutation in tissue [167], serum [168], and endoscopic ultrasound-guided needle aspirations [169] appears to be more useful than a search in pancreatic juice [170, 171]. Nevertheless, there is no validated screening strategy to monitor patients with chronic pancreatitis. A better knowledge of the genetic alterations and of the carcinogenesis of pancreatic tissue could be helpful in establishing earlier diagnosis of pancreatic cancer in these patients. Germ line mutations in the cationic trypsinogen gene (*PRSS1*) have been associated with an autosomal dominant form of hereditary pancreatitis, while germ line mutations in the serine protease inhibitor gene (*SPINK1*) have been associated with an autosomal recessive form of hereditary pancreatitis [172]. *PRSS1* gene mutations in hereditary pancreatitis have been extensively studied. Multiple mutation sites have been identified, most of which cluster in the N-terminal half of the molecule encoded by exons 2 and 3. The most common mutations are R122H and N29I. Some *PRSS1* gene mutations appear to increase the stability of trypsin by eliminating a trypsin autodegradation site while other PRSS1 gene mutations appear to enhance trypsinogen autoactivation, both of which eventually result in chronic pancreatitis [173–176]. Klöppel and colleagues [177, 178] have carefully examined pancreatic specimens from patients with hereditary pancreatitis and they made the hypothesis that hereditary pancreatitis begins with necrosis of the duct-lining cells and periductal tissue, and gradually progresses to dilation of the involved ducts, periductal fibrosis, and in advanced cases, intralobular fibrosis.

Microscopically, in the early stages, the involved ducts are characterized by epithelial injury and/or necrosis and inflammatory cell infiltration. Periductal fibrosis is more prominent than intralobular fibrosis and the pancreatic parenchyma away from the involved ducts is relatively well preserved. In the advanced stages of hereditary pancreatitis, there is extensive periductal as well as intralobular fibrosis, and the lobular parenchyma is eventually completely replaced by sclerotic tissue containing metaplastic acini and aggregates of islets of Langerhans. The ducts can be dilated or very irregular in shape, and some ducts contain protein plugs and calculi. Individuals with hereditary pancreatitis have an approximately 53-fold increased risk for pancreatic cancer after the age of 50 years compared with the general population. Cumulative rates of pancreatic adenocarcinoma in patients with hereditary pancreatitis reach 30–40% by the age of 70 years [179, 180]. Smoking, early onset of pancreatitis, and diabetes mellitus are associated risk factors for the development of pancreatic cancer in these patients [161] and smokers tend to develop disease 20 years before non-smokers [180]. No specific histopathologic phenotype of pancreatic cancer has been associated with hereditary pancreatitis. Instead, most patients have a classic tubular type of infiltrating ductal adenocarcinoma.

Hereditary Nonpolyposis Colorectal Cancer Syndrome

Hereditary nonpolyposis colorectal cancer syndrome (HNPCC) is an autosomal dominant hereditary disease characterized by early onset of colon cancer with a predilection for the right colon [181]. Patients with HNPCC have germ line mutations in genes coding for proteins associated with DNA mismatch repair. These genes include *hMSH2*, *hMLH1*, *hPMS1*, *hPMS2*, and *hMSH6/GTBP*. Adenocarcinomas of the colon in patients with HNPCC show microsatellite instability (MSI+) and a distinct medullary histopathology. In addition, patients with HNPCC are at increased risk for a spectrum of extracolonic neoplasms, including carcinomas of the endometrium, ovary, stomach, bile duct, kidney, bladder, ureter, and skin [181]. While some studies have suggested individuals with HNPCC may also have an increased risk for pancreatic cancer [138, 182], additional studies are needed to accurately quantify this risk. Lynch and colleagues [182] first reported pancreatic carcinoma in kindreds with HNPCC. Further evidence linking HNPCC and pancreatic cancer comes from a study of medullary carcinomas of the pancreas by Wilentz and colleagues [139]. The pancreatic cancers that arise in patients with HNPCC often have a distinctive medullary appearance. Medullary carcinoma of the pancreas is a rare variant of pancreatic adenocarcinoma. As with medullary carcinoma of the colon, it is associated with a better prognosis than conventional ductal adenocarcinoma [138, 139]. The morphology of pancreatic medullary carcinoma is very similar to that of medullary carcinoma of the colon. Unlike conventional ductal adenocarcinoma of the pancreas, most medullary carcinomas do not harbor *KRAS2* gene mutations. Instead, medullary carcinomas of the pancreas often harbor *BRAF* gene mutations and are MSI+ [139, 141, 183]. As one would expect in a neoplasm with genetic inactivation of a DNA mismatch repair gene, medullary carcinomas of the pancreas often show loss of expression of one of the DNA mismatch repair proteins (Mlh1 and Msh2) as recently reported for a patient with HNPCC due to a mutation of the hMSH2 mismatch repair gene [140]. The presence of medullary phenotype in a pancreatic cancer may suggest inherited susceptibility to HNPCC.

Familial Atypical Multiple Mole Melanoma

Familial atypical multiple mole melanoma (FAMMM) is an autosomal dominant inherited syndrome with incomplete penetrance. It is characterized by greater than normal numbers of melanocytic nevi, multiple atypical melanocytic nevi, and an increased risk of cutaneous malignant melanoma [184, 185]. Germ line mutations in the *CDKN2A/p16* gene are responsible for a portion of FAMMM cases [134, 185, 186]. A variety of cancers, other than melanoma, have been documented in kindreds with familial melanoma, including carcinoma of the lung, pancreas, and breast as well as sarcoma. A subset of FAMMMs is associated with pancreatic cancer. Kindreds with FAMMM have a 13- to 22-fold increased risk for pancreatic cancer [187] and the risk for pancreatic cancer among mutation carriers is 38-fold higher than that of the general population. Lynch and colleagues [188] studied 159 families with familial pancreatic carcinoma and identified 19 families with FAMMM. DNA testing revealed a germ line *CDKN2A/p16* gene mutation in every case.

Familial Adenomatous Polyposis

Familial adenomatous polyposis (FAP) is an autosomal, dominantly inherited disorder characterized by the development of hundreds to thousands of colonic adenomatous polyps at an early age. Some of the adenomas can progress to invasive adenocarcinoma, and, if untreated, invasive adenocarcinoma of the colon will develop in almost all patients by the age of 40 years [181]. Germ line mutations in adenomatous polyposis coli (*APC*) gene, a tumor suppressor gene, are responsible for the development of FAP [189, 190]. Patients with FAP are at increased risk for other neoplasms, including thyroid tumors, gastric, duodenal, and ampullary adenocarcinoma. Although the association of pancreatic cancer and FAP is not as strong as the association of FAP with other cancer types, several lines of evidence suggest that patients with FAP are also at increased risk for the development of pancreatic neoplasms. Pancreatic adenocarinoma has been described in individuals with germ line *APC* gene mutations, and patients with FAP may have a fourfold increase in risk for pancreatic adenocarcinoma [191]. Furthermore, in a report of a patient with FAP and an IPMN with high-grade dysplasia, it was noted that the IPMN showed biallelic inactivation of the *APC* gene, a fact that supports the genetic link between FAP and IPMN for this patient [192]. In addition to the association of FAP with pancreatic adenocarcinoma, Abraham and colleagues [137] reported a rare pancreatic neoplasm, pancreatoblastoma, arising in a patient with FAP. Pancreatoblastoma is a malignant epithelial neoplasm with acinar differentiation and squamoid nests. In contrast to conventional ductal adenocarcinoma of the pancreas, both sporadic and FAP-associated pancreatoblastomas lack KRAS2 and TP53 gene mutations. Instead, most cases harbor alterations in the APC/beta-catenin pathway [137].

Germ line DNA Mutations that Increase Susceptibility to Develop Pancreatic Adenocarcinoma

Most pancreatic cancers are sporadic. In sporadic forms, association with polymorphic somatic mutations, spontaneous or generated by environmental factors is of paramount importance. It is well known that sporadic forms of cancers result from the accumulation of genetic modifications or mutations (see above). Besides particular syndromes whose association with pancreatic cancer is well known, a lot of work remains to identify genetic variations associated with cancer predisposition. Recent advances in large-scale genotyping (SNP chips) increased considerably the number of markers that can be simultaneously genotyped. As a consequence, haplotype blocks can be determined much more precisely than with microsatellite markers. These methodologies allowed recent studies of associations covering all intragenic regions of the whole human genome, these studies being called "genome wide association studies" (GWAS). Several GWAS led to the identification of susceptibility markers of several types of cancers (review in New England Journal of Medicine in 2010) [193]. More than 700 GWAS are presently registered (http://gwas.nih.gov/index.html) [194]; among them, only four concern pancreatic cancer [195–198].

Two GWAS for pancreatic cancer [196, 197], have been conducted within the framework of the "Pancreatic Cancer Cohort Consortium," or PCCC, an international consortium comprising Centers from Europe, China, and North America. In these studies, 12 cohorts representing more than 3,800 cases of cancer and 3,900 controls allowed detection of four new genes associated with pancreatic cancer. The first locus, harboring several SNPs strongly associated with cancer occurrence, was localized on chromosome 9q34 and contains the first intron of the *ABO* gene. That gene encodes the glycosyl-transferase which catalyzes the transfer of sugars to antigen H (Histo-blood group ABO system transferase). By this mechanism, antigen H becomes antigen A or B, depending on the nature of the sugar transferred by the ABO enzyme. In individuals from group O, the gene is mutated and generates a nonfunctional protein. Association of the ABO gene with pancreatic cancer has already been suggested 50 years ago in studies showing a higher frequency of pancreatic or gastric cancer in individuals from groups A, B, or AB than from group O [199, 200]. That gene is also found altered in primary tumors and metastases of pancreatic cancer [201]. More recently, the PCCC added three other regions strongly associated with cancer [196]. The first one is an intergenic region of 600 kb in chromosome 13q22.1, located between genes *KLF5* and *KLF12*. These two members of the Kruppel transcription factor family are involved in regulating cell growth and transformation [202, 203]. The same region has already been found deleted in other cancers [204, 205], or associated with an increase in breast cancer susceptibility in families at risk for that cancer because of high incidence in relatives but negative for *BRCA1* and *BRCA2* mutations [206]. The second region is located on 1q32.1. That region harbors gene nuclear receptor subfamily 5 group A member 2 (*NR5A2*), at 91 kb upstream of the gene. The receptor encoded by gene *NR5A2* is present, in adults, in the exocrine pancreas, liver, gut, and ovaries. Its function is not completely understood. In liver, it would be a key regulator of CYP7A, an enzyme involved in the homeostasis of cholesterol and bile salts, and in steroidogenesis (reviewed in Frontier Bioscience in 2008) [207]. Also, HBV could use it to enhance the expression of its genome [208]. In the pancreas, NR5A2 would contribute to the regulation of the expression of several genes [209]. Finally, the importance of NR5A2 in embryogenesis was demonstrated in mice by showing that its knock-out results in embryonic lethality [210]. The last locus identified by this consortium is localized at 5p15.33. SNP markers that border this locus are within intron 13 of the *CLPTM1L* gene (cleft lip and palate transmembrane 1-like). *CLPTM1L* is poorly described. It is overexpressed in cell lines resistant to cisplatin and could play a role in apoptosis [211]. In the same locus rs401681 and only distant by 23 kb is located the TERT gene, which encodes the catalytic subunit of telomerase, an enzyme required to maintain the telomeres with a correct size. TERT activity is almost undetectable in normal cells but is increased in 90% of cancers [212]. However, a more recent study has shown that association of that marker (i.e., C–T shift at locus rs401681) with the increase risk of cancer occurrence is not due to a decrease in TERT activity [213].

One of the critical factors in the interpretation of GWAS is the ethnicity of the studied group. PCCC studies mentioned above have been conducted on Caucasian populations and do not necessarily reflect risk factors for other ethnic groups. In fact, a GWAS conducted by Biobank Japan (BJJ) has evidenced a different set of markers of pancreatic cancer [195]. Among loci strongly associated with the disease, three are specific to the Japanese population. The first one, in 6p25.3, is a 75 bp linkage disequilibrium block containing the *FOXQ1* gene (Forkhead-box (Fox) Q1), one of the 43 members of the Fox transcription factor family. *FOXQ1* had already been associated with pancreatic cancer [214] and colorectal cancer [215]. Finding this new marker 25 kb upstream from the beginning of the *FOXQ1* gene suggests a change in its promoter region that could alter its expression [195]. The second SNP showing significant association is located within gene Bicaudal-D homolog 1 (BICD1). The BICD1 protein is involved in vacuolar trafficking and associated with a shortening of telomeres [216]. Telomere shortening has been associated with pancreatic cancer [37, 217, 218], which makes BICD1 an excellent candidate to explain these findings. The last locus with a high odds ratio (OR = 3.73) corresponds to several SNPs that fall in the first intron of the *DPP6* gene. That gene encodes dipeptidyl-peptidase 6, a member of the DPP family devoid of the catalytic residues required for enzymatic activity but which acts through protein–protein interactions. The major activity known for that protein is the modulation of the activity of voltage-dependent potassium KCND2 channels in the CNS (reviewed by the Federation of European Biochemical Societies Journal in 2010) [219]. However, that gene also appears in one of the 12 pathways and metabolic processes identified by Jones et al. [66] In this study, DPP6, which is involved in the regulation of tumor invasion, shows somatic mutations in 3 of 24 pancreatic cancers.

The enormous wealth of information generated by the successive GWAS and by the other high-throughput studies provides the first opportunity to study at the genomic scale the genetic anomalies associated with pancreatic cancer. Their contribution to the understanding of cancer genetics, to risk prediction, and to the monitoring of cancer evolution opens-up new ways toward individualized treatments and, hopefully, cancer prevention. The major challenges of GWAS lie mostly in the management of studies, toward a standardization of results allowing easier interpretation of results. Analyses will have to take into account disease heterogeneity and ethnic factors (effect of stratification of probation) and the complex interactions between multiple genetic and environmental factors. Another limiting factor of GWAS is cost. There is a correlation between the number of samples required to establish an association and the degree of penetrance of a gene or the frequency of involved alleles [193]. For instance, to detect with a reasonable statistical significance a single locus with a moderate risk (odds ratio between 1.5 and 5), at least 1,000 cases and 1,000 controls are required. If the risk is lower (OR < 1.3), 10 times more cases and controls will be necessary, which requires multicenter international collaborations.

Two studies have extended the results obtained by analyzing tumor exome, including a study comparing the exomes of the primary tumor and corresponding metastases [220, 221]. In the first one, Yachida and colleagues [220] gathered data on the exomes of seven tumors previously characterized by Jones and colleagues [66] and compared them to data on the exomes of the corresponding metastases. The second study was conducted on 13 tumors and corresponding metastases [221]. Information produced by these studies allowed to fill in a high resolution framework of the various modifications and rearrangements that occur in the DNA of metastases, of the various underlying mechanisms, on the genes potentially involved in the metastatic process and finally on the time between the appearance of the cancerous cell and the occurrence of a clone with a metastatic profile.

Genomic instability is apparently the main characteristic of pancreatic cancer [221]. DNA sequencing in tumors and in cells at different stages of progression toward metastases revealed that the major chromosomal rearrangement is a duplication of fragments and their insertion in opposite

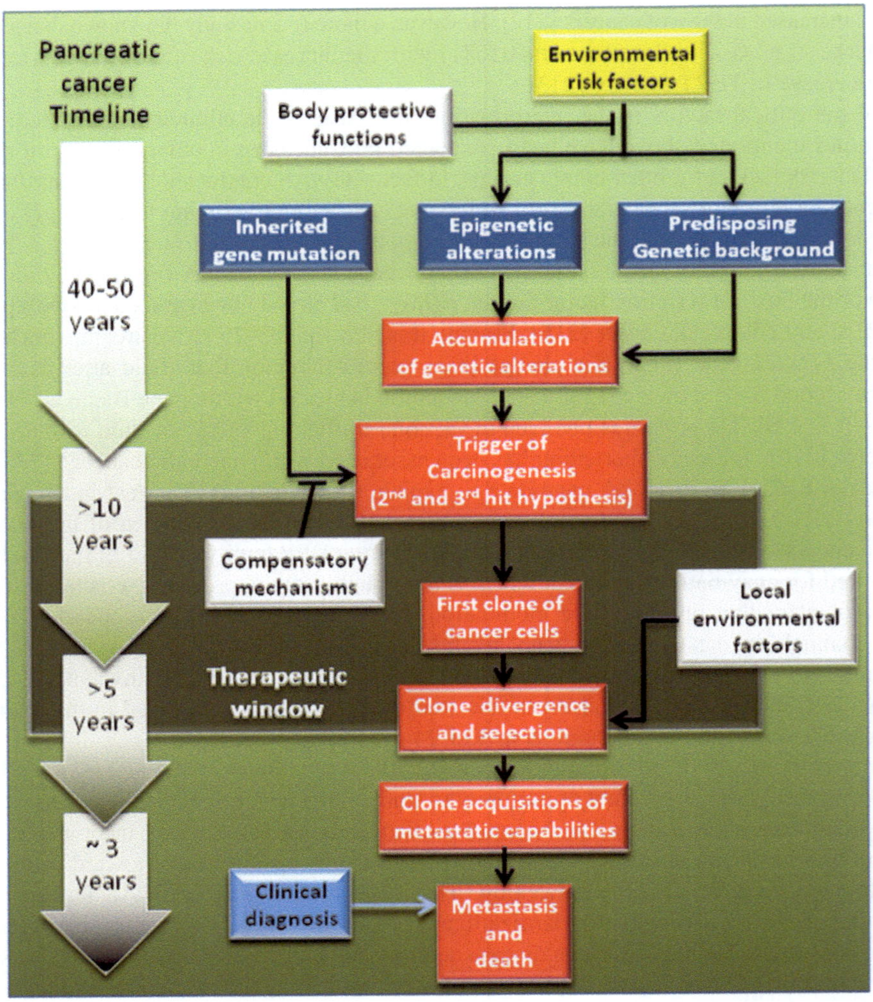

Fig. 5.2 Development of the pancreatic cancer. Development of sporadic pancreas cancer results from the combination of a genetic background of susceptibility, which remain largely to be characterized, with the protumoral environmental factors. In the familial forms of pancreas cancer, mutations on specific genes are responsible for the accumulation of genetics alteration. The role of the environmental factors is not clearly demonstrated in these forms of pancreas cancer. Accumulation of genetic alterations results in direct cancer development or, more probably, through development of the precancerous lesions PanIN and/or IPMN. At *left*, an estimated pancreatic cancer timescale associated with tumor progression. The *successive arrows* represent the estimated times from a healthy asymptomatic state to tumor initiation, acquisition of metastatic capabilities, and to the production of metastasis and the patient death. Therapeutic window: Period of more than 10 years from the birth of the founding cell of the parental, nonmetastatic, subclone, and the acquisition for one of the subclones of the metastatic profile

directions, a phenomenon called "fold-back inversions" by the authors. The origin of fold-back inversions is not known. A possible hypothesis is that early in tumor progression telomerase activity is altered or inhibited [42] generating a progressive shortening of telomeres and successive cycles of breaking and abnormal fusion of chromatides by a mechanism called "breakage-fusion-bridge" [222]. These rearrangements will determine the gain or loss of DNA and, as consequence, genomic instability. Interestingly, these results suggest that fold-back inversions arise early in tumor evolution and would trigger progression toward metastasis [221]. These results have shown unequivocally, for the first time, that the pancreatic tumor evolves from a first (parental) clone which generates progressively other subclones. Each subclone evolves independently by accumulating specific genetic modifications and, possibly, a tissue-specific metastatic affinity profile. Establishment of a phylogenetic correlation between genomic alterations of tumors and their metastases allowed drawing a kinetic of tumor progression [220]. Unexpectedly, the mean time between arising of the founding cell and the occurrence of the first parent clone (nonmetastasis) is 11.8 years. Another 6.8 years will be required for one of the subclones to acquire the metastatic profile, which occurs about 3 years before the patient dies. These findings, if confirmed, have major clinical consequences. They show that, if early detection of pancreatic cancer is possible, there is a large window of about 10 years for therapeutic intervention. A diagram summarizing these data is represented in Fig. 5.2.

Summary and Perspectives

The advent of genomic analysis, at both levels of gene expression and alternative splicing, as well as the increased in knowledge of the DNA structure with the detection of millions of SNPs and CNVs, or more recently, with the sequencing of the entire genome of coding regions using second generation sequencing techniques led us to rethink how the pathophysiology of pancreatic cancer must be analyzed and interpreted.

However, a really profitable use of all these genomic data still requires to standardize the different types of protocols and platforms and to build consensus in quality control and methods to validate the results. The interpretation of studies has been complicated by the lack of standards used for publications, the absence of a single gene annotation, and a delay in consolidation analysis. Furthermore, there is no fully established functional network allowing meta-analysis of high-throughput data that can be considered to date as a "golden standard" and, as consequence, a lot of work remains to better characterize the affected pathways in pancreatic cancer.

In fact, obtaining a critical mass of data easily connected, followed by the optimization and standardization in the use and interpretation of data from the powerful informatics tools, and finally, the validation of their clinical interest using standardized protocols, will eventually give us powerful tools for diagnosis, prognosis, and personalized treatment of this terrible cancer.

References

1. Greenlee RT, Hill-Harmon MB, Murray T, Thun M. Cancer statistics, 2001. CA Cancer J Clin. 2001;51(1): 15–36.
2. Devesa SS, Blot WJ, Stone BJ, Miller BA, Tarone RE, Fraumeni Jr JF. Recent cancer trends in the United States. J Natl Cancer Inst. 1995;87(3):175–82.
3. Yeo CJ, Abrams RA, Grochow LB, et al. Pancreaticoduodenectomy for pancreatic adenocarcinoma: postoperative adjuvant chemoradiation improves survival. A prospective, single-institution experience. Ann Surg. 1997;225(5):621–33; discussion 633–6.
4. Warshaw AL, Gu ZY, Wittenberg J, Waltman AC. Preoperative staging and assessment of resectability of pancreatic cancer. Arch Surg. 1990;125(2):230–3.
5. Lillemoe KD, Yeo CJ, Cameron JL. Pancreatic cancer: state-of-the-art care. CA Cancer J Clin. 2000;50(4): 241–68.
6. Cubilla AL, Fitzgerald PJ. Morphological lesions associated with human primary invasive nonendocrine pancreas cancer. Cancer Res. 1976;36(7 PT 2):2690–8.
7. Klimstra DS, Longnecker DS. K-ras mutations in pancreatic ductal proliferative lesions. Am J Pathol. 1994;145(6):1547–50.
8. Kozuka S, Sassa R, Taki T, et al. Relation of pancreatic duct hyperplasia to carcinoma. Cancer. 1979;43(4): 1418–28.

9. Hruban RH, Adsay NV, Albores-Saavedra J, et al. Pancreatic intraepithelial neoplasia: a new nomenclature and classification system for pancreatic duct lesions. Am J Surg Pathol. 2001;25(5):579–86.
10. Takaori K, Hruban RH, Maitra A, Tanigawa N. Current topics on precursors to pancreatic cancer. Adv Med Sci. 2006;51:23–30.
11. Grippo PJ, Nowlin PS, Demeure MJ, Longnecker DS, Sandgren EP. Preinvasive pancreatic neoplasia of ductal phenotype induced by acinar cell targeting of mutant Kras in transgenic mice. Cancer Res. 2003;63(9):2016–9.
12. Guerra C, Mijimolle N, Dhawahir A, et al. Tumor induction by an endogenous K-ras oncogene is highly dependent on cellular context. Cancer Cell. 2003;4(2):111–20.
13. Hingorani SR, Petricoin EF, Maitra A, et al. Preinvasive and invasive ductal pancreatic cancer and its early detection in the mouse. Cancer Cell. 2003;4(6):437–50.
14. Aguirre AJ, Bardeesy N, Sinha M, et al. Activated Kras and Ink4a/Arf deficiency cooperate to produce metastatic pancreatic ductal adenocarcinoma. Genes Dev. 2003;17(24):3112–26.
15. Bardeesy N, Aguirre AJ, Chu GC, et al. Both p16(Ink4a) and the p19(Arf)-p53 pathway constrain progression of pancreatic adenocarcinoma in the mouse. Proc Natl Acad Sci USA. 2006;103(15):5947–52.
16. Hingorani SR, Wang L, Multani AS, et al. Trp53R172H and KrasG12D cooperate to promote chromosomal instability and widely metastatic pancreatic ductal adenocarcinoma in mice. Cancer Cell. 2005;7(5):469–83.
17. Ijichi H, Chytil A, Gorska AE, et al. Aggressive pancreatic ductal adenocarcinoma in mice caused by pancreas-specific blockade of transforming growth factor-beta signaling in cooperation with active Kras expression. Genes Dev. 2006;20(22):3147–60.
18. Izeradjene K, Combs C, Best M, et al. Kras(G12D) and Smad4/Dpc4 haploinsufficiency cooperate to induce mucinous cystic neoplasms and invasive adenocarcinoma of the pancreas. Cancer Cell. 2007;11(3):229–43.
19. Kojima K, Vickers SM, Adsay NV, et al. Inactivation of Smad4 accelerates Kras(G12D)-mediated pancreatic neoplasia. Cancer Res. 2007;67(17):8121–30.
20. Sharpless NE, Ramsey MR, Balasubramanian P, Castrillon DH, DePinho RA. The differential impact of p16(INK4a) or p19(ARF) deficiency on cell growth and tumorigenesis. Oncogene. 2004;23(2):379–85.
21. Guerra C, Schuhmacher AJ, Canamero M, et al. Chronic pancreatitis is essential for induction of pancreatic ductal adenocarcinoma by K-Ras oncogenes in adult mice. Cancer Cell. 2007;11(3):291–302.
22. Day JD, Digiuseppe JA, Yeo C, et al. Immunohistochemical evaluation of HER-2/neu expression in pancreatic adenocarcinoma and pancreatic intraepithelial neoplasms. Hum Pathol. 1996;27(2):119–24.
23. Caldas C, Kern SE. K-ras mutation and pancreatic adenocarcinoma. Int J Pancreatol. 1995;18(1):1–6.
24. Hingorani SR, Tuveson DA. Ras redux: rethinking how and where Ras acts. Curr Opin Genet Dev. 2003;13(1):6–13.
25. Lohr M, Kloppel G, Maisonneuve P, Lowenfels AB, Luttges J. Frequency of K-ras mutations in pancreatic intraductal neoplasias associated with pancreatic ductal adenocarcinoma and chronic pancreatitis: a meta-analysis. Neoplasia. 2005;7(1):17–23.
26. Laghi L, Orbetegli O, Bianchi P, et al. Common occurrence of multiple K-RAS mutations in pancreatic cancers with associated precursor lesions and in biliary cancers. Oncogene. 2002;21(27):4301–6.
27. Sherr CJ. Cell cycle control and cancer. Harvey Lect. 2000;96:73–92.
28. Caldas C, Hahn SA, da Costa LT, et al. Frequent somatic mutations and homozygous deletions of the p16 (MTS1) gene in pancreatic adenocarcinoma. Nat Genet. 1994;8(1):27–32.
29. Schutte M, Hruban RH, Geradts J, et al. Abrogation of the Rb/p16 tumor-suppressive pathway in virtually all pancreatic carcinomas. Cancer Res. 1997;57(15):3126–30.
30. Ueki T, Toyota M, Sohn T, et al. Hypermethylation of multiple genes in pancreatic adenocarcinoma. Cancer Res. 2000;60(7):1835–9.
31. Wilentz RE, Geradts J, Maynard R, et al. Inactivation of the p16 (INK4A) tumor-suppressor gene in pancreatic duct lesions: loss of intranuclear expression. Cancer Res. 1998;58(20):4740–4.
32. Redston MS, Caldas C, Seymour AB, et al. p53 mutations in pancreatic carcinoma and evidence of common involvement of homocopolymer tracts in DNA microdeletions. Cancer Res. 1994;54(11):3025–33.
33. Maitra A, Adsay NV, Argani P, et al. Multicomponent analysis of the pancreatic adenocarcinoma progression model using a pancreatic intraepithelial neoplasia tissue microarray. Mod Pathol. 2003;16(9):902–12.
34. Hahn SA, Schutte M, Hoque AT, et al. DPC4, a candidate tumor suppressor gene at human chromosome 18q21.1. Science. 1996;271(5247):350–3.
35. Truty MJ, Urrutia R. Basics of TGF-beta and pancreatic cancer. Pancreatology. 2007;7(5–6):423–35.
36. Wilentz RE, Iacobuzio-Donahue CA, Argani P, et al. Loss of expression of Dpc4 in pancreatic intraepithelial neoplasia: evidence that DPC4 inactivation occurs late in neoplastic progression. Cancer Res. 2000;60(7):2002–6.
37. van Heek NT, Meeker AK, Kern SE, et al. Telomere shortening is nearly universal in pancreatic intraepithelial neoplasia. Am J Pathol. 2002;161(5):1541–7.
38. Gisselsson D. Chromosome instability in cancer: how, when, and why? Adv Cancer Res. 2003;87:1–29.

39. Klein WM, Hruban RH, Klein-Szanto AJ, Wilentz RE. Direct correlation between proliferative activity and dysplasia in pancreatic intraepithelial neoplasia (PanIN): additional evidence for a recently proposed model of progression. Mod Pathol. 2002;15(4):441–7.
40. Gansauge S, Gansauge F, Ramadani M, et al. Overexpression of cyclin D1 in human pancreatic carcinoma is associated with poor prognosis. Cancer Res. 1997;57(9):1634–7.
41. Hruban RH, Goggins M, Parsons J, Kern SE. Progression model for pancreatic cancer. Clin Cancer Res. 2000;6(8):2969–72.
42. Bardeesy N, DePinho RA. Pancreatic cancer biology and genetics. Nat Rev Cancer. 2002;2(12):897–909.
43. Tanaka M. Intraductal papillary mucinous neoplasm of the pancreas: diagnosis and treatment. Pancreas. 2004;28(3):282–8.
44. Kobari M, Egawa S, Shibuya K, et al. Intraductal papillary mucinous tumors of the pancreas comprise 2 clinical subtypes: differences in clinical characteristics and surgical management. Arch Surg. 1999;134(10):1131–6.
45. Kimura W, Sasahira N, Yoshikawa T, Muto T, Makuuchi M. Duct-ectatic type of mucin producing tumor of the pancreas–new concept of pancreatic neoplasia. Hepatogastroenterology. 1996;43(9):692–709.
46. Furukawa T, Kloppel G, Volkan Adsay N, et al. Classification of types of intraductal papillary-mucinous neoplasm of the pancreas: a consensus study. Virchows Arch. 2005;447(5):794–9.
47. Satoh K, Shimosegawa T, Moriizumi S, Koizumi M, Toyota T. K-ras mutation and p53 protein accumulation in intraductal mucin-hypersecreting neoplasms of the pancreas. Pancreas. 1996;12(4):362–8.
48. Yoshizawa K, Nagai H, Sakurai S, et al. Clonality and K-ras mutation analyses of epithelia in intraductal papillary mucinous tumor and mucinous cystic tumor of the pancreas. Virchows Arch. 2002;441(5):437–43.
49. La Rosa S, Uccella S, Billo P, Facco C, Sessa F, Capella C. Immunohistochemical localization of alpha- and betaA-subunits of inhibin/activin in human normal endocrine cells and related tumors of the digestive system. Virchows Arch. 1999;434(1):29–36.
50. Semba S, Moriya T, Kimura W, Yamakawa M. Phosphorylated Akt/PKB controls cell growth and apoptosis in intraductal papillary-mucinous tumor and invasive ductal adenocarcinoma of the pancreas. Pancreas. 2003;26(3):250–7.
51. Furukawa T, Fujisaki R, Yoshida Y, et al. Distinct progression pathways involving the dysfunction of DUSP6/MKP-3 in pancreatic intraepithelial neoplasia and intraductal papillary-mucinous neoplasms of the pancreas. Mod Pathol. 2005;18(8):1034–42.
52. House MG, Guo M, Iacobuzio-Donahue C, Herman JG. Molecular progression of promoter methylation in intraductal papillary mucinous neoplasms (IPMN) of the pancreas. Carcinogenesis. 2003;24(2):193–8.
53. Sessa F, Solcia E, Capella C, et al. Intraductal papillary-mucinous tumours represent a distinct group of pancreatic neoplasms: an investigation of tumour cell differentiation and K-ras, p53 and c-erbB-2 abnormalities in 26 patients. Virchows Arch. 1994;425(4):357–67.
54. Hahn SA, Seymour AB, Hoque AT, et al. Allelotype of pancreatic adenocarcinoma using xenograft enrichment. Cancer Res. 1995;55(20):4670–5.
55. Inoue H, Furukawa T, Sunamura M, Takeda K, Matsuno S, Horii A. Exclusion of SMAD4 mutation as an early genetic change in human pancreatic ductal tumorigenesis. Genes Chromosomes Cancer. 2001;31(3):295–9.
56. Kimura M, Abe T, Sunamura M, Matsuno S, Horii A. Detailed deletion mapping on chromosome arm 12q in human pancreatic adenocarcinoma: identification of a I-cM region of common allelic loss. Genes Chromosomes Cancer. 1996;17(2):88–93.
57. Iacobuzio-Donahue CA, Klimstra DS, Adsay NV, et al. Dpc-4 protein is expressed in virtually all human intraductal papillary mucinous neoplasms of the pancreas: comparison with conventional ductal adenocarcinomas. Am J Pathol. 2000;157(3):755–61.
58. Forcet C, Etienne-Manneville S, Gaude H, et al. Functional analysis of Peutz-Jeghers mutations reveals that the LKB1 C-terminal region exerts a crucial role in regulating both the AMPK pathway and the cell polarity. Hum Mol Genet. 2005;14(10):1283–92.
59. Sato N, Rosty C, Jansen M, et al. STK11/LKB1 Peutz-Jeghers gene inactivation in intraductal papillary-mucinous neoplasms of the pancreas. Am J Pathol. 2001;159(6):2017–22.
60. Schena M, Shalon D, Davis RW, Brown PO. Quantitative monitoring of gene expression patterns with a complementary DNA microarray. Science. 1995;270(5235):467–70.
61. Buchholz M, Boeck W, Fensterer H, et al. Use of DNA arrays/microarrays in pancreatic research. Pancreatology. 2001;1(6):581–6.
62. Ranganathan P, Harsha HC, Pandey A. Molecular alterations in exocrine neoplasms of the pancreas. Arch Pathol Lab Med. 2009;133(3):405–12.
63. Hirono S, Yamaue H, Hoshikawa Y, et al. Molecular markers associated with lymph node metastasis in pancreatic ductal adenocarcinoma by genome-wide expression profiling. Cancer Sci. 2010;101(1):259–66.
64. Stratford JK, Bentrem DJ, Anderson JM, et al. A six-gene signature predicts survival of patients with localized pancreatic ductal adenocarcinoma. PLoS Med. 2010;7(7):e1000307.

65. Hezel AF, Kimmelman AC, Stanger BZ, Bardeesy N, Depinho RA. Genetics and biology of pancreatic ductal adenocarcinoma. Genes Dev. 2006;20(10):1218–49.
66. Jones S, Zhang X, Parsons DW, et al. Core signaling pathways in human pancreatic cancers revealed by global genomic analyses. Science. 2008;321(5897):1801–6.
67. Jones S, Hruban RH, Kamiyama M, et al. Exomic sequencing identifies PALB2 as a pancreatic cancer susceptibility gene. Science. 2009;324(5924):217.
68. Harsha HC, Kandasamy K, Ranganathan P, et al. A compendium of potential biomarkers of pancreatic cancer. PLoS Med. 2009;6(4):e1000046.
69. Xu K, Cui J, Olman V, Yang Q, Puett D, Xu Y. A comparative analysis of gene-expression data of multiple cancer types. PLoS One. 2010;5(10):e13696.
70. Cutts RJ, Gadaleta E, Hahn SA, Crnogorac-Jurcevic T, Lemoine NR, Chelala C. The Pancreatic Expression database: 2011 update. Nucleic Acids Res. 2011;39:D1023–8.
71. Lee RC, Feinbaum RL, Ambrose V. C. elegans heterochronic gene lin-4 encodes small RNAs with antisense complementarity to lin-14. Cell. 1993;75(5):843–54.
72. Wightman B, Ha I, Ruvkun G. Posttranscriptional regulation of the heterochronic gene lin-14 by lin-4 mediates temporal pattern formation in *C. elegans*. Cell. 1993;75(5):855–62.
73. Zeng Y. Principles of micro-RNA production and maturation. Oncogene. 2006;25(46):6156–62.
74. Lee CT, Risom T, Strauss WM. Evolutionary conservation of microRNA regulatory circuits: an examination of microRNA gene complexity and conserved microRNA-target interactions through metazoan phylogeny. DNA Cell Biol. 2007;26(4):209–18.
75. Bentwich I, Avniel A, Karov Y, et al. Identification of hundreds of conserved and nonconserved human microRNAs. Nat Genet. 2005;37(7):766–70.
76. Calin GA, Dumitru CD, Shimizu M, et al. Frequent deletions and down-regulation of micro-RNA genes miR15 and miR16 at 13q14 in chronic lymphocytic leukemia. Proc Natl Acad Sci USA. 2002;99(24):15524–9.
77. Carthew RW. Gene regulation by microRNAs. Curr Opin Genet Dev. 2006;16(2):203–8.
78. Lagos-Quintana M, Rauhut R, Lendeckel W, Tuschl T. Identification of novel genes coding for small expressed RNAs. Science. 2001;294(5543):853–8.
79. Rodriguez A, Griffiths-Jones S, Ashurst JL, Bradley A. Identification of mammalian microRNA host genes and transcription units. Genome Res. 2004;14(10A):1902–10.
80. Yi R, Qin Y, Macara IG, Cullen BR. Exportin-5 mediates the nuclear export of pre-microRNAs and short hairpin RNAs. Genes Dev. 2003;17(24):3011–6.
81. Lund E, Guttinger S, Calado A, Dahlberg JE, Kutay U. Nuclear export of microRNA precursors. Science. 2004;303(5654):95–8.
82. Bernstein E, Caudy AA, Hammond SM, Hannon GJ. Role for a bidentate ribonuclease in the initiation step of RNA interference. Nature. 2001;409(6818):363–6.
83. Hammond SM, Boettcher S, Caudy AA, Kobayashi R, Hannon GJ. Argonaute2, a link between genetic and biochemical analyses of RNAi. Science. 2001;293(5532):1146–50.
84. Khvorova A, Reynolds A, Jayasena SD. Functional siRNAs and miRNAs exhibit strand bias. Cell. 2003;115(2):209–16.
85. Hwang HW, Mendell JT. MicroRNAs in cell proliferation, cell death, and tumorigenesis. Br J Cancer. 2006;94(6):776–80.
86. Engels BM, Hutvagner G. Principles and effects of microRNA-mediated post-transcriptional gene regulation. Oncogene. 2006;25(46):6163–9.
87. Hobert O. Common logic of transcription factor and microRNA action. Trends Biochem Sci. 2004;29(9):462–8.
88. Eulalio A, Rehwinkel J, Stricker M, et al. Target-specific requirements for enhancers of decapping in miRNA-mediated gene silencing. Genes Dev. 2007;21(20):2558–70.
89. Koorstra JB, Hustinx SR, Offerhaus GJ, Maitra A. Pancreatic carcinogenesis. Pancreatology. 2008;8(2):110–25.
90. Singh M, Maitra A. Precursor lesions of pancreatic cancer: molecular pathology and clinical implications. Pancreatology. 2007;7(1):9–19.
91. Wong T, Howes N, Threadgold J, et al. Molecular diagnosis of early pancreatic ductal adenocarcinoma in high-risk patients. Pancreatology. 2001;1(5):486–509.
92. Lu J, Getz G, Miska EA, et al. MicroRNA expression profiles classify human cancers. Nature. 2005;435(7043):834–8.
93. Szafranska AE, Davison TS, John J, et al. MicroRNA expression alterations are linked to tumorigenesis and non-neoplastic processes in pancreatic ductal adenocarcinoma. Oncogene. 2007;26(30):4442–52.
94. Roldo C, Missiaglia E, Hagan JP, et al. MicroRNA expression abnormalities in pancreatic endocrine and acinar tumors are associated with distinctive pathologic features and clinical behavior. J Clin Oncol. 2006;24(29):4677–84.
95. Volinia S, Calin GA, Liu CG, et al. A microRNA expression signature of human solid tumors defines cancer gene targets. Proc Natl Acad Sci USA. 2006;103(7):2257–61.

96. Jiang J, Lee EJ, Gusev Y, Schmittgen TD. Real-time expression profiling of microRNA precursors in human cancer cell lines. Nucleic Acids Res. 2005;33(17):5394–403.
97. Olson P, Lu J, Zhang H, et al. MicroRNA dynamics in the stages of tumorigenesis correlate with hallmark capabilities of cancer. Genes Dev. 2009;23(18):2152–65.
98. Lee EJ, Gusev Y, Jiang J, et al. Expression profiling identifies microRNA signature in pancreatic cancer. Int J Cancer. 2007;120(5):1046–54.
99. Habbe N, Koorstra JB, Mendell JT, et al. MicroRNA miR-155 is a biomarker of early pancreatic neoplasia. Cancer Biol Ther. 2009;8(4):340–6.
100. Sood P, Krek A, Zavolan M, Macino G, Rajewsky N. Cell-type-specific signatures of microRNAs on target mRNA expression. Proc Natl Acad Sci USA. 2006;103(8):2746–51.
101. Watanabe S, Ueda Y, Akaboshi S, Hino Y, Sekita Y, Nakao M. HMGA2 maintains oncogenic RAS-induced epithelial-mesenchymal transition in human pancreatic cancer cells. Am J Pathol. 2009;174(3):854–68.
102. Lodygin D, Tarasov V, Epanchintsev A, et al. Inactivation of miR-34a by aberrant CpG methylation in multiple types of cancer. Cell Cycle. 2008;7(16):2591–600.
103. Gironella M, Seux M, Xie MJ, et al. Tumor protein 53-induced nuclear protein 1 expression is repressed by miR-155, and its restoration inhibits pancreatic tumor development. Proc Natl Acad Sci USA. 2007;104(41):16170–5.
104. Okamura S, Arakawa H, Tanaka T, et al. p53DINP1, a p53-inducible gene, regulates p53-dependent apoptosis. Mol Cell. 2001;8(1):85–94.
105. Tomasini R, Samir AA, Pebusque MJ, et al. P53-dependent expression of the stress-induced protein (SIP). Eur J Cell Biol. 2002;81(5):294–301.
106. Tomasini R, Samir AA, Carrier A, et al. TP53INP1s and homeodomain-interacting protein kinase-2 (HIPK2) are partners in regulating p53 activity. J Biol Chem. 2003;278(39):37722–9.
107. Gommeaux J, Cano C, Garcia S, et al. Colitis and colitis-associated cancer are exacerbated in mice deficient for tumor protein 53-induced nuclear protein 1. Mol Cell Biol. 2007;27(6):2215–28.
108. Ito Y, Motoo Y, Yoshida H, et al. Decreased expression of tumor protein p53-induced nuclear protein 1 (TP53INP1) in breast carcinoma. Anticancer Res. 2006;26(6B):4391–5.
109. Jiang PH, Motoo Y, Garcia S, Iovanna JL, Pebusque MJ, Sawabu N. Down-expression of tumor protein p53-induced nuclear protein 1 in human gastric cancer. World J Gastroenterol. 2006;12(5):691–6.
110. Torrisani J, Bournet B, du Rieu MC, et al. let-7 MicroRNA transfer in pancreatic cancer-derived cells inhibits in vitro cell proliferation but fails to alter tumor progression. Hum Gene Ther. 2009;20(8):831–44.
111. Bloomston M, Frankel WL, Petrocca F, et al. MicroRNA expression patterns to differentiate pancreatic adenocarcinoma from normal pancreas and chronic pancreatitis. JAMA. 2007;297(17):1901–8.
112. Croce CM. Causes and consequences of microRNA dysregulation in cancer. Nat Rev Genet. 2009;10(10):704–14.
113. Moriyama T, Ohuchida K, Mizumoto K, et al. MicroRNA-21 modulates biological functions of pancreatic cancer cells including their proliferation, invasion, and chemoresistance. Mol Cancer Ther. 2009;8:1067.
114. Mendell JT. miRiad roles for the miR-17-92 cluster in development and disease. Cell. 2008;133(2):217–22.
115. Ventura A, Young AG, Winslow MM, et al. Targeted deletion reveals essential and overlapping functions of the miR-17 through 92 family of miRNA clusters. Cell. 2008;132(5):875–86.
116. Lee KH, Lotterman C, Karikari C, et al. Epigenetic silencing of MicroRNA miR-107 regulates cyclin-dependent kinase 6 expression in pancreatic cancer. Pancreatology. 2009;9(3):293–301.
117. Burk U, Schubert J, Wellner U, et al. A reciprocal repression between ZEB1 and members of the miR-200 family promotes EMT and invasion in cancer cells. EMBO Rep. 2008;9(6):582–9.
118. Weiler J, Hunziker J, Hall J. Anti-miRNA oligonucleotides (AMOs): ammunition to target miRNAs implicated in human disease? Gene Ther. 2006;13(6):496–502.
119. Cheng AM, Byrom MW, Shelton J, Ford LP. Antisense inhibition of human miRNAs and indications for an involvement of miRNA in cell growth and apoptosis. Nucleic Acids Res. 2005;33(4):1290–7.
120. Chan JA, Krichevsky AM, Kosik KS. MicroRNA-21 is an antiapoptotic factor in human glioblastoma cells. Cancer Res. 2005;65(14):6029–33.
121. Krutzfeldt J, Rajewsky N, Braich R, et al. Silencing of microRNAs in vivo with 'antagomirs'. Nature. 2005;438(7068):685–9.
122. Felli N, Fontana L, Pelosi E, et al. MicroRNAs 221 and 222 inhibit normal erythropoiesis and erythroleukemic cell growth via kit receptor down-modulation. Proc Natl Acad Sci USA. 2005;102(50):18081–6.
123. Takamizawa J, Konishi H, Yanagisawa K, et al. Reduced expression of the let-7 microRNAs in human lung cancers in association with shortened postoperative survival. Cancer Res. 2004;64(11):3753–6.
124. Coughlin SS, Calle EE, Patel AV, Thun MJ. Predictors of pancreatic cancer mortality among a large cohort of United States adults. Cancer Causes Control. 2000;11(10):915–23.
125. Petersen GM, de Andrade M, Goggins M, et al. Pancreatic cancer genetic epidemiology consortium. Cancer Epidemiol Biomarkers Prev. 2006;15(4):704–10.

126. Pogue-Geile KL, Chen R, Bronner MP, et al. Palladin mutation causes familial pancreatic cancer and suggests a new cancer mechanism. PLoS Med. 2006;3(12):e516.
127. Earl J, Yan L, Vitone LJ, et al. Evaluation of the 4q32-34 locus in European familial pancreatic cancer. Cancer Epidemiol Biomarkers Prev. 2006;15(10):1948–55.
128. Salaria SN, Illei P, Sharma R, et al. Palladin is overexpressed in the non-neoplastic stroma of infiltrating ductal adenocarcinomas of the pancreas, but is only rarely overexpressed in neoplastic cells. Cancer Biol Ther. 2007;6(3):324–8.
129. Klein AP, de Andrade M, Hruban RH, et al. Linkage analysis of chromosome 4 in families with familial pancreatic cancer. Cancer Biol Ther. 2007;6(3):320–3.
130. Slater E, Amrillaeva V, Fendrich V, et al. Palladin mutation causes familial pancreatic cancer: absence in European families. PLoS Med. 2007;4(4):e164.
131. Zogopoulos G, Rothenmund H, Eppel A, et al. The P239S palladin variant does not account for a significant fraction of hereditary or early onset pancreas cancer. Hum Genet. 2007;121(5):635–7.
132. Wang W, Chen S, Brune KA, Hruban RH, Parmigiani G, Klein AP. PancPRO: risk assessment for individuals with a family history of pancreatic cancer. J Clin Oncol. 2007;25(11):1417–22.
133. Giardiello FM, Brensinger JD, Tersmette AC, et al. Very high risk of cancer in familial Peutz-Jeghers syndrome. Gastroenterology. 2000;119(6):1447–53.
134. Goldstein AM, Fraser MC, Struewing JP, et al. Increased risk of pancreatic cancer in melanoma-prone kindreds with p16INK4 mutations. N Engl J Med. 1995;333(15):970–4.
135. Lowenfels AB, Maisonneuve P, Cavallini G, et al. Pancreatitis and the risk of pancreatic cancer. International Pancreatitis Study Group. N Engl J Med. 1993;328(20):1433–7.
136. Whitcomb DC, Gorry MC, Preston RA, et al. Hereditary pancreatitis is caused by a mutation in the cationic trypsinogen gene. Nat Genet. 1996;14(2):141–5.
137. Abraham SC, Wu TT, Klimstra DS, et al. Distinctive molecular genetic alterations in sporadic and familial adenomatous polyposis-associated pancreatoblastomas: frequent alterations in the APC/beta-catenin pathway and chromosome 11p. Am J Pathol. 2001;159(5):1619–27.
138. Yamamoto H, Itoh F, Nakamura H, et al. Genetic and clinical features of human pancreatic ductal adenocarcinomas with widespread microsatellite instability. Cancer Res. 2001;61(7):3139–44.
139. Wilentz RE, Goggins M, Redston M, et al. Genetic, immunohistochemical, and clinical features of medullary carcinoma of the pancreas: a newly described and characterized entity. Am J Pathol. 2000;156(5):1641–51.
140. Banville N, Geraghty R, Fox E, et al. Medullary carcinoma of the pancreas in a man with hereditary nonpolyposis colorectal cancer due to a mutation of the MSH2 mismatch repair gene. Hum Pathol. 2006;37(11):1498–502.
141. Goggins M, Offerhaus GJ, Hilgers W, et al. Pancreatic adenocarcinomas with DNA replication errors (RER+) are associated with wild-type K-ras and characteristic histopathology. Poor differentiation, a syncytial growth pattern, and pushing borders suggest RER+. Am J Pathol. 1998;152(6):1501–7.
142. Su GH, Hruban RH, Bansal RK, et al. Germline and somatic mutations of the STK11/LKB1 Peutz-Jeghers gene in pancreatic and biliary cancers. Am J Pathol. 1999;154(6):1835–40.
143. Furukawa T. Molecular genetics of intraductal papillary-mucinous neoplasms of the pancreas. J Hepatobiliary Pancreat Surg. 2007;14(3):233–7.
144. Sinilnikova OM, Mazoyer S, Bonnardel C, Lynch HT, Narod SA, Lenoir GM. BRCA1 and BRCA2 mutations in breast and ovarian cancer syndrome: reflection on the Creighton University historical series of high risk families. Fam Cancer. 2006;5(1):15–20.
145. Couch FJ, Johnson MR, Rabe KG, et al. The prevalence of BRCA2 mutations in familial pancreatic cancer. Cancer Epidemiol Biomarkers Prev. 2007;16(2):342–6.
146. Murphy KM, Brune KA, Griffin C, et al. Evaluation of candidate genes MAP2K4, MADH4, ACVR1B, and BRCA2 in familial pancreatic cancer: deleterious BRCA2 mutations in 17%. Cancer Res. 2002;62(13):3789–93.
147. Lal G, Liu G, Schmocker B, et al. Inherited predisposition to pancreatic adenocarcinoma: role of family history and germ-line p16, BRCA1, and BRCA2 mutations. Cancer Res. 2000;60(2):409–16.
148. Hahn SA, Greenhalf B, Ellis I, et al. BRCA2 germline mutations in familial pancreatic carcinoma. J Natl Cancer Inst. 2003;95(3):214–21.
149. van der Heijden MS, Brody JR, Dezentje DA, et al. In vivo therapeutic responses contingent on Fanconi anemia/BRCA2 status of the tumor. Clin Cancer Res. 2005;11(20):7508–15.
150. McCabe N, Lord CJ, Tutt AN, Martin NM, Smith GC, Ashworth A. BRCA2-deficient CAPAN-1 cells are extremely sensitive to the inhibition of Poly (ADP-Ribose) polymerase: an issue of potency. Cancer Biol Ther. 2005;4(9):934–6.
151. Bryant HE, Schultz N, Thomas HD, et al. Specific killing of BRCA2-deficient tumours with inhibitors of poly(ADP-ribose) polymerase. Nature. 2005;434(7035):913–7.
152. Brose MS, Rebbeck TR, Calzone KA, Stopfer JE, Nathanson KL, Weber BL. Cancer risk estimates for BRCA1 mutation carriers identified in a risk evaluation program. J Natl Cancer Inst. 2002;94(18):1365–72.
153. Thompson D, Easton DF. Cancer Incidence in BRCA1 mutation carriers. J Natl Cancer Inst. 2002;94(18):1358–65.

154. Skudra S, Staka A, Pukitis A, et al. Association of genetic variants with pancreatic cancer. Cancer Genet Cytogenet. 2007;179(1):76–8.
155. Boardman LA, Thibodeau SN, Schaid DJ, et al. Increased risk for cancer in patients with the Peutz-Jeghers syndrome. Ann Intern Med. 1998;128(11):896–9.
156. Hizawa K, Iida M, Matsumoto T, et al. Cancer in Peutz-Jeghers syndrome. Cancer. 1993;72(9):2777–81.
157. Spigelman AD, Murday V, Phillips RK. Cancer and the Peutz-Jeghers syndrome. Gut. 1989;30(11):1588–90.
158. Giardiello FM, Welsh SB, Hamilton SR, et al. Increased risk of cancer in the Peutz-Jeghers syndrome. N Engl J Med. 1987;316(24):1511–4.
159. Sahin F, Maitra A, Argani P, et al. Loss of Stk11/Lkb1 expression in pancreatic and biliary neoplasms. Mod Pathol. 2003;16(7):686–91.
160. Le Bodic L, Bignon JD, Raguenes O, et al. The hereditary pancreatitis gene maps to long arm of chromosome 7. Hum Mol Genet. 1996;5(4):549–54.
161. Rebours V, Boutron-Ruault MC, Schnee M, et al. Risk of pancreatic adenocarcinoma in patients with hereditary pancreatitis: a national exhaustive series. Am J Gastroenterol. 2008;103(1):111–9.
162. Rosty C, Geradts J, Sato N, et al. p16 Inactivation in pancreatic intraepithelial neoplasias (PanINs) arising in patients with chronic pancreatitis. Am J Surg Pathol. 2003;27(12):1495–501.
163. Cylwik B, Nowak HF, Puchalski Z, Barczyk J. Epithelial anomalies in chronic pancreatitis as a risk factor of pancreatic cancer. Hepatogastroenterology. 1998;45(20):528–32.
164. Ardengh JC, Lopes CV, Campos AD, Pereira de Lima LF, Venco F, Modena JL. Endoscopic ultrasound and fine needle aspiration in chronic pancreatitis: differential diagnosis between pseudotumoral masses and pancreatic cancer. J Pancreas. 2007;8(4):413–21.
165. Iordache S, Saftoiu A, Cazacu S, et al. Endoscopic ultrasound approach of pancreatic cancer in chronic pancreatitis patients in a tertiary referral centre. J Gastrointestin Liver Dis. 2008;17(3):279–84.
166. Varadarajulu S, Tamhane A, Eloubeidi MA. Yield of EUS-guided FNA of pancreatic masses in the presence or the absence of chronic pancreatitis. Gastrointest Endosc. 2005;62(5):728–736; quiz 751, 753.
167. Talar-Wojnarowska R, Gasiorowska A, Smolarz B, et al. Clinical significance of K-ras and c-erbB-2 mutations in pancreatic adenocarcinoma and chronic pancreatitis. Int J Gastrointest Cancer. 2005;35(1):33–41.
168. Maire F, Micard S, Hammel P, et al. Differential diagnosis between chronic pancreatitis and pancreatic cancer: value of the detection of KRAS2 mutations in circulating DNA. Br J Cancer. 2002;87(5):551–4.
169. Urgell E, Puig P, Boadas J, et al. Prospective evaluation of the contribution of K-ras mutational analysis and CA 19.9 measurement to cytological diagnosis in patients with clinical suspicion of pancreatic cancer. Eur J Cancer. 2000;36(16):2069–75.
170. Queneau PE, Adessi GL, Thibault P, et al. Early detection of pancreatic cancer in patients with chronic pancreatitis: diagnostic utility of a K-ras point mutation in the pancreatic juice. Am J Gastroenterol. 2001;96(3):700–4.
171. Uehara H, Nakaizumi A, Tatsuta M, et al. Diagnosis of pancreatic cancer by detecting telomerase activity in pancreatic juice: comparison with K-ras mutations. Am J Gastroenterol. 1999;94(9):2513–8.
172. Witt H, Luck W, Hennies HC, et al. Mutations in the gene encoding the serine protease inhibitor, Kazal type 1 are associated with chronic pancreatitis. Nat Genet. 2000;25(2):213–6.
173. Teich N, Rosendahl J, Toth M, Mossner J, Sahin-Toth M. Mutations of human cationic trypsinogen (PRSS1) and chronic pancreatitis. Hum Mutat. 2006;27(8):721–30.
174. Teich N, Nemoda Z, Kohler H, et al. Gene conversion between functional trypsinogen genes PRSS1 and PRSS2 associated with chronic pancreatitis in a six-year-old girl. Hum Mutat. 2005;25(4):343–7.
175. Sahin-Toth M. Human cationic trypsinogen. Role of Asn-21 in zymogen activation and implications in hereditary pancreatitis. J Biol Chem. 2000;275(30):22750–5.
176. Sahin-Toth M, Toth M. Gain-of-function mutations associated with hereditary pancreatitis enhance autoactivation of human cationic trypsinogen. Biochem Biophys Res Commun. 2000;278(2):286–9.
177. Kloppel G, Detlefsen S, Feyerabend B. Fibrosis of the pancreas: the initial tissue damage and the resulting pattern. Virchows Arch. 2004;445(1):1–8.
178. Kloppel G. Chronic pancreatitis, pseudotumors and other tumor-like lesions. Mod Pathol. 2007;20 Suppl 1:S113–131.
179. Lowenfels AB, Maisonneuve P, DiMagno EP, et al. Hereditary pancreatitis and the risk of pancreatic cancer. International Hereditary Pancreatitis Study Group. J Natl Cancer Inst. 1997;89(6):442–6.
180. Lowenfels AB, Maisonneuve P, Whitcomb DC, Lerch MM, DiMagno EP. Cigarette smoking as a risk factor for pancreatic cancer in patients with hereditary pancreatitis. JAMA. 2001;286(2):169–70.
181. Rustgi AK. The genetics of hereditary colon cancer. Genes Dev. 2007;21(20):2525–38.
182. Lynch HT, Voorhees GJ, Lanspa SJ, McGreevy PS, Lynch JF. Pancreatic carcinoma and hereditary nonpolyposis colorectal cancer: a family study. Br J Cancer. 1985;52(2):271–3.
183. Calhoun ES, Jones JB, Ashfaq R, et al. BRAF and FBXW7 (CDC4, FBW7, AGO, SEL10) mutations in distinct subsets of pancreatic cancer: potential therapeutic targets. Am J Pathol. 2003;163(4):1255–60.

184. Lynch HT, Frichot 3rd BC, Lynch JF. Familial atypical multiple mole-melanoma syndrome. J Med Genet. 1978;15(5):352–6.
185. Goldstein AM, Goldin LR, Dracopoli NC, Clark Jr WH, Tucker MA. Two-locus linkage analysis of cutaneous malignant melanoma/dysplastic nevi. Am J Hum Genet. 1996;58(5):1050–6.
186. Zhu G, Duffy DL, Eldridge A, et al. A major quantitative-trait locus for mole density is linked to the familial melanoma gene CDKN2A: a maximum-likelihood combined linkage and association analysis in twins and their sibs. Am J Hum Genet. 1999;65(2):483–92.
187. Lynch HT, Fusaro RM, Lynch JF, Brand R. Pancreatic cancer and the FAMMM syndrome. Fam Cancer. 2008;7(1):103–12.
188. Lynch HT, Brand RE, Hogg D, et al. Phenotypic variation in eight extended CDKN2A germline mutation familial atypical multiple mole melanoma-pancreatic carcinoma-prone families: the familial atypical mole melanoma-pancreatic carcinoma syndrome. Cancer. 2002;94(1):84–96.
189. Groden J, Thliveris A, Samowitz W, et al. Identification and characterization of the familial adenomatous polyposis coli gene. Cell. 1991;66(3):589–600.
190. Kinzler KW, Nilbert MC, Su LK, et al. Identification of FAP locus genes from chromosome 5q21. Science. 1991;253(5020):661–5.
191. Giardiello FM, Offerhaus GJ, Lee DH, et al. Increased risk of thyroid and pancreatic carcinoma in familial adenomatous polyposis. Gut. 1993;34(10):1394–6.
192. Maire F, Hammel P, Terris B, et al. Intraductal papillary and mucinous pancreatic tumour: a new extracolonic tumour in familial adenomatous polyposis. Gut. 2002;51(3):446–9.
193. Manolio TA. Genomewide association studies and assessment of the risk of disease. N Engl J Med. 2010;363(2):166–76.
194. Hindorff LA, Sethupathy P, Junkins HA, et al. Potential etiologic and functional implications of genome-wide association loci for human diseases and traits. Proc Natl Acad Sci USA. 2009;106(23):9362–7.
195. Low SK, Kuchiba A, Zembutsu H, et al. Genome-wide association study of pancreatic cancer in Japanese population. PLoS One. 2010;5(7):e11824.
196. Petersen GM, Amundadottir L, Fuchs CS, et al. A genome-wide association study identifies pancreatic cancer susceptibility loci on chromosomes 13q22.1, 1q32.1 and 5p15.33. Nat Genet. 2010;42(3):224–8.
197. Amundadottir L, Kraft P, Stolzenberg-Solomon RZ, et al. Genome-wide association study identifies variants in the ABO locus associated with susceptibility to pancreatic cancer. Nat Genet. 2009;41(9):986–90.
198. Diergaarde B, Brand R, Lamb J, et al. Pooling-based genome-wide association study implicates gamma-glutamyltransferase 1 (GGT1) gene in pancreatic carcinogenesis. Pancreatology. 2010;10(2–3):194–200.
199. Aird I, Bentall HH, Roberts JA. A relationship between cancer of stomach and the ABO blood groups. Br Med J. 1953;1(4814):799–801.
200. Marcus DM. The ABO and Lewis blood-group system. Immunochemistry, genetics and relation to human disease. N Engl J Med. 1969;280(18):994–1006.
201. Itzkowitz SH, Yuan M, Ferrell LD, et al. Cancer-associated alterations of blood group antigen expression in the human pancreas. J Natl Cancer Inst. 1987;79(3):425–34.
202. Dong JT, Chen C. Essential role of KLF5 transcription factor in cell proliferation and differentiation and its implications for human diseases. Cell Mol Life Sci. 2009;66(16):2691–706.
203. Nakamura Y, Migita T, Hosoda F, et al. Kruppel-like factor 12 plays a significant role in poorly differentiated gastric cancer progression. Int J Cancer. 2009;125(8):1859–67.
204. Chen C, Brabham WW, Stultz BG, et al. Defining a common region of deletion at 13q21 in human cancers. Genes Chromosomes Cancer. 2001;31(4):333–44.
205. Baudis M, Cleary ML. Progenetix.net: an online repository for molecular cytogenetic aberration data. Bioinformatics. 2001;17(12):1228–9.
206. Kainu T, Juo SH, Desper R, et al. Somatic deletions in hereditary breast cancers implicate 13q21 as a putative novel breast cancer susceptibility locus. Proc Natl Acad Sci USA. 2000;97(17):9603–8.
207. Lee YK, Moore DD. Liver receptor homolog-1, an emerging metabolic modulator. Front Biosci. 2008;13:5950–8.
208. Li M, Xie YH, Kong YY, Wu X, Zhu L, Wang Y. Cloning and characterization of a novel human hepatocyte transcription factor, hB1F, which binds and activates enhancer II of hepatitis B virus. J Biol Chem. 1998;273(44):29022–31.
209. Fayard E, Schoonjans K, Annicotte JS, Auwerx J. Liver receptor homolog 1 controls the expression of carboxyl ester lipase. J Biol Chem. 2003;278(37):35725–31.
210. Pare JF, Malenfant D, Courtemanche C, et al. The fetoprotein transcription factor (FTF) gene is essential to embryogenesis and cholesterol homeostasis and is regulated by a DR4 element. J Biol Chem. 2004;279(20):21206–16.
211. Yamamoto K, Okamoto A, Isonishi S, Ochiai K, Ohtake Y. A novel gene, CRR9, which was up-regulated in CDDP-resistant ovarian tumor cell line, was associated with apoptosis. Biochem Biophys Res Commun. 2001;280(4):1148–54.

212. Kim NW, Piatyszek MA, Prowse KR, et al. Specific association of human telomerase activity with immortal cells and cancer. Science. 1994;266(5193):2011–5.
213. Pooley KA, Tyrer J, Shah M, et al. No association between TERT-CLPTM1L single nucleotide polymorphism rs401681 and mean telomere length or cancer risk. Cancer Epidemiol Biomarkers Prev. 2010;19(7):1862–5.
214. Landi MT, Chatterjee N, Yu K, et al. A genome-wide association study of lung cancer identifies a region of chromosome 5p15 associated with risk for adenocarcinoma. Am J Hum Genet. 2009;85(5):679–91.
215. Kaneda H, Arao T, Tanaka K, et al. FOXQ1 is overexpressed in colorectal cancer and enhances tumorigenicity and tumor growth. Cancer Res. 2010;70(5):2053–63.
216. Mangino M, Brouilette S, Braund P, et al. A regulatory SNP of the BICD1 gene contributes to telomere length variation in humans. Hum Mol Genet. 2008;17(16):2518–23.
217. Kobitsu K, Tsutsumi M, Tsujiuchi T, et al. Shortened telomere length and increased telomerase activity in hamster pancreatic duct adenocarcinomas and cell lines. Mol Carcinog. 1997;18(3):153–9.
218. Buchler P, Conejo-Garcia JR, Lehmann G, et al. Real-time quantitative PCR of telomerase mRNA is useful for the differentiation of benign and malignant pancreatic disorders. Pancreas. 2001;22(4):331–40.
219. Yu DM, Yao TW, Chowdhury S, et al. The dipeptidyl peptidase IV family in cancer and cell biology. FEBS J. 2010;277(5):1126–44.
220. Yachida S, Jones S, Bozic I, et al. Distant metastasis occurs late during the genetic evolution of pancreatic cancer. Nature. 2010;467(7319):1114–7.
221. Campbell PJ, Yachida S, Mudie LJ, et al. The patterns and dynamics of genomic instability in metastatic pancreatic cancer. Nature. 2010;467(7319):1109–13.
222. McClintock B. The stability of broken ends of chromosomes in *Zea mays*. Genetics. 1941;26(2):234–82.

Chapter 6
Diagnostic, Prognostic, and Therapeutic Value of Gene Signatures in Non-Small Cell Lung Cancer

Rafael Rosell, Miquel Taron, Christian Diego Rolfo, Delvys Rodriguez-Abreu, and Jia Wei

Introduction

Lung cancer is the primary cause of cancer mortality in developed countries. In the majority of cases, lung cancer is metastatic at the time of diagnosis. Although low-dose spiral computed tomography (CT) has proven to be effective in the early detection of lung cancer, providing higher resectability and higher long-term survival rates, the capacity of annual CT screening to reduce lung cancer mortality in heavy smokers has yet to be demonstrated. Numerous ongoing large-scale randomized trials are under way in high-risk individuals, with different study designs. The initial results should be available in the next 2 years [1]. Biomarker research in CT screening trials, combining noninvasive genomic and proteomic analyses, could lead to a significant improvement in early detection, offering a potential contribution to diagnostic algorithms, assessment of individual risk, and management of CT-detected cancers [1]. Surgical resection for early (stage I–II) non-small cell lung cancer (NSCLC) remains the only reliable treatment for cure. Patients with early disease who do not undergo surgery have a median survival of less than 1 year, while those who undergo appropriate surgery have a median survival of more than 4 years. A decision not to undergo surgery by patients with newly diagnosed early NSCLC has been associated with perceptions of communication and prognosis, older age, multiple comorbidities, and black race [2]. Even in patients without lymph node involvement, tumor cells were found in bone marrow of 54% of the patients [3]. MicroRNAs (miRNAs) are present in human plasma in a remarkably stable form that is protected from endogenous RNase activity. miRNAs could be an ideal class of blood-based biomarkers for cancer detection [4]. Serum levels of four miRNAs (miR-486, miR-30d, miR-1, and miR-499) were significantly associated with survival in stage I–IIIA NSCLC patients who were resected and received adjuvant chemotherapy [5]. Although intense efforts have been dedicated to the development of useful biomarkers, many have not been established for diagnosis or decision-making in lung cancer. Blood-based biomarkers could have a great advantage because only a small tumor specimen

R. Rosell (✉) • M. Taron
Department of Medical Oncology, Germans Trias I Pujol, Catalan Institute of Oncology, Badalona, Spain
e-mail: rrosell@iconcologia.net

C.D. Rolfo
Medical Oncology, Oncology Unit, Clinica Rotger, Palma de Mallorca, Spain

D. Rodriguez-Abreu
Department of Medical Oncology, Universitario Insular de Gran Canaria, Las Palmas de Gran Canaria, Spain

J. Wei
Medical School of Nanjing University and Clinical Cancer Institute of Nanjing University, The Comprehensive Cancer Centre of Drum Tower Hospital, Nanjing, China

A. Russo et al. (eds.), *Diagnostic, Prognostic and Therapeutic Value of Gene Signatures*, Current Clinical Pathology, DOI 10.1007/978-1-61779-358-5_6, © Springer Science+Business Media, LLC 2012

can be obtained by usual diagnostic modalities, such as fine-needle aspiration and transbronchial biopsy. Circulating tumor cell (CTC) count (using the CellSearch System, Veridex) was significantly higher in NSCLC patients than in nonmalignant patients [6]. However, the CTC test showed a moderate diagnostic performance. The CTC test has been approved by the US Food and Drug Administration (FDA) for monitoring of blood from metastatic breast and colon cancer patients [6]. A microfluidic platform for detecting CTCs has also been developed, and CTCs were detected in all blood samples taken from NSCLC patients, permitting the identification of EGFR mutations in 92% of the cases [7]. Genomic assays for the prediction of clinical outcome are being used in breast cancer, including MammaPrint, and a 21-gene recurrence score (Oncotype DX) [8].

Prognosis in Early NSCLC: The Role of Gene Signatures

A recent meta-analysis based on individual data based on 11,107 resected NSCLC patients showed that the survival benefit of adjuvant chemotherapy was 4% at 5 years and the survival benefit of adjuvant chemotherapy plus radiotherapy was 5% at 5 years [9]. Two new adjuvant chemotherapy trials with paclitaxel plus carboplatin have shown a trend toward improved survival in early NSCLC [10, 11]. However, the benefit of adjuvant treatment remains suboptimal. In early resected NSCLC, there are three main issues to be resolved. Firstly, we must be able to distinguish two groups of patients: those who will not relapse (low-risk patients) – and who can thus be spared adjuvant treatment – and those who will (high-risk patients). Secondly, we must personalize adjuvant treatment in the group of high-risk patients, since a majority of these patients will be resistant to cisplatin-based chemotherapy or postoperative radiotherapy. Finally, we must predict the pattern of metastases, since this can help design-specific targeted treatments. In breast cancer patients, gene signatures predicting lung [12], bone [13] or brain [14] metastases have been reported. The gene expression analyses for predicting brain metastases include the cyclooxygenase (COX2), the epidermal growth factor receptor (EGFR) ligand HBEGF, and the a2,6-sialyltransferase ST6GALNAC5 [14]. A novel strategy will be the use of PCR-based quantitative analysis for personalized tumor monitoring of plasma samples. An almost universal feature of cancer is the widespread rearrangement of chromosomes as a result of chromosome instability. The consequences of chromosomal instability include copy number alterations (duplications, amplifications, deletions), inversions, insertions, and translocations. Tumor-specific chromosomal rearrangements could be used as biomarkers for monitoring tumor response and detecting residual disease after surgery. A method called personalized analysis of rearranged ends can identify translocations in solid tumors. PCR with primers spanning the breakpoints was able to detect mutant DNA molecules present at levels lower than 0.001% and identified mutated circulating DNA in patient plasma samples [15].

A retrospective analysis of 787 patients, mainly early-stage NSCLC, performed at Duke University, identified low- and high-risk patients. In patients younger than 70 years, high-risk patients (with the shortest recurrence-free survival) had increased activation of the Src and tumor necrosis factor pathways compared to low-risk patients. High-risk patients older than 70 years demonstrated increased activation of the wound healing and invasiveness pathways compared to low-risk patients. In women, high-risk patients demonstrated increased activation of the invasiveness and STAT3 pathways, while high-risk men demonstrated increased activation of the STAT3, tumor necrosis factor, EGFR, and wound healing pathways [16].

Tumor growth encompasses many aspects of normal wound healing, and a wound response (WR) gene expression signature is reactivated in many types of human cancers, including breast and lung [17]. The WR signature is composed of 512 genes that define the transcriptional response of fibroblasts to serum, the soluble fraction of clotted blood. In early breast cancer and lung adenocarcinoma, the WR signature provides prognostic risk stratification of metastasis development (Fig. 6.1) [17].

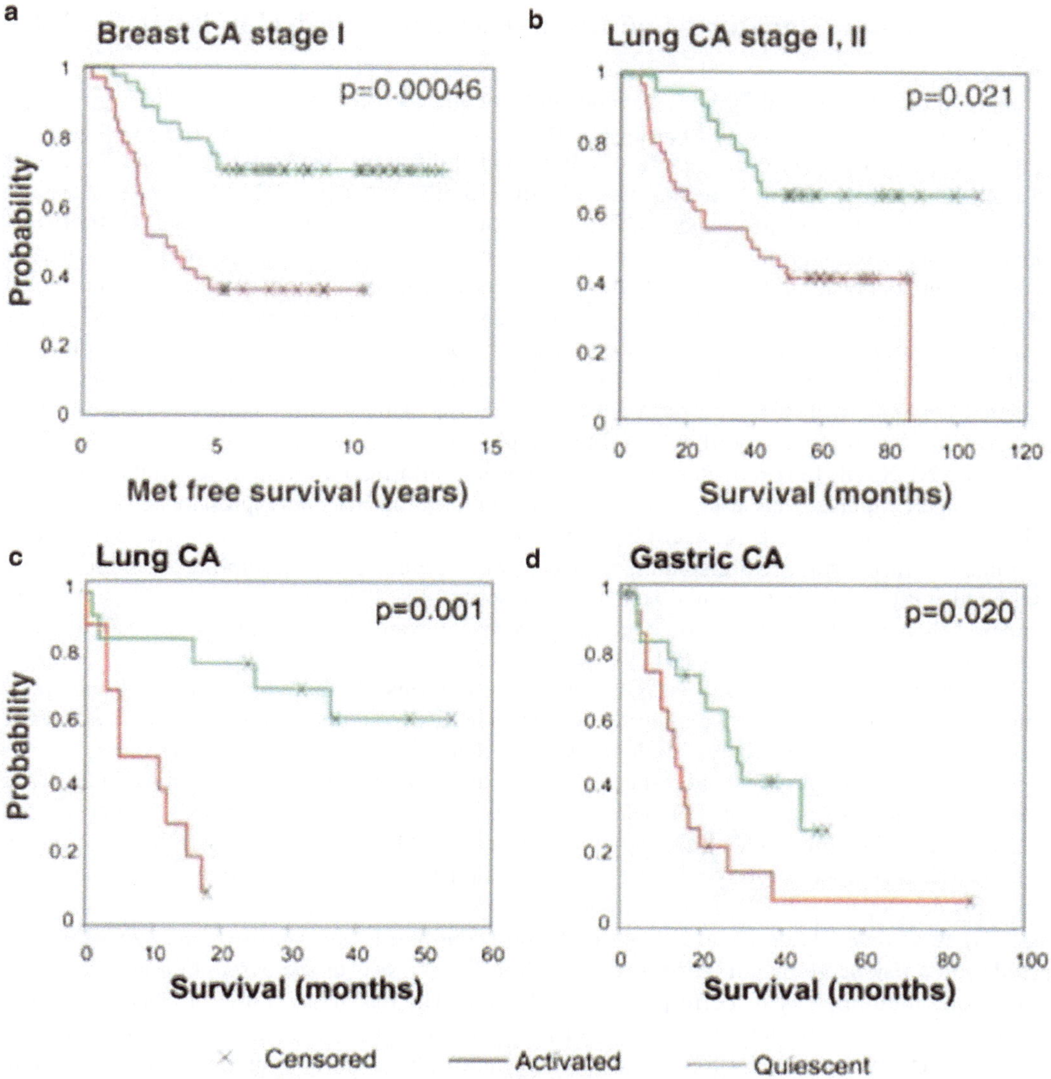

Fig. 6.1 The wound response signature integrates 512 genes and is also defined as the fibroblast core serum response. Kaplan–Meier survival curves are shown for early breast, lung, and gastric cancer according to the activation of the wound response signature [17]

Several gene expression signatures, generally containing nonoverlapping genes, provide similar predictive information on clinical outcome, and a model combining several signatures did not perform better than did each of the signatures separately. These signatures may be largely different from one another as regards gene identity, but they occupy overlapping prognostic space [8, 18]. The invasiveness gene signature (IGS), containing 186 genes, is prognostic not only in breast, but also in other tumors, including lung [19]. The IGS includes genes involved in the nuclear factor-kB pathway, the RAS-mitogen-activated protein kinase pathway, and epigenetic control of gene expression [19]. Of the 186 genes in the IGS, only six overlap with the WR signature [19].

Gene Expression Signatures and Recurrence-Free Survival in Early NSCLC

Among completely resected NSCLC patients, 40% of stage I, 66% of stage II, and 75% of stage IIIA patients die within 5 years of resection [20], mainly due to the development of distant metastases. In some clinical trials, the benefit of adjuvant chemotherapy has been negligible in stage IB [21, 22], although a recent meta-analysis has identified a moderate benefit [23]. Although at present there are no reliable clinical predictors of relapse after surgery in early-stage NSCLC, transcriptional analysis of primary tumors has identified gene expression profiles strongly related to disease recurrence in adenocarcinoma [24–30] and, to a lesser extent, in squamous cell carcinoma (SCC) [27–30]. The lung metagene model is a gene expression profile that predicts recurrence in early NSCLC (including stage IA) with an overall accuracy of 72% [30]. A meta-analysis [29] of data sets from seven microarray studies [24, 26, 31, 32] identified a 64-gene-expression signature that predicted survival with 85% accuracy. However, the use of these gene signatures has not been implemented due to practical difficulties. Importantly, the simultaneous coactivation of two genes (TTF1 and NKX2–8) [33] provides the same prognostic information of short recurrence-free survival in stage I patients, similar to the metagene model [30], the 64-gene-expression signature [29], and the 50-gene signature (Fig. 6.2) [26]. The coupled overexpression of TTF1 and NKX2–8 is associated with resistance to cisplatin, taxanes, gemcitabine, and vinorelbine but is positively correlated with response to pemetrexed [33]. However, the majority of predictive models are hampered by the difficulty of reproducing them. A recent example demonstrated the lack of value of the lung metagene model [30, 101].

The quantitative PCR (QPCR) assay is convenient in terms of laboratory work-load and applicable for large-scale routine use, making it a viable alternative to more complex microarrays. QPCR also allows for accurate and reproducible RNA quantification. The expression pattern of eight genes determined by QPCR correlated with survival in lung adenocarcinoma [34]. Similarly, QPCR-based three- [35], four- [36], and five- [37] gene signatures and a five-miRNA signature [38] correlated with metastasis-free survival and overall survival in early NSCLC. A three-gene prognostic model [39] includes a key gene, hypoxia-inducible factor 1a (HIF1a). The construction of the small gene signatures developed with QPCR is based on the prognostic value of each gene as determined in a multivariate analysis. Each gene that is significant according to the multivariate analysis is then included in a risk score model, generated by adding the z-scores of the expression levels of each of the genes multiplied by its corresponding coefficient. The risk score is used to classify patients into high or low risk of metastasis and death [37]. The five-gene signature [37] comprises dual-specificity phosphatase 6 (DUSP6), monocyte-to-macrophage differentiation-associated protein (MMD), signal transducer and activator of transcription 1 (STAT1), HER3/neu receptor tyrosine kinase (ERBB3), and lymphocyte-specific protein tyrosine kinase (LCK).

Intriguingly, special AT-rich binding protein 1 (SATB1), originally identified as a protein that recognized double-stranded DNA with a high degree of base-unpairing [40], is a genome organizer that upregulates metastasis-associated genes, including genes involved in EGF signaling, such as ERBB1, ERBB2, ERBB3, and ERBB4 [41]. In addition, SATB1 upregulates multiple other genes that stimulate invasion and mediate angiogenesis and bone metastasis, such as connective tissue growth factor. SATB1 nuclear staining significantly correlates with survival in 985 patients with ductal breast carcinoma stratified by SATB1 expression level [41]. Importantly, since SATB1 tethers multiple genomic loci and regulates chromatin structure and gene expression [42], the analysis of SATB1 mRNA or protein expression could provide important prognostic information that merits testing in NSCLC in the clinical setting. Fibronectin is also upregulated by SATB1 [41] and has been identified in a six-gene expression signature that predicted survival in diffuse large-B-cell lymphoma [43]. Fibronectin, an extracellular matrix glycoprotein, is highly expressed in tobacco-related lung disease and stimulates lung cancer growth [44]. In our experience, median survival of resected SCC patients with low levels of fibronectin mRNA was not reached while it was 31 months for those with high levels ($P = 0.002$) [35].

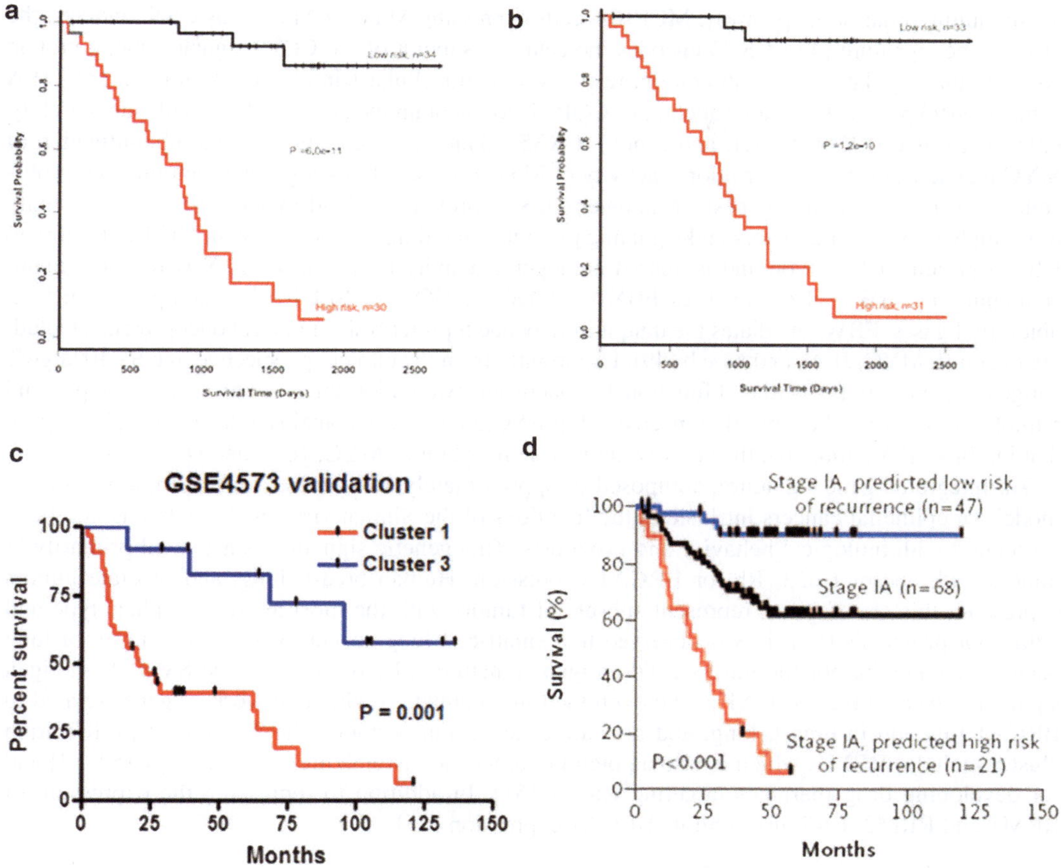

Fig. 6.2 Kaplan–Meier survival curves for early resected NSCLC according to different gene signatures: (**a**) 64 genes [29] (**b**) 50 genes [26] (**c**) the two-gene signature of TTF1 and NKX2-8 [33] and (**d**) the metagene signature [30, 101]. (Fig. 6.2c reprinted with permission, Copyright 2009 National Academy of Sciences, USA); (Fig. 6.2d reprinted with permission, Copyright Massachusetts Medical Society 2006)

miRNAs are attractive candidates as upstream regulators of metastatic progression because miRNAs can posttranscriptionally regulate entire sets of genes. Quantitative stem-loop PCR of five miRNAs showed that patients with high-risk scores in their miRNA signatures had poor overall and disease-free survival compared to patients with low-risk scores [38]. The five-miRNA signature includes two protective miRNAs (let-7a and miR-221) and three miRNAs indicative of poor survival (miR-137, miR-372, and miR-182) [38]. Interestingly, miR-335 regulates a set of metastasis genes and has been shown to predict bone and lung metastases in breast cancer [45].

The WR signature predicts survival in several tumors, including NSCLC (Fig. 6.1) [17]. It is important to know that the coordinate amplification of CSN5 (also known as JAB1 or COPS5, residing on 8q13) and MYC (8q24) regulate WR signature activation in breast cancer. Coexpression of CSN5 with MYC is sufficient to induce the WR signature [46]. A high expression level of both CSN5 and MYC was a significant predictor of poor patient survival in breast tumors, with efficacy equivalent to that observed for the WR signature [46].

The induction of a proteasome signature was associated with an activated WR signature. MCF10A cells with the activated WR signature were more susceptible to death by drugs that inhibit the ubiquitin–proteasome pathway. MCF10A cells expressing MYC or MYC plus CSN5 induced the proteasome signature [47]. CSN5 encodes the catalytic subunit of the COP9 signalosome, a protein complex that regulates cell proliferation, response to extracellular stimuli, cell migration, and DNA damage checkpoints. The main function of COP9 is to maintain the activity of the multisubunit ubiquitin ligase SCF (SKP1, CUL1, and F-box). CNS5 enhances the cotranscriptional ubiquitination of MYC that activates the transcriptional activity of MYC on a set of target genes promoting cell proliferation, invasion, and angiogenesis. Monomeric CSN5 protein can bind to and modulate the activity of multiple transcription factors and signaling proteins, including interactions with HIF1-a, leading to HIF-1a protein stabilization and increased angiogenic activity [48]. F-box and WD repeat domain-containing 7 (FBW7, also known as FBXW7, CDC4, AGO, and SEL10) is a component of SCF ubiquitin ligases. FBW7 mediates the ubiquitin-dependent proteolysis of several oncoproteins, including cyclin E, MYC, JUN, and Notch [49]. Importantly from the clinical perspective, the SV40 large T antigen oncoprotein binds to and functionally inactivates two major tumor suppressor genes, p53 and retinoblastoma (Rb), that are often inactivated in NSCLC. Large T contains a decoy phospho-degron that inhibits FBW7 function, thus repressing its role in cyclin E, MYC, JUN, and Notch [50].

An integrated gene signature, composed of approximately 150 genes, from multiple transgenic models of epithelial cancers intrinsic to the functions of the Simian virus 40 T/t antigens, has been associated with biological behavior and prognosis. This genetic signature is activated primarily in tumors with aberrant p53, Rb, or BRCA1 expression. Human breast, lung, and prostate tumors expressing this set of genes represent subsets of tumors with the most aggressive phenotype and with poor prognosis [51]. It was observed that small-cell lung cancer, SCC, and a subset of lung adenocarcinoma harbor the intrinsic T/t-antigen signature [51]. Analysis of the SV40 T/t-antigen signature revealed that BRCA1 is overexpressed in conjunction with a network of genes related to BRCA1 function in breast, lung, and prostate cancers. Furthermore, the T/t-antigen proliferation cluster includes RRM1, which is also a potential target for customizing chemotherapy [52–54] and for developing drug therapies targeting RRM1 [51]. In addition to repressing the expression of RRM1 and RRM2, let-7 also inhibits BRCA1 expression [55].

Prognostic and Predictive Roles of BRCA1

We performed QPCR in frozen lung cancer tissue specimens from 126 early NSCLC patients who had undergone surgical resection and evaluated the association between survival and expression levels of nine genes involved in DNA repair pathways and in invasion and metastasis. For validation, we used paraffin-embedded specimens from 58 other NSCLC patients. A strong intergene correlation was observed between expression levels of all nine genes except nuclear factor of activated T cells (NFAT), for example, between ERCC1, RRM1, and BRCA1. Along with disease stage (stage I vs II vs III), BRCA1 mRNA expression significantly correlated with overall survival (HR, 1.98; $P=0.02$). In the independent cohort of 58 patients, BRCA1 mRNA expression also significantly correlated with survival (HR, 2.4; $P=0.04$) [56]. When only stage I patients were examined, median survival was significantly different according to expression levels of ERCC1, MZF1, Twist, and BRCA1 [56]. Our findings indicate that although BRCA1 is closely related to ERCC1, RRM1, and other genes like MZF1, it stands out as the most significant prognostic marker of relapse. Patients whose tumors had high BRCA1 expression had significantly worse survival and should be candidates for adjuvant chemotherapy. Intriguingly, in vitro studies have shown that BRCA1 can regulate differential sensitivity to different classes of chemotherapy agents [57, 58]. The absence of BRCA1 results in high sensitivity to cisplatin, whereas its presence increases sensitivity to antimicrotubule agents [57, 58]. Therefore, it is plausible that patients with the highest expression levels would receive more benefit from antimicrotubule, nonplatinum-based chemotherapy.

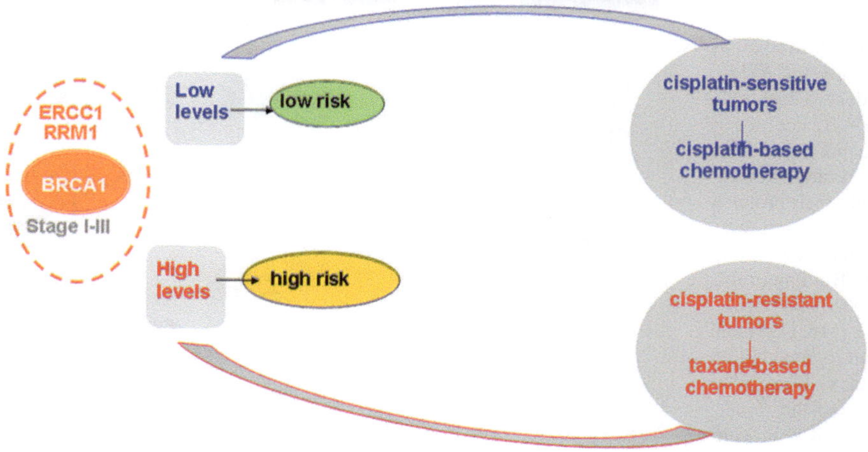

Fig. 6.3 The BRCA1 model shows the potential prognostic and predictive relevance in resected NSCLC. Low levels of BRCA1 mRNA have a lower risk of disease recurrence. These tumors could be sensitive to platinum combinations. In contrast, patients with high levels of BRCA1 mRNA have a higher risk of relapse and could be more resistant to platinum combinations. Alternatively, these tumors could be sensitive to tubulin-binding agents, such as taxanes [56]

The fact that high levels of ERCC1 or RRM1 transcripts conferred a higher risk of relapse [56] provides further evidence for the role of the loss of let-7 in upregulation of ERCC1 and RRM1, as well as BRCA1 [55] and for the upregulation of BRCA1 and RRM1 in the SV40 T/t-antigen signature [51]. Paradoxically, contradictory findings [59, 60], leading to opposed strategies of customizing adjuvant chemotherapy, have reported that the lack of ERCC1 protein implies a higher risk of relapse and a greater sensitivity to cisplatin-based chemotherapy [59]. Nevertheless, the clinical evidence that overexpression of ERCC1, RRM1, and especially BRCA1 confers poor survival in early NSCLC patients indicates the high risk involved in adjuvant chemotherapy. Against the current standard of cisplatin-based chemotherapy, noncisplatin-based chemotherapy, including antimicrotubule drugs, may be the proper treatment for the majority of patients with a high risk of relapse (Fig. 6.3) [56].

It has been further demonstrated that genes belonging to DNA repair (including BRCA1) and replication pathways are overexpressed in resected NSCLC patients and associated with a more aggressive phenotype [61].

BRCA1: A Potential Biomarker for Cisplatin-Based Chemotherapy

Experimental findings suggest that both DNA double-strand breaks (DSBs) and a DNA damage response (DDR) can be induced by ionizing radiation [62], hypoxia [63], DNA-damaging agents, and activated oncogenes [64]. In precancerous lesions, p53 binding protein 1 (53BP1) localized at foci, and histone H2AX, ataxia telangiectasia (ATM), and checkpoint kinase 2 (Chk2) were phosphorylated, suggesting the presence of DNA DSBs [65, 66]. In lung cancer, there is evidence for the presence of DSBs when phosphorylated histone H2AX and 53BP1 foci are present with a high proliferation index but low levels of apoptosis [64, 65]. More than 50% of surgically resected lung cancers show phosphorylation of Chk2 [66]. Many proteins, including ATM, g-H2AX, mediator of DNA damage checkpoint protein (MDC1), BRCA1, Chk1, and Chk2, are involved in the ionizing radiation-induced DDR pathway [62]. Under nonirradiated normoxic conditions, g-H2AX and 53BP1 are not activated; however, under nonirradiated anoxic conditions, g-H2AX can be induced

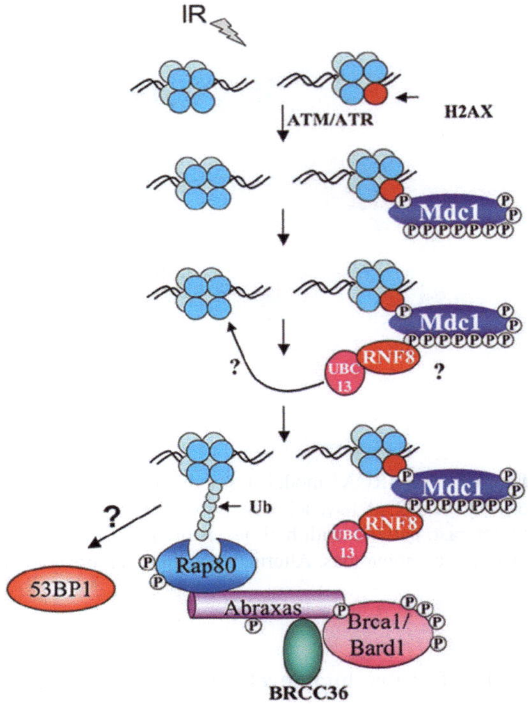

Fig. 6.4 In response to DNA DSBs induced by irradiation or chemotherapy, the ATM/ATR-dependent phosphorylation of H2AX creates gH2AX, the initial signal for subsequent accumulation of signaling and repair proteins to DNA breaks to form the so-called ionizing radiation-induced foci. Binding of the mediator of DNA damage checkpoint 1 (Mdc1) protein to the phosphorylated tail of H2AX at sites of DNA breakage recruits the ubiquitin ligase RNF8, which generates ubiquitin chains bound by RAP80 and Abraxas, which in turn recruit BRCA1 [71] (Copyright 2007 National Academy of Sciences, USA)

through the chromatin [63]. At the core of DDR signaling, ATM is central, activating g-H2AX. A large-scale proteomic analysis of proteins phosphorylated in DDR identified multiple super-complexes, including BRCA1, the COP9 signalosome, and the AKT-insulin pathway [67].

A proposed model for DDR to irradiation involves the formation of a BRCA1 complex [68, 69]. In DDR, ATM and ATR phosphorylate H2AX on Ser-139 [70], which serves to recruit the MDC1 protein to chromatin, where it is also phosphorylated. RNF8/UBC13 complexes go to sites of DNA damage through their forkhead domain and initiate the synthesis of K63 polyubiquitin chains on chromatin that recruit the BRCA1 complex through the ubiquitin-interacting motif domains (UIM) of RAP80 (Fig. 6.4) [71, 72]. RAP80 targets a complex containing Abraxas, BRCA1–BARD1 (BRCA1-associated ring domain protein 1) and BRCC36 [68, 69]. BRCC36 is frequently overexpressed in breast cancer, and its depletion disrupts irradiation-induced phosphorylation of BRCA1, thereby sensitizing breast cancer cells to irradiation-induced apoptosis [73]. Cells lacking MDC1 are also sensitive to ionizing irradiation [74].

The Potential Relevance of BRCA1 Sumoylation

The limited efficacy of current chemotherapy approaches is epitomized in metastatic (stage IV) NSCLC, where median survival is only 10–11 months with either noncustomized platinum-based chemotherapy [75, 76] or customized cisplatin-based chemotherapy based on ERCC1 mRNA expression [77], and the 2-year survival rate is only 14–21% [75–77].

Receptor-associated protein 80 (RAP80) or ubiquitin-interacting motif containing 1 (UIMC1) is a nuclear protein containing two functional UIMs at its amino terminus. It has been shown that RAP80 plays a critical role in DDR signaling [68, 69, 78, 79]. These studies report that RAP80 translocates

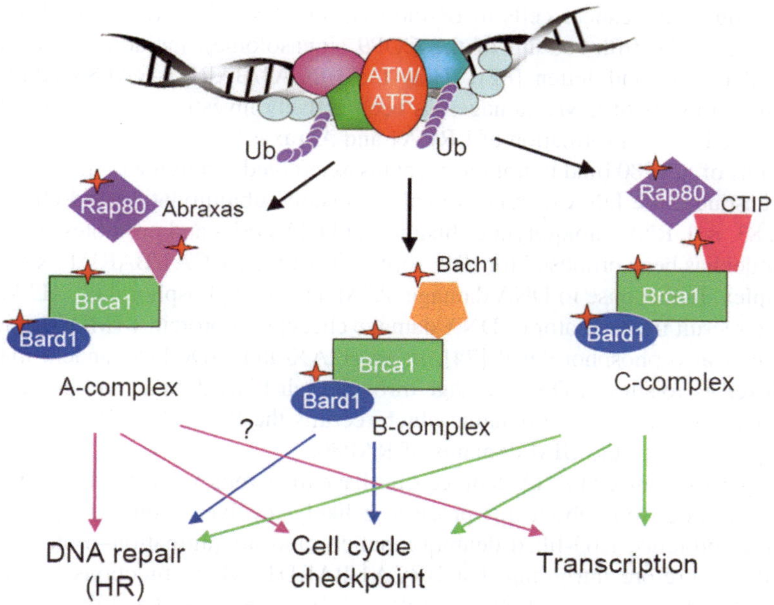

Fig. 6.5 RAP80 interacts with BRCA1 through binding to Abraxas. The RAP80–BRCA1 interaction decreases when Abraxas – but not BACH1 or CtIP – is depleted [68] (Reprinted with permission from American Association for the Advancement of Science)

to ionizing radiation (IR)-induced foci (IRIF) after IR and that the UIMs are essential for relocalization. It was further found that RAP80 forms a complex with BRCA1 and that this association is dependent on the BRCA1 COOH-terminal (BRCT) repeats of BRCA1. BRCA1 plays a critical role in DNA repair and activation of cell cycle checkpoints. RAP80 depletion disrupts the translocation of BRCA1 to IRIF and causes defects in G2-M checkpoint activation after IR [68, 69, 78]. In addition, knockdown of RAP80 expression by small interfering RNA reduces DSB-induced homology-directed recombination and increases the sensitivity of cells to IR-induced cytotoxicity [68, 79].

Abraxas binds BRCA1 to the mutual exclusion of BRCA1-associated C-terminal helicase (BACH)/BRCA1-interacting protein (Brip1) and CtBP-interacting protein (CTIP) through the pSer–X–X–Phe motif. The BRCA1–RAP80–Abraxas complex (Fig. 6.5) is clearly involved in DDR [68]. It is likely that different BRCA1 complexes play redundant roles or promote multiple distinct steps in DDR. For example, the three complexes (Fig. 6.5) are required for homologous recombination repair [68]. Furthermore, both the BRCA1–RAP80–Abraxas complex and the BRCA1–RAP80–CTIP complex are required for the G2-M checkpoint. These two complexes are also involved in transcription, through their association with RAP80 (Fig. 6.5) [68]. It is also interesting to note that RAP80 binds the estrogen receptor, suggesting that the BRCA1–RAP80 complexes might mediate the BRCA1 role in estrogen signaling in breast cancer [80]. Similar to our findings of poor prognosis in NSCLC patients with elevated BRCA1 expression [56], high BACH1/Brip1 transcript levels found in more aggressive breast cancers with an estrogen receptor-negative, progesterone receptor-negative or HER-2-positive status [81] seem to contradict their function as tumor suppressors.

In addition to BRCA1, RAP 80 and Abraxas, a deubiquitinating enzyme – BRCC36 – has also been shown to be present in the RAP80–BRCA1 complex [69]. Interestingly, BRCC36 is also aberrantly expressed in many breast cancers. Along the same lines, downregulation of BRCC36 expression impairs the DNA repair pathway activated in response to IR by inhibiting BRCA1 activation, thereby sensitizing breast cancer cells to IR-induced apoptosis [73]. BRCC36 displays sequence homology to CSN5 (the fifth subunit of the COP9 signasolome), including a conserved JAMM motif (reviewed in Yan and Jetten [82]). How the BRCA1/BARD1–RAP80–Abraxas–BRCC36 complex localizes to sites of DNA damage has been actively investigated. The two UIM motifs of RAP80 are required for foci formation of BRCA1 and Abraxas [68, 78]. It has been also shown that the UIM domains of RAP80 bind to ubiquitin chains assembled through K63 linkages [69]. In addition, RNF8 transduces the DNA-damage signal via histone ubiquitylation and checkpoint protein assembly [71, 83, 84]. RNF8 ubiquitylates histones at DNA DSBs and promotes assembly of repair proteins. A model has been proposed for IRIF formation of the BRCA1/BARD1–RAP80–Abraxas–BRCC36 complex. In response to DNA damage, ATM and ATR phosphorylates H2AX on Ser-139, which serves to recruit the mediator of DNA damage checkpoint protein 1 (MDC1) protein to chromatin, where it is also phosphorylated [74]. In an H2AX-and MDC1-dependent manner, RNF8–Ubc13 complexes go to sites of DNA damage through their FHA domain and initiate the synthesis of K63 polyubiquitin chains on chromatin, which recruits the BRCA1/BARD1–RAP80–Abraxas–BRCC36 complex through the UIM domains of RAP80.

Protein complexes formed at DNA damage sites have to be removed after the damage is repaired. Although little is understood about this process, it likely involves dephosphorylation of gamma-H2AX and other proteins, K63-liked deubiquitination, and ubiquitination–proteasome-dependent degradation. It is therefore intriguing that BRCA1/BARD1, which functions as an E3 ligase, is associated with the deubiquitinase BRCC36 [69, 85]. It can be speculated that the deubiquitinase activity of BRCC36 might play a role in terminating the DNA repair signaling at a later stage of DDR [82].

In addition, the SUMO pathways components (UBC9-protein inhibitor of activated STAT (PIAS)4 and UBC9–PIAS1) also accumulate at DSBs, where they catalyze the SUMOylation of BRCA1 [86, 87]. SUMOylation stimulates BRCA1 E3 ubiquitin ligase activity, leading to ubiquitylation of target proteins at DSBs, including the H2AX. Depletion of PIAS1 and PIAS4 impaired recruitment of BRCA1 to DSBs, significantly impaired ubiquitylation at DSBs, and reduced ubiquitylation of H2AX. PIAS SUMO ligases are required for homologous recombination and nonhomologous end-joining. PIAS1 and PIAS4 depletion resulted in ionizing radiation hypersensitivity [86, 87]. Therefore, these could be essential components – together with BRCA1 – for predicting response to radiotherapy and cisplatin-based chemotherapy.

BRCA1 as a Differential Modulator of Chemosensitivity

A growing body of evidence indicates that BRCA1 confers sensitivity to apoptosis induced by antimicrotubule drugs (paclitaxel and vincristine) but induces resistance to DNA-damaging agents (cisplatin and etoposide) and radiotherapy [88–91]. These preclinical findings are supported by a variety of experimental models in breast and ovarian cancer cells: inducible expression of BRCA1-enhanced paclitaxel sensitivity [92]; a short interfering RNA-mediated inactivation of endogenous BRCA1 led to paclitaxel and docetaxel resistance [57, 58, 93], and reconstitution of BRCA1-deficient cells with wild-type BRCA1 enhanced sensitivity to paclitaxel and vinorelbine [57]. This differential modulating effect of BRCA1 mRNA expression was also observed in tumor cells isolated from malignant effusions of NSCLC and gastric cancer patients, where BRCA1 mRNA levels correlated negatively with cisplatin sensitivity and positively with docetaxel sensitivity [94]. Five retrospective studies – in NSCLC [95, 96], ovarian [58, 97] and bladder [98] cancer patients – found that low or intermediate BRCA1 mRNA levels correlated with a significantly longer survival following platinum-based chemotherapy [58, 95] while survival in patients with higher BRCA1 expression increased following taxane-based chemotherapy [58, 96].

NSCLC with EGFR Mutations

We evaluated the feasibility of large-scale screening for EGFR mutations in advanced NSCLC patients and the impact of EGFR mutations on clinical outcome to customized erlotinib [99]. EGFR mutations were found in 350 of 2,105 patients (16.6%). Mutations were found more frequently in women (69.7%; $P<0.001$), in never-smokers (66.6%; $P<0.001$), and in adenocarcinomas (80.9%; $P<0.001$). Deletions in exon 19 were found in 62.3% and L858R mutations in 37.8% of patients. Metastatic NSCLC patients with EGFR mutations were eligible for erlotinib treatment. Median progression-free and overall survival for 217 patients receiving erlotinib was 14 months and 27 months, respectively. Large-scale screening of patients for EGFR mutations, with subsequent customization of erlotinib, is feasible and leads to impressive outcomes and is becoming standard practice [99].

However, the treatment approach in early resected NSCLC patients with EGFR mutations is still unclear although the majority of cases display favorable outcomes [100]. Other genetic alterations, such as activation of complementary signaling pathways, could influence outcome in patients with EGFR mutations and should be kept in mind for targeted therapy approaches.

References

1. Pastorino U. Lung cancer screening. Br J Cancer. 2010;102(12):1681–6.
2. Cykert S, Dilworth-Anderson P, Monroe MH, et al. Factors associated with decisions to undergo surgery among patients with newly diagnosed early-stage lung cancer. JAMA. 2010;303(23):2368–76.
3. Pantel K, Izbicki J, Passlick B, et al. Frequency and prognostic significance of isolated tumour cells in bone marrow of patients with non-small-cell lung cancer without overt metastases. Lancet. 1996;347(9002):649–53.
4. Mitchell PS, Parkin RK, Kroh EM, et al. Circulating microRNAs as stable blood-based markers for cancer detection. Proc Natl Acad Sci USA. 2008;105(30):10513–8.
5. Hu Z, Chen X, Zhao Y, et al. Serum microRNA signatures identified in a genome-wide serum microRNA expression profiling predict survival of non-small-cell lung cancer. J Clin Oncol. 2010;28(10):1721–6.
6. Tanaka F, Yoneda K, Kondo N, et al. Circulating tumor cell as a diagnostic marker in primary lung cancer. Clin Cancer Res. 2009;15(22):6980–6.
7. Maheswaran S, Sequist LV, Nagrath S, et al. Detection of mutations in EGFR in circulating lung-cancer cells. N Engl J Med. 2008;359(4):366–77.
8. Sotiriou C, Pusztai L. Gene-expression signatures in breast cancer. N Engl J Med. 2009;360(8):790–800.
9. Arriagada R, Auperin A, Burdett S, et al. Adjuvant chemotherapy, with or without postoperative radiotherapy, in operable non-small-cell lung cancer: two meta-analyses of individual patient data. Lancet. 2010;375(9722):1267–77.
10. Pisters KM, Vallieres E, Crowley JJ, et al. Surgery with or without preoperative paclitaxel and carboplatin in early-stage non-small-cell lung cancer: Southwest Oncology Group Trial S9900, an intergroup, randomized, phase III trial. J Clin Oncol. 2010;28(11):1843–9.
11. Felip E, Rosell R, Maestre JA, et al. Preoperative chemotherapy plus surgery versus surgery plus adjuvant chemotherapy versus surgery alone in early-stage non-small-cell lung cancer. J Clin Oncol. 2010;28(19):3138–45.
12. Minn AJ, Gupta GP, Siegel PM, et al. Genes that mediate breast cancer metastasis to lung. Nature. 2005;436(7050):518–24.
13. Kang Y, Siegel PM, Shu W, et al. A multigenic program mediating breast cancer metastasis to bone. Cancer Cell. 2003;3(6):537–49.
14. Bos PD, Zhang XH, Nadal C, et al. Genes that mediate breast cancer metastasis to the brain. Nature. 2009;459(7249):1005–9.
15. Leary RJ, Kinde I, Diehl F, et al. Development of personalized tumor biomarkers using massively parallel sequencing. Sci Transl Med. 2010. doi:10.1126/scitranslmed.3000702.
16. Mostertz W, Stevenson M, Acharya C, et al. Age- and sex-specific genomic profiles in non-small cell lung cancer. JAMA. 2010;303(6):535–43.
17. Chang HY, Sneddon JB, Alizadeh AA, et al. Gene expression signature of fibroblast serum response predicts human cancer progression: similarities between tumors and wounds. PLoS Biol. 2004;2(2):E7.

18. Massague J. Sorting out breast-cancer gene signatures. N Engl J Med. 2007;356(3):294–7.
19. Liu R, Wang X, Chen GY, et al. The prognostic role of a gene signature from tumorigenic breast-cancer cells. N Engl J Med. 2007;356(3):217–26.
20. Strauss GM. Adjuvant chemotherapy of lung cancer: methodologic issues and therapeutic advances. Hematol Oncol Clin North Am. 2005;19(2):263–81, vi.
21. Pignon JP, Tribodet H, Scagliotti GV, et al. Lung adjuvant cisplatin evaluation: a pooled analysis by the LACE Collaborative Group. J Clin Oncol. 2008;26(21):3552–9.
22. Douillard JY, Rosell R, De Lena M, et al. Adjuvant vinorelbine plus cisplatin versus observation in patients with completely resected stage IB-IIIA non-small-cell lung cancer (Adjuvant Navelbine International Trialist Association [ANITA]): a randomised controlled trial. Lancet Oncol. 2006;7(9):719–27.
23. Felip E, Rosell R, Maestre J, et al. Preoperative chemotherapy plus surgery versus surgery plus adjuvant chemotherapy versus surgery alone in early-stage non-small cell lung cancer. J Clin Oncol. 2010. doi:10.1200/JCO.2009.27.6204.
24. Bhattacharjee A, Richards WG, Staunton J, et al. Classification of human lung carcinomas by mRNA expression profiling reveals distinct adenocarcinoma subclasses. Proc Natl Acad Sci USA. 2001;98(24):13790–5.
25. Garber ME, Troyanskaya OG, Schluens K, et al. Diversity of gene expression in adenocarcinoma of the lung. Proc Natl Acad Sci USA. 2001;98(24):13784–9.
26. Beer DG, Kardia SL, Huang CC, et al. Gene-expression profiles predict survival of patients with lung adenocarcinoma. Nat Med. 2002;8(8):816–24.
27. Wigle DA, Jurisica I, Radulovich N, et al. Molecular profiling of non-small cell lung cancer and correlation with disease-free survival. Cancer Res. 2002;62(11):3005–8.
28. Raponi M, Zhang Y, Yu J, et al. Gene expression signatures for predicting prognosis of squamous cell and adenocarcinomas of the lung. Cancer Res. 2006;66(15):7466–72.
29. Lu Y, Lemon W, Liu PY, et al. A gene expression signature predicts survival of patients with stage I non-small cell lung cancer. PLoS Med. 2006;3(12):e467.
30. Potti A, Mukherjee S, Petersen R, et al. A genomic strategy to refine prognosis in early-stage non-small-cell lung cancer. N Engl J Med. 2006;355(6):570–80.
31. Borczuk AC, Shah L, Pearson GD, et al. Molecular signatures in biopsy specimens of lung cancer. Am J Respir Crit Care Med. 2004;170(2):167–74.
32. Bild AH, Yao G, Chang JT, et al. Oncogenic pathway signatures in human cancers as a guide to targeted therapies. Nature. 2006;439(7074):353–7.
33. Hsu DS, Acharya CR, Balakumaran BS, et al. Characterizing the developmental pathways TTF-1, NKX2-8, and PAX9 in lung cancer. Proc Natl Acad Sci USA. 2009;106(13):5312–7.
34. Endoh H, Tomida S, Yatabe Y, et al. Prognostic model of pulmonary adenocarcinoma by expression profiling of eight genes as determined by quantitative real-time reverse transcriptase polymerase chain reaction. J Clin Oncol. 2004;22(5):811–9.
35. Skrzypski M, Jassem E, Taron M, et al. Three-gene expression signature predicts survival in early-stage squamous cell carcinoma of the lung. Clin Cancer Res. 2008;14(15):4794–9.
36. Raz DJ, Ray MR, Kim JY, et al. A multigene assay is prognostic of survival in patients with early-stage lung adenocarcinoma. Clin Cancer Res. 2008;14(17):5565–70.
37. Chen HY, Yu SL, Chen CH, et al. A five-gene signature and clinical outcome in non-small-cell lung cancer. N Engl J Med. 2007;356(1):11–20.
38. Yu SL, Chen HY, Chang GC, et al. MicroRNA signature predicts survival and relapse in lung cancer. Cancer Cell. 2008;13(1):48–57.
39. Lau SK, Boutros PC, Pintilie M, et al. Three-gene prognostic classifier for early-stage non small-cell lung cancer. J Clin Oncol. 2007;25(35):5562–9.
40. Cai S, Han HJ, Kohwi-Shigematsu T. Tissue-specific nuclear architecture and gene expression regulated by SATB1. Nat Genet. 2003;34(1):42–51.
41. Han HJ, Russo J, Kohwi Y, Kohwi-Shigematsu T. SATB1 reprogrammes gene expression to promote breast tumour growth and metastasis. Nature. 2008;452(7184):187–93.
42. Yasui D, Miyano M, Cai S, Varga-Weisz P, Kohwi-Shigematsu T. SATB1 targets chromatin remodelling to regulate genes over long distances. Nature. 2002;419(6907):641–5.
43. Lossos IS, Czerwinski DK, Alizadeh AA, et al. Prediction of survival in diffuse large-B-cell lymphoma based on the expression of six genes. N Engl J Med. 2004;350(18):1828–37.
44. Han S, Ritzenthaler JD, Sitaraman SV, Roman J. Fibronectin increases matrix metalloproteinase 9 expression through activation of c-Fos via extracellular-regulated kinase and phosphatidylinositol 3-kinase pathways in human lung carcinoma cells. J Biol Chem. 2006;281(40):29614–24.
45. Tavazoie SF, Alarcon C, Oskarsson T, et al. Endogenous human microRNAs that suppress breast cancer metastasis. Nature. 2008;451(7175):147–52.

46. Adler AS, Lin M, Horlings H, Nuyten DS, van de Vijver MJ, Chang HY. Genetic regulators of large-scale transcriptional signatures in cancer. Nat Genet. 2006;38(4):421–30.
47. Wong DJ, Nuyten DS, Regev A, et al. Revealing targeted therapy for human cancer by gene module maps. Cancer Res. 2008;68(2):369–78.
48. Adler AS, Littlepage LE, Lin M, et al. CSN5 isopeptidase activity links COP9 signalosome activation to breast cancer progression. Cancer Res. 2008;68(2):506–15.
49. Akhoondi S, Sun D, von der Lehr N, et al. FBXW7/hCDC4 is a general tumor suppressor in human cancer. Cancer Res. 2007;67(19):9006–12.
50. Welcker M, Clurman BE. FBW7 ubiquitin ligase: a tumour suppressor at the crossroads of cell division, growth and differentiation. Nat Rev Cancer. 2008;8(2):83–93.
51. Deeb KK, Michalowska AM, Yoon CY, et al. Identification of an integrated SV40 T/t-antigen cancer signature in aggressive human breast, prostate, and lung carcinomas with poor prognosis. Cancer Res. 2007;67(17):8065–80.
52. Rosell R, Scagliotti G, Danenberg KD, et al. Transcripts in pretreatment biopsies from a three-arm randomized trial in metastatic non-small-cell lung cancer. Oncogene. 2003;22(23):3548–53.
53. Rosell R, Danenberg KD, Alberola V, et al. Ribonucleotide reductase messenger RNA expression and survival in gemcitabine/cisplatin-treated advanced non-small cell lung cancer patients. Clin Cancer Res. 2004;10(4):1318–25.
54. Souglakos J, Boukovinas I, Taron M, et al. Ribonucleotide reductase subunits M1 and M2 mRNA expression levels and clinical outcome of lung adenocarcinoma patients treated with docetaxel/gemcitabine. Br J Cancer. 2008;98(10):1710–5.
55. Johnson CD, Esquela-Kerscher A, Stefani G, et al. The let-7 microRNA represses cell proliferation pathways in human cells. Cancer Res. 2007;67(16):7713–22.
56. Rosell R, Skrzypski M, Jassem E, et al. BRCA1: a novel prognostic factor in resected non-small-cell lung cancer. PLoS ONE. 2007;2(11):e1129.
57. Quinn JE, Kennedy RD, Mullan PB, et al. BRCA1 functions as a differential modulator of chemotherapy-induced apoptosis. Cancer Res. 2003;63(19):6221–8.
58. Quinn JE, James CR, Stewart GE, et al. BRCA1 mRNA expression levels predict for overall survival in ovarian cancer after chemotherapy. Clin Cancer Res. 2007;13(24):7413–20.
59. Olaussen KA, Dunant A, Fouret P, et al. DNA repair by ERCC1 in non-small-cell lung cancer and cisplatin-based adjuvant chemotherapy. N Engl J Med. 2006;355(10):983–91.
60. Zheng Z, Chen T, Li X, Haura E, Sharma A, Bepler G. DNA synthesis and repair genes RRM1 and ERCC1 in lung cancer. N Engl J Med. 2007;356(8):800–8.
61. Saviozzi S, Ceppi P, Novello S, et al. Non-small cell lung cancer exhibits transcript overexpression of genes associated with homologous recombination and DNA replication pathways. Cancer Res. 2009;69(8):3390–6.
62. Su TT. Cellular responses to DNA damage: one signal, multiple choices. Annu Rev Genet. 2006;40:187–208.
63. Bristow RG, Hill RP. Hypoxia and metabolism, Hypoxia, DNA repair and genetic instability. Nat Rev Cancer. 2008;8(3):180–92.
64. Halazonetis TD, Gorgoulis VG, Bartek J. An oncogene-induced DNA damage model for cancer development. Science. 2008;319(5868):1352–5.
65. Gorgoulis VG, Vassiliou LV, Karakaidos P, et al. Activation of the DNA damage checkpoint and genomic instability in human precancerous lesions. Nature. 2005;434(7035):907–13.
66. DiTullio Jr RA, Mochan TA, Venere M, et al. 53BP1 functions in an ATM-dependent checkpoint pathway that is constitutively activated in human cancer. Nat Cell Biol. 2002;4(12):998–1002.
67. Matsuoka S, Ballif BA, Smogorzewska A, et al. ATM and ATR substrate analysis reveals extensive protein networks responsive to DNA damage. Science. 2007;316(5828):1160–6.
68. Wang B, Matsuoka S, Ballif BA, et al. Abraxas and RAP80 form a BRCA1 protein complex required for the DNA damage response. Science. 2007;316(5828):1194–8.
69. Sobhian B, Shao G, Lilli DR, et al. RAP80 targets BRCA1 to specific ubiquitin structures at DNA damage sites. Science. 2007;316(5828):1198–202.
70. Rogakou EP, Pilch DR, Orr AH, Ivanova VS, Bonner WM. DNA double-stranded breaks induce histone H2AX phosphorylation on serine 139. J Biol Chem. 1998;273(10):5858–68.
71. Wang B, Elledge SJ. Ubc13/Rnf8 ubiquitin ligases control foci formation of the Rap80/Abraxas/Brca1/Brcc36 complex in response to DNA damage. Proc Natl Acad Sci USA. 2007;104(52):20759–63.
72. Harper JW, Elledge SJ. The DNA damage response: ten years after. Mol Cell. 2007;28(5):739–45.
73. Chen X, Arciero CA, Wang C, Broccoli D, Godwin AK. BRCC36 is essential for ionizing radiation-induced BRCA1 phosphorylation and nuclear foci formation. Cancer Res. 2006;66(10):5039–46.
74. Stewart GS, Wang B, Bignell CR, Taylor AM, Elledge SJ. MDC1 is a mediator of the mammalian DNA damage checkpoint. Nature. 2003;421(6926):961–6.

75. Fossella F, Pereira JR, von Pawel J, et al. Randomized, multinational, phase III study of docetaxel plus platinum combinations versus vinorelbine plus cisplatin for advanced non-small-cell lung cancer: the TAX 326 study group. J Clin Oncol. 2003;21(16):3016–24.
76. Scagliotti GV, Parikh P, von Pawel J, et al. Phase III study comparing cisplatin plus gemcitabine with cisplatin plus pemetrexed in chemotherapy-naive patients with advanced-stage non-small-cell lung cancer. J Clin Oncol. 2008;26(21):3543–51.
77. Cobo M, Isla D, Massuti B, et al. Customizing cisplatin based on quantitative excision repair cross-complementing 1 mRNA expression: a phase III trial in non-small-cell lung cancer. J Clin Oncol. 2007;25(19):2747–54.
78. Kim H, Chen J, Yu X. Ubiquitin-binding protein RAP80 mediates BRCA1-dependent DNA damage response. Science. 2007;316(5828):1202–5.
79. Yan J, Kim YS, Yang XP, et al. The ubiquitin-interacting motif containing protein RAP80 interacts with BRCA1 and functions in DNA damage repair response. Cancer Res. 2007;67(14):6647–56.
80. Yan J, Kim YS, Yang XP, Albers M, Koegl M, Jetten AM. Ubiquitin-interaction motifs of RAP80 are critical in its regulation of estrogen receptor alpha. Nucleic Acids Res. 2007;35(5):1673–86.
81. Eelen G, Vanden Bempt I, Verlinden L, et al. Expression of the BRCA1-interacting protein Brip1/BACH1/FANCJ is driven by E2F and correlates with human breast cancer malignancy. Oncogene. 2008;27(30):4233–41.
82. Yan J, Jetten AM. RAP80 and RNF8, key players in the recruitment of repair proteins to DNA damage sites. Cancer Lett. 2008;271(2):179–90.
83. Huen MS, Grant R, Manke I, et al. RNF8 transduces the DNA-damage signal via histone ubiquitylation and checkpoint protein assembly. Cell. 2007;131(5):901–14.
84. Mailand N, Bekker-Jensen S, Faustrup H, et al. RNF8 ubiquitylates histones at DNA double-strand breaks and promotes assembly of repair proteins. Cell. 2007;131(5):887–900.
85. Wu W, Koike A, Takeshita T, Ohta T. The ubiquitin E3 ligase activity of BRCA1 and its biological functions. Cell Div. 2008;3:1.
86. Morris JR, Boutell C, Keppler M, et al. The SUMO modification pathway is involved in the BRCA1 response to genotoxic stress. Nature. 2009;462(7275):886–90.
87. Galanty Y, Belotserkovskaya R, Coates J, Polo S, Miller KM, Jackson SP. Mammalian SUMO E3-ligases PIAS1 and PIAS4 promote responses to DNA double-strand breaks. Nature. 2009;462(7275):935–9.
88. Lafarge S, Sylvain V, Ferrara M, Bignon YJ. Inhibition of BRCA1 leads to increased chemoresistance to microtubule-interfering agents, an effect that involves the JNK pathway. Oncogene. 2001;20(45):6597–606.
89. Husain A, He G, Venkatraman ES, Spriggs DR. BRCA1 up-regulation is associated with repair-mediated resistance to cis-diamminedichloroplatinum(II). Cancer Res. 1998;58(6):1120–3.
90. Bhattacharyya A, Ear US, Koller BH, Weichselbaum RR, Bishop DK. The breast cancer susceptibility gene BRCA1 is required for subnuclear assembly of Rad51 and survival following treatment with the DNA cross-linking agent cisplatin. J Biol Chem. 2000;275(31):23899–903.
91. Abbott DW, Thompson ME, Robinson-Benion C, Tomlinson G, Jensen RA, Holt JT. BRCA1 expression restores radiation resistance in BRCA1-defective cancer cells through enhancement of transcription-coupled DNA repair. J Biol Chem. 1999;274(26):18808–12.
92. Mullan PB, Quinn JE, Gilmore PM, et al. BRCA1 and GADD45 mediated G2/M cell cycle arrest in response to antimicrotubule agents. Oncogene. 2001;20(43):6123–31.
93. Chabalier C, Lamare C, Racca C, Privat M, Valette A, Larminat F. BRCA1 downregulation leads to premature inactivation of spindle checkpoint and confers paclitaxel resistance. Cell Cycle. 2006;5(9):1001–7.
94. Wang L, Wei J, Qian X, et al. ERCC1 and BRCA1 mRNA expression levels in metastatic malignant effusions is associated with chemosensitivity to cisplatin and/or docetaxel. BMC Cancer. 2008;8:97.
95. Taron M, Rosell R, Felip E, et al. BRCA1 mRNA expression levels as an indicator of chemoresistance in lung cancer. Hum Mol Genet. 2004;13(20):2443–9.
96. Boukovinas I, Papadaki C, Mendez P, et al. Tumor BRCA1, RRM1 and RRM2 mRNA expression levels and clinical response to first-line gemcitabine plus docetaxel in non-small-cell lung cancer patients. PLoS ONE. 2008;3(11):e3695.
97. Weberpals J, Garbuio K, O'Brien A, et al. The DNA repair proteins BRCA1 and ERCC1 as predictive markers in sporadic ovarian cancer. Int J Cancer. 2009;124(4):806–15.
98. Font A, Taron M, Gago JL, et al. BRCA1 mRNA expression and outcome to neoadjuvant cisplatin-based chemotherapy in bladder cancer. Ann Oncol. 2010;22(1):139–44.
99. Rosell R, Moran T, Queralt C, et al. Screening for epidermal growth factor receptor mutations in lung cancer. N Engl J Med. 2009;361(10):958–67.
100. Marks JL, Broderick S, Zhou Q, et al. Prognostic and therapeutic implications of EGFR and KRAS mutations in resected lung adenocarcinoma. J Thorac Oncol. 2008;3(2):111–6.
101. Potti A, Mukherjee S. Retraction: A genomic strategy to refine prognosis in early-stage non-small-cell lung cancer. N Engl J Med 2006;355:570–80.

Chapter 7
Gene Signatures in Gastric Cancer

Laura Ottini, Mario Falchetti, and Gabriella Nesi

Introduction

Although the incidence and mortality of gastric cancer (GC) have fallen over the past 70 years, GC continues to be the second leading cause of cancer death and the fourth most common malignant neoplasia across the world [1]. GC still represents a major clinical challenge because it has a poor prognosis and limited treatment options due to its relative resistance to radiotherapy and chemotherapy. At present, tumour stage provides the major prognostic variables used in clinical management of GC patients. However, GC with similar morphology may display different biological aggressiveness, prognosis and response to therapy.

It is now widely accepted that GC develops through the accumulation of genetic and epigenetic alterations affecting oncogenes and tumour suppressor genes and that alterations in mechanisms that control genomic instability lay at the base of this process [2]. Current knowledge on the molecular mechanisms underlying gastric carcinogenesis indicate that two major genomic instability pathways are involved in the pathogenesis of GC, microsatellite instability (MSI) and chromosome instability (CIN) [3, 4]. From a molecular standpoint, there is evidence that gastric carcinogenesis is a long-term multistep process associated with alteration in genomic stability and accumulation of multiple gene abnormalities. According to a metaplasia–adenoma–carcinoma progression model [5], a sequence of molecular changes related to MSI and CIN phenotypes may be observed in gastric carcinogenesis.

Several attempts to classify GC have been made over the past decades. Most successful, and widely used, is the classification by Lauren, which, by microscopic morphology alone, distinguishes two main cancer phenotypes, diffuse and intestinal subtypes, which appear clearly as dissimilar clinical and epidemiological entities. Although most of the genetic alterations that have been reported are observed in both intestinal and diffuse GCs, it has become apparent that these two tumour types result from different genetic pathways [2, 6].

MSI, *p53* mutation, reduced p27 expression, cyclin E overexpression and c-met 6.0-kb transcripts are involved in malignant transformation from precancerous lesions to intestinal-type GC. In addition, *DCC* loss, *APC* mutations, 1q loss of heterozygosity (LOH), p27 loss, reduced expression of transforming growth factor (TGF)-beta type I receptor and *HER2* gene amplification

L. Ottini (✉) • M. Falchetti
Department of Molecular Medicine, Policlinico Umberto I, Sapienza University of Rome, Rome, Italy
e-mail: laura.ottini@uniroma1.it

G. Nesi
Department of Critical Care Medicine, Division of Pathological Anatomy, University of Florence, Florence, Italy

A. Russo et al. (eds.), *Diagnostic, Prognostic and Therapeutic Value of Gene Signatures*, Current Clinical Pathology, DOI 10.1007/978-1-61779-358-5_7,

are frequently associated with an advanced stage of intestinal-type gastric carcinomas. In comparison, LOH at chromosome 17p (p53) and mutation or loss of *E-cadherin* are more often implicated in the development of diffuse-type GC, while gene amplification of *k-sam* and *c-met*, and loss of *p27* lead to disease progression and metastatic spread. The two types of gastric carcinoma organize different patterns of interplay between neoplastic and stromal cells through the growth factor/cytokine receptor system that has a critical role in cell growth, apoptosis, morphogenesis, angiogenesis and metastasis.

In addition to genetic alterations, epigenetic alterations are also involved in carcinogenesis. In particular in gastric carcinogenesis, the CpG islands methylator phenotype (CIMP), characterized by abnormal degree of hypermethylation in the context of CpG islands localized in gene promoters, may lead to the transcriptional silencing of various genes including *E-cadherin*, *p16*, *p15* and *hMLH1* [7]. Interestingly, GC hypermethylation of gene promoters progressively increases with histopathology progression from chronic gastritis, intestinal metaplasia and adenoma to carcinoma [8]. Thus, CIMP may represent a distinct pathway in GC, although there is evidence of a high frequency of CIMP phenotype in GC displaying MSI phenotype.

Recently, the role of micro-RNA (miRNA) deregulation in gastric carcinogenesis has been pointed out. Micro-RNAs are involved in many important biological processes of cell growth control and proliferation, and alterations in these patterns are possibly crucial to cancer initiation, progression and treatment outcome.

Overall, a complete knowledge about GC molecular signatures may provide an accurate biological rationale for predicting clinical behaviour and to establish more appropriate therapeutic interventions. Here, we will focus on the role of genomic instability, genetic variations and miRNA expression in GC development and progression and their possible clinical relevance.

Genomic Instability

Loss of genomic stability represents a key molecular step in GC development and progression. Genomic instability occurs early in the carcinogenesis process and creates a permissive environment for the accumulation of genetic and epigenetic alterations in tumour suppressor genes and oncogenes. Two main phenotypes of genomic instability, including MSI and CIN, are recognized in GC. MSI, characterized by high frequency of mutations in simple repetitive sequences (microsatellite), is responsible for a well-defined subset of GCs. CIN, characterized by high frequency of allelic losses (LOHs), gene deletions/amplifications and aneuploidy, comprises heterogeneous subsets of GC.

In addition to MSI and CIN, the CIMP plays an important role in gastric carcinogenesis. CIMP, characterized by abnormal degree of hypermethylation in the context of CpG islands within gene promoters, may lead to the transcriptional silencing of various genes involved in GC development and progression. Evidence suggests that although MSI, CIN and CIMP phenotypes can be distinguished from one another, there might be some degree of overlap [2].

MSI

MSI is a common feature of GC due to a deficit in the DNA mismatch repair system (MMR) and derives from the presence of spontaneous DNA replication errors in simple repetitive sequences [4]. During cell replication, MMR recognizes base-pair mismatches which occur by addition or deletion of one base. A standard panel of microsatellite markers, including mononucleotide (*BAT25* and

BAT26) and dinucleotide (*D2S123*, *D5S346* and *D17S250*) repeats, has been recommended and guidelines for MSI testing (Bethesda Guidelines) have been drawn up [9]. Using the reference panel, three levels of MSI can be identified: high-level MSI (MSI-H), low-level MSI (MSI-L) and microsatellite stable (MSS). Recently, it has been established that mononucleotide repeats are instrumental in detecting MSI-H tumours because of their high sensitivity and specificity, and MSI-L has been defined as instability limited to dinucleotide loci [10]. After the adoption of the Bethesda panel, MSI-H phenotype was reported in a range of 5–50% of all gastric carcinomas with significant differences in various ethnic groups.

MSI-H appears to be a phenotypical marker of an underlying cellular defect involving MMR. Functional inactivation of MMR genes (including *hMLH1* and *hMSH2*) by mutations or epigenetic mechanisms is responsible for the MSI-H phenotype in GC. Abnormal loss of protein expression of either hMLH1 or hMSH2 has been observed in MSI-H gastric carcinomas [11]. In particular, altered expression of hMLH1 has been associated with gene inactivation by promoter hypermethylation.

MSI-H GCs follow a molecular pathway of tumour progression characterized by the presence of multiple frameshift mutations affecting mononucleotide tracts within genes involved in cancer-related molecular networks, which control cellular homeostasis at different levels. MSI-related mutations occur in many genes at variable frequencies. Genes regulating cell cycle and apoptotic signalling are frequently targeted in MSI-H gastric carcinomas and include *TGFbRII*, *IGFIIR*, *TCF4*, *RIZ*, *BAX*, *CASPASE5*, *FAS*, *BCL10* and *APAF1* [12]. Moreover, genes involved in genomic integrity maintenance, i.e. *hMSH6*, *hMSH3*, *MED1*, *RAD50*, *BLM*, *ATR* and *MRE11*, are also frequently altered in MSI-H tumours [13, 14]. Several studies indicate that, in most MSI-H GCs, multiple target genes are simultaneously mutated and multiple hits impact on different genes in the same pathway [15]. The occurrence of mutations in specific sets of cancer-related genes may confer distinctive clinico-pathological features to MSI-H gastric carcinomas, which tend to occur as expanding tumours of the distal stomach, usually displaying an intestinal histotype with prominent lymphoid cell infiltration, and do not often give rise to lymph node metastasis, regardless of the extent of wall invasion [15–19]. Moreover, MSI-H tumours generally show improved survival compared to MSS/MSI-L cases [15].

MSI genotyping may allow the identification of discrete molecular GC subtypes and MSI seems to represent one of the most promising molecular markers with prognostic but also predictive value for chemosensitivity. In fact, MSI testing could also result in more patients being assigned to proper treatment based on their disease profile. Since, an intact MMR system is a determinant of sensitivity to a variety of chemotherapeutic agents, including 5-fluorouracil (5-FU), it is therefore reasonable to consider MSI-H tumours as a separate entity when determining response to chemotherapy. At present, data regarding the relevance of MSI for predicting the prognosis and benefits of 5-FU-based chemotherapy in GC are scarce to draw conclusions [20, 21]. In colon cancer, MSI status is regarded as a predictive factor since a significant improved prognosis for MSS patients treated with 5-FU-based chemotherapy is observed compared to MSI-positive cases [22, 23].

CIN

CIN is the most common type of genomic instability observed in human cancers and it has been reported in at least the 60% of gastrointestinal tumours [4]. CIN is characterized by changes in chromosome copy number (aneuploidy) and alterations in chromosomal regions, including LOHs, gene deletions and/or amplifications [24]. All these alterations may lead to oncogene activation and/or tumour suppressor gene inactivation. The identification of specific patterns of chromosome gains/losses occurring during the progression from adenoma to carcinoma and the observation that CIN is an early event in tumour formation and increases with tumour progression are consistent with the idea that CIN is a relevant pathogenic process in GC. The allelotype of gastric carcinoma is similar to that of colorectal and oesophageal cancers, suggesting the presence of a common genetic pathway for tumour development. Some of these chromosomal segments include genes, which are strongly implicated in carcinogenesis, such as the *p53* gene on chromosome 17, *DCC*, *DPC4* and *SMAD2* genes on chromosome 18, and *APC* and *MCC* genes on chromosome 5 [25, 26].

By contrast with MSI, the mechanism underlying CIN is largely unknown. Mitotic chromosomal mis-aggregation and errors in the mitotic spindle checkpoint have been implicated. *APC* is one of the major genes involved in the regulation of chromosome segregation; cells carrying *APC* mutations may acquire structural alterations in chromosomes and aneuploidy [27]. *APC* mutations have been observed in about 10% of GC [28]. *p53* is one of the most important genes involved in the regulation of the mitotic checkpoint [29]. *p53* point mutations are observed in 30–50% of GCs and *p53* locus is targeted by LOH in 60% of GCs [28]. Inactivation of proteins involved in DNA damage checkpoints, chromosome metabolism and centrosome function, cell proliferation, apoptosis, cell adhesion and in neoangiogenesis has been shown also to be involved in CIN pathway [29]. The progression of CIN positive (CIN+) GC is characterized by frequent LOH at the *APC* locus (30–40%) and, at lower level (3–20%), by *K-ras* activating point mutations, specifically at codons 12, 13, 59 e 61 [30, 31].

CIN has been demonstrated to be a valuable prognostic factor and tumour stage indicator in GC. As with other tumours, aneuploidy is generally considered an unfavourable prognostic factor [32–35], though contrasting results have been reported [36–39]. High CIN levels have also been associated with a shorter survival in GC patients [40] and several studies have found that tumours with LOH at chromosome 5q, 18q or 17p had a poorer prognosis than tumours that did not show LOH at these sites [25, 26]. High LOH frequencies have been also identified at several chromosome arms, including 1p, 3p, 4p, 7p, 8p, 8q, 9p, 12p, 13q, 20q and 22q [41, 42] thus supporting the hypothesis that, in GC, tumour progression, and consequently survival, correlates with the accumulation of genomic instability. A few studies have shown an association between high CIN levels with a good response to cisplatin-based chemotherapy and poor survival [43–55].

CIMP

Epigenetic changes, such as aberrant methylation of CpG islands in promoter regions are commonly detected in human cancers and can permanently inactivate tumour-suppressor genes and affect important pathways of cell cycle regulation and proliferation. The methylation of CpG islands may be considered a third molecular phenotype of GC and the tumour-related genes more commonly methylated are *APC, E-cadherin, MHL1, CDKN2A, CDKN2B* and *RUNX3*. It has also been widely reported that *CDKN2A, E-cadherin* and *MLH1* are more frequently inactivated by promoter methylation rather than by mutations [56]. Hypermethylation of gene promoters progressively increases with histopathological progression from chronic gastritis, intestinal metaplasia, adenoma and carcinoma, suggesting a distinct pathway in gastric carcinogenesis and progression [48, 57].

A series of individual methylated genes has been related to prognosis in GC. Methylation of tumour-suppressor genes, such as *E-cadherin* [58], *DKK3* [59], *PTEN* [60] and *MGMT* [61] of putative tumour-suppressor genes, such as *TFPI2* [7] and *CACNA2D3* [62], and of other tumour-related genes, such as *PCDH10* [63] and *SOX2* [64], has been associated with shorter disease-free and/or overall survival. The combined use of *APC* and *E-cadherin* methylation markers has identified a subgroup of patients with a worse prognosis [43]. Conversely, methylation of single genes has been associated with a better prognosis in some cases. Patients showing methylation of *APC* [44], the M1 region of *MAL* promoter [65] and cyclooxygenase-2 (*COX2*) [46] showed prolonged survival, compared to patients without methylation of these genes.

In order to define CIMP tumours as a group characterized by distinct genetic, morphological or clinical characteristics compared to tumours with other predominant forms of genomic instability the definition of CIMP-positive as been suggested to be quantitative and a low CIMP level (CIMP-L) and high CIMP level (CIMP-H) can be identified if less than 50% and more than 50% of genes/loci are respectively methylated. CIMP status is associated with clinically useful information and patients with CIMP-negative phenotype have significantly shorter survival than those with high CIMP levels [48, 50]. Intriguingly, concurrent hypermethylation of gene promoters is associated with MSI-H phenotype in GC and concordant methylation of multiple genes/loci (CIMP-H) is associated with better survival but is not an independent predictor of prognosis in GC [48]. Overall, much overlap of MSI and CIMP has been noted in GC, suggesting that MSI is a confounding factor.

Irrespective of CIMP being a separate pathway in gastric carcinogenesis, the presence of hypermethylation of important genes could be clinically relevant, because methylation can be reverted by DNA methyltransferase inhibitors reactivating genes, thus CIMP is emerging as interesting candidate for the development of *ad hoc* therapeutic strategies [66].

Molecular Pathways in GC

Genetic and genomic variations occurring in genes and molecules that participate in proliferation, invasion and metastasis (e.g. growth factors and their receptors, signal transducers, cell cycle and apoptosis regulators, cell adhesion molecules, DNA repair genes and matrix metalloproteinases) are involved in GC development and progression and may influence the prognosis of patients with GC. In particular, deregulation of oncogenic and tumour suppressor pathways such as *HER2, K-ras, p53, p21, MYC* and *Wnt/b-catenin* are known to occur with varying frequencies in GC [67] indicating that GC is a molecularly heterogeneous disease. Indeed, experimental evidence indicates that most cancer phenotypes (uncontrolled growth, resistance to apoptosis, cell invasion, metastasis and angiogenesis) are largely controlled by complex interactions between multiple pro- and anti-oncogenic signalling circuits [68], suggesting that pathway interactions may play an important role in influencing disease behaviour.

Cell Growth and Proliferation

HER2 (HER2/neu or ErbB-2), a glycoprotein with tyrosine kinase activity, is a member of the ErbB family of receptors. HER2 is codified by a gene located on chromosome 17q21 and does not bind to any known ligand. Some studies demonstrated that overexpression of HER2 is selectively found in intestinal tumours and may serve as a prognostic marker for tumour invasion and lymph node metastasis. Overexpression of HER2 protein in GC has been reported to range from 7.4% to 38% [68–70]. The prognostic value of HER2 expression and/or amplification has been widely investigated with controversial findings. Although most available studies indicate that the HER2 overexpression is an independent prognostic factor associated with a shorter disease-free [71] and overall survival [72, 73], some studies failed to confirm its prognostic role on multivariate analysis [53] or to find a correlation between HER2 overexpression and survival parameters [70–73]. Also associated with poor survival is the presence of HER2 amplification [74, 75].

The epidermal growth factor receptor (EGFR or ErbB-1), also a member of the ErbB family of receptors, is a trans-membrane protein that homo- or heterodimerizes with other *EGFR* family members at the cell membrane. Receptor dimerization causes activation of the intrinsic cytoplasmic kinase domain, resulting in the phosphorylation of several tyrosine residues [76]. The active *EGFR* stimulates the *MAPK* cascade and *PI3K* survival pathways [77]. *EGFR* is overexpressed in a maximum frequency of 38% of GC [78–80] and very few cases were reported to harbour gene structural alterations like gene amplification or mutations [81–83]. EGFR mutations have been identified as predictors of response to EGFR tyrosine kinase inhibitors (TKIs), thus, a small subset of gastric adenocarcinoma patients may respond to EGFR TKIs [84].

The *K-ras* oncogene encodes a membrane-associated protein, $p21^{RAS}$, with intrinsic GTPase activity involved in cellular signal transduction. Point mutations at codons 12, 13 and 61 of *K-ras* result in a shift of *K-ras* protein toward the activated state, which constitutively activates the mitogenic signal transduction pathway. In GC, the *K-ras* gene mutation frequency varies between 3 and 8% and, whenever present, *K-ras* mutations normally cluster in the MSI subset (30% of MSI cases) [85].

The oncogene *c-met*, encoding for the hepatocyte growth factor receptor, is preferentially amplified in diffuse-type tumours and has been described to be well correlated with stage and prognosis [53]. Overexpression of c-met has also shown to be associated with lower survival probability [55].

k-sam oncogene, a member of the fibroblast growth factor receptor family, is more frequently activated in diffuse-type tumours. Overexpression of k-sam occurs in approximately 32% of diffuse-type GCs, and the prognosis of k-sam positive patients is poorer than that of k-sam negative patients [86].

Cell Cycle and Apoptosis

The p53 protein plays a fundamental role in cell cycle control. Mutations of *p53* are present in about 40% of early and advanced well-differentiated GC [87]. A lower incidence of *p53* mutations has been shown in young patients compared to older patients [88]. p53 can be investigated by immunohistochemical techniques, bearing in mind that the half-life of the p53 mutant protein is prolonged. Cells carrying the p53 mutant protein can be stained with antibodies against p53, whereas cells carrying normal p53 are negative. Sequencing of the gene after screening can also be performed in order to determine the mutation location within the gene [89]. Overexpression of p53 often occurs in the early stages of intestinal-type tumours, and there is no significant difference between early and advanced cancers. In contrast, p53 abnormalities are not often seen in the early stages of diffuse-type tumours, but tend to occur as the disease progresses [90].

p53 cell cycle regulatory function is mediated by different effectors. One of these is a cyclin-dependant kinase inhibitor (CDK I), the p21 protein. The cell cycle checkpoints are controlled by a cascade of phosphorylation. Protein kinases such as cyclin-dependent kinases are activated by cyclins and inhibited by CDK I, although p21 is up-regulated not only through a p53 pathway, but through a TGFbRII pathway, as well. Levels of p21 expression could indicate the absence of a functional p53 protein in neoplastic cells. It has been reported that the survival of GC patients with p21-positive tumours is significantly longer than that of patients with p21-negative tumours [91]. The expression of p21 is usually assessed in combination with p53 status and contributes in predicting the clinical outcome of GC patients [92, 93].

It has been suggested that the cyclin-dependent inhibitor p27, which controls the transition from G1 to S in the cell cycle, has prognostic relevance in GC. Reduced p27 expression is detected in approximately 40–50% of GCs [42]. Some studies have shown that tumours with a low expression of p27 protein are poorly differentiated and at an advanced stage [94, 95]. However, some authors have found no difference in overall survival of GC patients whether with high or low p27 expression [96]. p53, p21 and p27 have also been analysed in combination, confirming their role as prognostic markers [90].

c-myc encodes a multifunctional, nuclear phosphoprotein that plays a role in cell cycle progression, apoptosis and cellular transformation. It functions as a transcription factor that regulates transcription of specific target genes. The c-myc protein has been shown to have a significantly enhanced expression in well-differentiated GC and to be associated with a poor prognosis [97, 98]. Although c-myc is a short-lived protein in normal cells, its stability is increased in transformed cells through several mechanisms. One of these has recently been identified in the overexpression of a human oncoprotein, the cancerous inhibitor of protein phosphate 2A (CIP2A) that stabilizes c-myc [99]. Interestingly, the expression of CIP2A has been associated with reduced overall survival in GC patients [100].

A fundamental negative regulator of the cell cycle is *pRb,* a protein encoded by the tumour suppressor gene *RB.* Poor prognosis of GC patients with low levels of pRb expression has been reported [101].

BCL2 and *p53* are closely linked in the regulation of apoptosis. LOH at the *BCL2* locus is frequently observed in GC. The overexpression of BCL2 may have a role in the development of GCs. It has been shown that BCL2 overexpression reduces cellular proliferative activity and correlates with a less aggressive biological behaviour of the tumour. The prognostic role of BCL2 on its own or in association with p53 has not yet been elucidated [102].

BAX gene encodes a protein belonging to the BCL family members. Negative BAX protein expression has been associated with de-differentiation, lymph node metastasis and shorter survival, suggesting that BAX status may play a role in the development and differentiation of GC and tumour progression [103].

Invasion and Metastasis

Invasion and metastasis are complex processes that require multiple genetic alterations and, more importantly, their coordinated interactions. Among the metastasis-associated factors, the Wnt-frizzled-b-catenin signalling pathway is frequently activated in GC leading to poor differentiation and increased tumour invasiveness. Activation of Wnt signalling pathway leads to inhibition of GSK-3b activity, resulting in deregulated accumulation of cytoplasmic b-catenin that can shift to nucleus and activates transcription of a series of genes. In addition to Wnt activation, mutational events targeting APC, Axin or b-catenin itself also result in b-catenin deregulation [104]. However, the incidences of APC mutation are from 4 to 20% in intestinal GCs and only 0–5% in diffuse GCs [28, 105]. To date, no Axin gene mutation has been reported in adenocarcinomas of the stomach, and frequent loss of Axin locus found in other cancer systems does not lead to imbalanced b-catenin distribution [106]. The data of b-catenin mutations in the hotspot codons of exon 3 are highly variable from 0 to 26% in GCs. On the other hand, more frequent b-catenin nuclear translocalization was found in 26–42% of GCs from different ethnic populations [107]. Overall, Wnt/b-catenin signalling pathway is activated in most of the GCs, and since it plays essential roles in cancer dissemination has potential value in prospecting the metastasis risk of GC patients.

E-cadherin, is a trasmembrane glycoprotein and is located at the adherens junctions. Its cytoplasmic tail binds either b-catenin or g-catenin, which in turn mediates interaction with the actin cytoskeleton via a-catenin as a linker. The intact function of E-cadherin is crucial for the establishment and maintenance of epithelial tissue polarity and structural integrity. Around 25–40% of hereditary diffuse GCs are caused by heterozygous E-cadherin. The inactivation of the second allele occurs by mutation and methylation events, and this results in the complete inactivation of the protein [108]. Reduced expression of E-cadherin correlates with infiltrative and metastatic ability in GC [58]. Patients with E-cadherin positive GCs showed statistically significant prolonged 3- and 5-year survival rates, compared to patients with E-cadherin negative tumours [109]. It has been shown that serum soluble E-cadherin is increased in several non-neoplastic diseases and also in various cancers including gastric tumours [110]. E-cadherin may be a potentially useful prognostic marker, and high levels of soluble E-cadherin correlate with the depth of tumour invasion, as well as inoperability [111]. In addition, levels higher than 10,000 ng/mL predict a survival of less than 3 years in more than 90% of patients [112].

Mucins (MUC) are high-molecular weight glycoproteins containing oligosaccharides. These glycoproteins constitute the major components of the mucus that protects the gastric epithelium. Overexpression of MUC1 has been linked to poor prognosis in GC patients [113]. It has been reported that MUC1 may accelerate tumour invasion by the impairment of E-cadherin [114]. The combined expression of MUC1 and E-cadherin shows that survival for GC patients with abnormal E-cadherin/MUC-positive expression was shorter than for patients with other expression patterns [110].

Overexpression of cyclin E, a member of the cyclin family required for the transition from G1 to S phase, correlates with invasiveness and proliferation and may be a marker of tumour aggressiveness. Although somatic mutations of the cell cycle inhibitor *p16MTSI* are rare, its reduced expression is associated with depth of invasion and metastatic potential in both diffuse- and intestinal-type gastric carcinomas. However, recent data show that the survival of GC patients with cyclin E positive tumours is not significantly shorter than that of negative patients [91].

Tumour-associated proteases and their inhibitors play a central role in tumour invasion and metastasis. The positive correlation of histological data with the urokinase-type plasminogen activator (uPA) and the plasminogen activator inhibitor type I (PAI-1) has been reported. Moreover, the independent prognostic impact of both uPA and PAI-1 on the survival of GC patients has been demonstrated. Elevated uPA and PAI-1 levels have been shown to be associated with shorter survival [115]. A trend towards poor prognosis has also been observed in patients with high expression of the u-PA receptor (u-PAR) [116] and the uPA system may therefore be a target for novel therapeutic agents. The prognostic role of some uPA genotypes has recently been investigated and an association between the exon 6 C/T polymorphism with invasive phenotype, but not with susceptibility or survival, was demonstrated [117].

Angiogenesis

Vascular endothelial growth factor (VEGF) is a pro-angiogenic factor, frequently overexpressed in tumours. Mutations of *p53*, which under physiological conditions downregulates VEGF, may be responsible for its overexpression. A correlation of the expression of VEGF with lymph node and liver metastasis has been described [118] and patients with VEGF-positive tumours have a rather worse prognosis than those with VEGF-negative tumours [119, 120].

RUNX3, a gene that codifies for a member of the runt domain-containing family of transcription factors, frequently shows loss of expression due to hemizygous deletion and hypermethylation in GC. This gene, generally expressed in 45–50% of GC patients [121] positively regulates the expression of the proapoptotic protein BIM and p21, and negatively regulates VEGF, thus affecting apoptosis, cell growth arrest and angiogenesis [122]. The loss or substantial decrease of RUNX3 protein expression in GC has been significantly associated with shorter survival [123].

The hypoxia inducible factor, HIF-1a, is a transcription factor that plays an essential role in cellular and systemic homeostatic responses to hypoxia. Its upregulation (high HIF-1a mRNA levels) has been found to be positively correlated with VEGF protein expression in GC patients and overall survival of patients with high mRNA levels of HIF-1a and VEGF was shorter compared to patients with different features [124–126].

Signal transducers and activators of transcription (STAT) behave as signal transducers in the cytoplasm and as transcription factors in the nucleus. Thus, an aberrant activation of STATs, especially STAT3, is often associated with cell survival, proliferation, and transformation. Dysregulated STAT3 activation has been linked to the development and progression of gastric adenocarcinoma via induction of VEGF overexpression leading to an elevated angiogenic phenotype [127]. STAT3 may be used as a molecular staging biomarker predicting poor prognosis of GC [128]. Several reports indicate that constitutively activated STAT3 is a target for antitumour drug discovery [129].

The secreted protein acidic and rich in cysteine (SPARC) or osteonectin is a member of a family of matricellular proteins that modulates cell–matrix interactions and cell function without participating in the structural scaffolding of the extracellular matrix. Since SPARC alters membrane permeability, cell shape, proliferation, migration and attachment, it may play a role in angiogenesis. It has been reported that its overexpression correlates with poor prognosis [130].

Micro-RNA and GC

Micro-RNAs are regulatory RNAs involved in the pathogenesis of many types of cancer [131–136], including GC [137]. miRNA carries out its biological functions by repressing the expression of its target genes, which often belong to the same metabolic or signalling pathway, through base-pairing with endogenous mRNAs. Partial or perfect base-pairing allows translational repression and mRNA degradation, respectively. It is estimated that up to 30% of genes in the human genome are regulated by miRNA [138]. miRNA transcription is similar to that of protein-coding genes, suggesting that the regulation of miRNA expression could also be controlled by transcriptional factors [139]. Moreover, the majority of miRNAs is located within introns and their expression is linked to the regulation of their host genes [140]. Alteration of genes involved in miRNA processing could also contribute to miRNA deregulation [141]. Finally, epigenetic alteration of DNA, such as DNA promoter hypermethylation and histone modification, may have critical roles in miRNA deregulation [142].

The biogenesis and the function of miRNA may be likely altered in GC. Cytogenetic and molecular analyses have revealed frequent chromosomal abnormalities in GC including gains at 20q, 8q, 20p, 7q, 17q, 5p and 13q and deletions at 19p, 18q, 5q, 21q, 4p, 4q, 15q and 17p [143]. Several miRNAs are hosted in these regions, currently investigated in GC, suggesting a role of chromosomal abnormality in miRNA deregulation [144].

Molecular Pathways and miRNA in GC

miRNAs are involved in important biological processes related to proliferation, apoptosis, differentiation, metastasis, angiogenesis and immune response, deregulation of which is possibly crucial to cancer initiation, progression and treatment outcome [145–149].

Aberrant miRNA expression may enhance cell cycle progression through downregulating the expression of CDK inhibitors in GC. For example, TGFb is known to suppress GC cell proliferation through transcriptional upregulation of p21 [150]. In this respect, miR-106b and miR-93, both of which are upregulated in GC and are downstream targets of the oncogenic transcription factor E2F-1, directly target p21 and thus impair the tumour-suppressive activity of TGFb [139]. The miRNAs in two clusters (miR-106b-93-25 and miR-222-221) have also been reported to suppress the p21 family of CDK inhibitors and, consistently with their biological roles, both clusters are upregulated in GC [151]. In addition to direct targeting of CDK inhibitors, miRNA has been shown to affect the expression of a CDK inhibitor-interacting protein known as anion exchanger-1 (AE1), which is expressed in GC cells. AE1 sequesters p16 in the cytoplasm and thus promotes cell proliferation [152]. The expression of AE1 is modulated by miR-24, and transfection of miR-24 leads the return of AE1-sequestered p16 to the nucleus and the inhibition of cell proliferation [153].

Regarding apoptosis control, miRNA deregulation alters the expression of Bcl-2 family members in GC. For instance, TGFb is known to induce RUNX3, which interacts with FoxO3a/FKHRL1 to activate the proapoptotic protein BIM and induce apoptosis in GC cells [154, 155]. In this regard, miR-106b and miR-93 impair TGFb-induced apoptosis in GC cells by inhibiting Bim expression [139]. Overexpression of miR-130b has also been reported to suppress TGFb-induced Bim expression and apoptosis by targeting RUNX3 in GC cells [156]. Moreover, several miRNAs, including miR-15b, miR-16, miR-34 and miR-181b, have been shown to target directly the antiapoptotic protein Bcl-2 and positively regulate apoptosis [157, 158]. miR-34 is a downstream target of p53 [159]. Overexpression of miR-34 increases caspase-3 activation and impairs growth in p53-mutant GC cells [160].

Macrophage migration inhibitory factor (MIF) is a lymphokine whose expression is increased in *H. pylori*-infected mucosa and GC [161]. MIF may bind to CD74 to reduce apoptosis in gastric epithelial cells by downregulating p53 phosphorylation and stimulating Bcl-2 expression [162]. MIF also promotes gastric epithelial cell proliferation through transactivation of epidermal growth factor receptor and the PI3K/Akt pathway [163, 164]. MIF expression may be reduced by miR-451, whose expression is correspondently reduced in GC. Restoration of miR-451 expression decreases expression of MIF target genes in GC cells, leading to cell proliferation reduction and cell death enhancement in response to irradiation. Moreover, there is a significant inverse correlation between miR-451 and MIF expression in GC biopsies, suggesting that miR-451 functions as a tumour suppressor by repressing MIF [165].

miRNA deregulation has been demonstrated to alter apoptosis by regulating pro-survival signalling: the PI3K/Akt pathway and NF-kB signalling. miR-375 suppresses the activity of PI3K/Akt pathway through direct targeting PDK1, a kinase that phosphorylates Akt. miR-375 is one of the most downregulated miRNAs in GC in which ectopic expression of miR-375 substantially reduces cell viability through induction of caspase-dependent apoptotic pathway [166].

NF-kB signalling inhibits apoptosis and, importantly, is induced during *H. pylori*-associated gastritis and is constitutively active in GC [167, 168]. miR-218 induces apoptosis in GC by targeting a positive regulator of NF-kB transcriptional activity and also by inhibiting NF-kB-regulated proliferative and anti-apoptotic genes, such as cyclooxygenase-2 [169]. NF-kB can be directly targeted by other miRNAs in GC like miR-9 [170].

miRNA deregulation may contribute to GC cell invasiveness. Enforced expression of miR-21 increases the invasiveness of cultured GC cells by directly targeting RECK a tumour-suppressor gene, which inhibits tumour metastasis and angiogenesis by modulating MMPs, including MMP9, MMP2 and MMP14 [171]. Activation of PI3K/Akt pathway by miR-21 may also confer the ability to avoid detachment-induced anoikis to GC cells. In addition, miR-21 may target PDCD4, a tumour suppressor. Low levels of PDCD4 mRNA are correlated with lymph node metastasis and venous invasion in GC [171, 172]. Similarly, the oncogenic miRNA miR-106a is upregulated and its expression correlates with invasion as well as lymphatic and distant metastasis [173]. Downregulation of miR-218 is also implicated in GC metastasis by activation of Slit/Robo1 signalling pathway [174].

miRNA deregulation seems to modulate transcription through alteration of chromatin architecture in GC. In particular, the high mobility group A2 (HMGA2) is a non-histone chromosomal protein, which can promote the assembly of regulatory protein complexes at sites of transcription [175]. This gene is expressed abundantly during early development, but at very low levels in adult tissues. HMGA2 overexpression is a hallmark of various benign and malignant tumours including GC, in which it is associated with serosal invasion and is evaluated as an independent prognostic factor for poor clinical outcome. The let-7 miRNA family has been demonstrated to negatively regulate HMGA2 and an inverse relationship between the expression of let-7 and HMGA2 has been observed in GC [176]. These findings suggest that the loss of inhibition by let-7 contributes to HMGA2 overexpression and probably enhances transcription in GC. Alterations of miRNA expression may affect transcriptional activities of various oncogenic/tumour-suppressing transcription factors by regulating the expression of prohibition [177]. This transcriptional co-regulator has been reported to enhance the transcriptional activity of p53, but have opposite effects on E2F-1 and NF-kB [178–180]. MiR-27a, an miRNA upregulated in GC, directly binds to prohibitin mRNA, thus switching off protein expression. Suppression of miR-27a also impairs GC cell proliferation, suggesting that miR-27a may act as an oncogene by targeting prohibitin [177]. miRNA deregulation has been reported to alter the expression of MeCP2, a methyl-CpG-binding protein involved in gene silencing by promoter hypermethylation. MeCP2 has been identified as the direct target of miR-212, an miRNA frequently downregulated in GC, and the restoration of MeCP2 expression by downregulation of miR-212 may promote carcinogenesis by enhancing epigenetic silencing of tumour-suppressor genes [181].

miRNA and Clinical Implication in GC

The development of sensitive and specific biomarkers may improve current management of GC, including cancer early detection, differentiation, progression and recurrence monitoring and treatment response evaluation. miRNAs have the following advantages as biomarkers (a) miRNAs are involved in tumorigenesis; (b) miRNAs are tissue, tumour, or even pathology specific; and (c) some miRNAs are related to treatment response or patients' survival [149].

Some recent studies suggest that polymorphisms in the miRNA genes may serve as novel risk predictors for GC. A polymorphism of miR-27a genome region is associated with a higher risk for the development of gastric mucosal atrophy in Japanese men [182]. An association of miRNA-196a-2 gene polymorphism with GC risk has been reported in a Chinese population [183]. Circulating miRNA profiling has been suggested as useful tool for non-invasive early diagnosis of GC. Plasma concentrations of various miRNAs, such as miR-17-5p, miR-21, miR-106a, miR-106b, are higher whereas let-7a is lower in GC patients compared to healthy people [184]. High levels of miR-17 and miR-106a in peripheral blood of GC patients have also been confirmed in another study [185]. Probably, the inclusion of miRNAs into the panels of biomarkers may enhance the sensitivity and specificity of diagnostic tests for GC.

Prognosis of GC patients is heterogeneous, with overall 5-year survival rates of about 20% [186]. Efforts have been put forth to predict disease outcome and response to treatment. For example, a seven-miRNA signature (miR-10b, miR-21, miR-223, miR-338, let-7a, miR-30a-5p and miR-126) is closely associated with relapse-free and overall survival among GC patients [187]. High expression levels of miR-20b or miR-150 [188] or downregulation of miR-451 [165] or miR-218 [174] are also associated with poor survival, whereas there is a correlation between miR-27a and lymph node metastasis [188]. In addition, miR-125b, miR-199a and miR-100 may represent a progression-related signature, whereas low expression of let-7 g and high expression of miR-214 are associated with shorter overall survival, independently of depth of invasion, lymph node metastasis and stage [137]. These prognostic miRNAs could be applicable to future decisions concerning treatment.

The association of miRNA deregulation with malignant diseases indicates the great potential of miRNAs as therapeutic targets. The basic strategy of current miRNA-based treatment studies is either to antagonize the expression of target miRNAs with antisense technology, to restore or else to strengthen the function of given miRNAs in order to inhibit the expression of specific protein-coding genes. There are several types of modifications for anti-miRNA antisense oligoribonucleotides (AMOs), locked nucleic acid, phosphorothioate backbone and cholesterol conjugation, which show difference in stability to nuclease degradation and in affinity and specificity to target miRNAs [189]. Several studies have evaluated the effects of reducing or knocking down the expression of overexpressed miRNAs with AMOs in GCs. For example, knockdown of miR-21 by AMOs caused a significant reduction in cell proliferation and a significant increase in apoptosis [171]. Loss-of-function of miR-27a inhibits GC cell growth in vitro [190]. Conversely, the induction of tumour suppressor miRNA with viral or liposomal delivery of miRNA mimics may taper the growth of GCs. Functional restoration of miR-34 by miR-34 mimics or lentiviral miR-34a could impair cell growth, accumulate the cells in G1 phase and increase caspase-3 activation in p53-mutant GC cells [160]. Chemotherapy resistance remains one of the major problems in improving overall survival and quality of life in GC patients. miRNAs-based therapy could be used to modulate the response of cancer cells to chemotherapy. For instance, miR-15b and miR-16 were downregulated in the multi-drug resistant GC cell line SGC7901/VCR and an enforced overexpression of miR-15b or miR-16 could sensitize these GC cells to apoptosis induced by vincristine [157]. The chemotherapy sensitizing effect of the miRNAs was partly mediated by modulation of apoptosis via targeting BCL2.

Summary

Gastric carcinomas are histologically and genetically heterogeneous and are influenced by gene–environment interactions resulting in the activation of multiple molecular pathways. The evidence of different patterns of genomic instability in GC may allow the identification of specific GC subsets characterized by peculiar molecular alterations and clinico-pathological features and this information may indicate improved management for patients care. In this respect, MSI is a promising tool. Indeed, MSI-H GCs identify a well-defined subset of GC characterized by unique clinico-pathological characteristics and, importantly, MSI-H GC cases show a relatively improved long-term survival compared to MSS/MSI-L counterparts. Thus, a role for MSI as prognostic marker is currently being evaluated. MSI can be also used as a marker for early diagnosis of GC considering that it can be detected in both gastric adenoma and intestinal metaplasia, which are precancerous lesions associated with well-differentiated GC. In this regard, it is noteworthy that the presence and the extent of MSI, but also CIN phenotype, assessed in endoscopic biopsy specimens from GC patients, was shown to provide valuable information for making a preoperative genetic diagnosis of GC [54].

Since GC is a heterogeneous disease, it may be amenable to different therapeutic treatment depending upon tumour mutational profiles, particularly whether they display MSI, CIN or CIMP phenotype. Currently, there is no consensus whether adjuvant therapy is differentially beneficial for patients with MSI+or MSI-GC. However, in vitro data suggest that MSI+and CIN+cancers differ in their response to therapy-induced DNA damage, and MSI+cancers are relatively insensitive to 5-fluorouracil. From a clinical standpoint, DNA methylation changes in GC represent an attractive therapeutic target, as epigenetic alterations are, in theory, reversible by using DNA methylation inhibitors, which have been demonstrated to restore gene expression and to exert antitumour effects in vitro and in vivo laboratory models. Similarly, the development of novel in vivo RNA delivery systems may enhance the importance of miRNA as cancer therapeutic drugs.

Recently, it has become apparent that gene expression profile may provide an accurate biologic rationale for predicting clinical behaviour and outcome [189, 190]. By comparing gene expression in normal and abnormal gastric cells, sets of genes have been identified as putative markers. In addition, comparison of expression profiles between malignant gastric histological subtypes has allowed the identification of additional genes that may represent a distinct molecular gene expression signature. The identification of genes with differential expression in specific gastric tumours may provide insight into the initiating mechanisms predicting progression and facilitate precise diagnosis.

At present, a complete understanding of the basis of genomic instability, aberrant methylation of cancer genome, miRNAs and gene expression profiling has to be achieved in GC but the translation of molecular genetics to new diagnostic, prognostic and therapeutic modalities has already shown its relevant role in GC patients' management.

References

1. Jemal A, Siegel R, Ward E, et al. Cancer statistics, 2009. CA Cancer J Clin. 2009;59:225–49.
2. Hiyama T, Tanaka S, Yoshihara M, et al. Chromosomal and microsatellite instability in sporadic GC. J Gastroenterol Hepatol. 2004;19:756–60.
3. Imai K, Yamamoto H. Carcinogenesis and microsatellite instability: the interrelationship between genetics and epigenetics. Carcinogenesis. 2008;29:673–80.
4. Sugai T, Habano W, Jiao YF, et al. Analysis of genetic alterations associated with DNA diploidy, aneuploidy and multiploidy in GCs. Oncology. 2005;68:548–57.
5. Correa P. Human gastric carcinogenesis: a multistep and multifactorial process–First American Cancer Society Award Lecture on Cancer Epidemiology and Prevention. Cancer Res. 1992;52:6735–40.
6. Tahara E. Genetic pathways of two types of GC. IARC Sci Publ. 2004;157:327–49.
7. Jee CD, Kim MA, Jung EJ, et al. Identification of genes epigenetically silenced by CpG methylation in human gastric carcinoma. Eur J Cancer. 2009;45:1282–93.
8. Lee JH, Park SJ, Abraham SC, et al. Frequent CpG island methylation in precursor lesions and early gastric adenocarcinomas. Oncogene. 2004;23:4646–54.
9. Boland CR, Thibodeau SN, Hamilton SR, et al. A National Cancer Institute Workshop on microsatellite instability for cancer detection and familial predisposition: development of international criteria for the determination of microsatellite instability in colorectal cancer. Cancer Res. 1998;58:5248–57.
10. Gologan A, Krasinskas A, Hunt J, et al. Performance of the revised Bethesda guidelines for identification of colorectal carcinomas with a high level of microsatellite instability. Arch Pathol Lab Med. 2005;129:1390–7.
11. Simpson AJ, Caballero OL, Pena SD. Microsatellite instability as a tool for the classification of GC. Trends Mol Med. 2001;7:76–80.
12. Iacopetta BJ, Soong R, House AK, et al. Gastric carcinomas with microsatellite instability: clinical features and mutations to the TGF-beta type II receptor, IGFII receptor, and BAX genes. J Pathol. 1999;187:428–32.
13. Ottini L, Falchetti M, Saieva C, et al. MRE11 expression is impaired in GC with microsatellite instability. Carcinogenesis. 2004;25:2337–43.
14. Kim MS, Kim SS, Ahn CH, et al. Frameshift mutations of Wnt pathway genes AXIN2 and TCF7L2 in gastric carcinomas with high microsatellite instability. Hum Pathol. 2009;40:58–64.
15. Falchetti M, Saieva C, Lupi R, et al. GC with high-level microsatellite instability: target gene mutations, clinicopathologic features, and long-term survival. Hum Pathol. 2008;39:925–32.
16. Hayden JD, Cawkwell L, Quirke P, et al. Prognostic significance of microsatellite instability in patients with gastric carcinoma. Eur J Cancer. 1997;33:2342–6.
17. Yamamoto H, Perez-Piteira J, Yoshida T, et al. GCs of the microsatellite mutator phenotype display characteristic genetic and clinical features. Gastroenterology. 1999;116:1348–57.
18. Choi SW, Choi JR, Chung YJ, et al. Prognostic implications of microsatellite genotypes in gastric carcinoma. Int J Cancer. 2000;89:378–83.
19. Corso G, Pedrazzani C, Marrelli D, et al. Correlation of microsatellite instability at multiple loci with long-term survival in advanced gastric carcinoma. Arch Surg. 2009;144:722–7.
20. Yashiro M, Inoue T, Nishioka N, Matsuoka T, Boland CR, Hirakawa K. Allelic imbalance at p53 and microsatellite instability are predictive markers for resistance to chemotherapy in gastric carcinoma. Ann Surg Oncol. 2009;16(10):2926–35.
21. Oki E, Kakeji Y, Zhao Y, Yoshida R, Ando K, Masuda T, et al. Chemosensitivity and survival in GC patients with microsatellite instability. Ann Surg Oncol. 2009;16(9):2510–5.
22. Sargent DJ, Marsoni S, Monges G, Thibodeau SN, Labianca R, Hamilton SR, et al. Defective mismatch repair as a predictive marker for lack of efficacy of fluorouracil-based adjuvant therapy in colon cancer. J Clin Oncol. 2010;28(20):3219–26.
23. Guastadisegni C, Colafranceschi M, Ottini L, Dogliotti E. Microsatellite instability as a marker of prognosis and response to therapy: a meta-analysis of colorectal cancer survival data. Eur J Cancer. 2010;46(15):2788–98.
24. Sanchez-Perez I, Garcia Alonso P, Belda Iniesta C. Clinical impact of aneuploidy on GC patients. Clin Transl Oncol. 2009;11:493–8.
25. Sanz-Ortega J, Sanz-Esponera J, Caldes T, et al. LOH at the APC/MCC gene (5Q21) in GC and preneoplastic lesions.Prognostic implications. Pathol Res Pract. 1996;192:1206–10.
26. Bamias AT, Bai MC, Agnantis NJ, et al. Prognostic significance of the deleted in colorectal cancer gene protein expression in high-risk resected gastric carcinoma. Cancer Invest. 2003;21:333–40.
27. Fodde R, Kuipers J, Rosenberg C, et al. Mutations in the APC tumour suppressor gene cause chromosomal instability. Nat Cell Biol. 2001;3:433–8.
28. Tamura G. Alterations of tumor suppressor and tumor-related genes in the development and progression of GC. World J Gastroenterol. 2006;12:192–8.

29. Tarapore P, Fukasawa K. Loss of p53 and centrosome hyperamplification. Oncogene. 2002;21:6234–40.
30. El-Rifai W, Powell SM. Molecular and biologic basis of upper gastrointestinal malignancy. Gastric carcinoma. Surg Oncol Clin N Am. 2002;11:273–91.
31. Nardone G. Review article: molecular basis of gastric carcinogenesis. Aliment Pharmacol Ther. 2003;17:75–81.
32. Yonemura Y, Ooyama S, Sugiyama K, et al. Retrospective analysis of the prognostic significance of DNA ploidy patterns and S-phase fraction in gastric carcinoma. Cancer Res. 1990;50:509–14.
33. Rugge M, Sonego F, Panozzo M, et al. Pathology and ploidy in the prognosis of GC with no extranodal metastasis. Cancer. 1994;73:1127–33.
34. Danesi DT, Spanò M, Fabiano A, et al. Flow cytometric DNA ploidy, p53, PCNA, and c-erbB-2 protein expressions as predictors of survival in surgically resected GC patients. Cytometry. 2000;42:27–34.
35. Wiksten JP, Lundin J, Nordling S, et al. Comparison of the prognostic value of a panel of tissue tumor markers and established clinicopathological factors in patients with GC. Anticancer Res. 2008;28:2279–87.
36. Ohyama S, Yonemura Y, Miyazaki I. Proliferative activity and malignancy in human GCs. Significance of the proliferation rate and its clinical application. Cancer. 1992;69:314–21.
37. Lee KH, Lee JS, Lee JH, et al. Prognostic value of DNA flow cytometry in stomach cancer: a 5-year prospective study. Br J Cancer. 1999;79:1727–35.
38. Sendler A, Gilbertz KP, Becker I, et al. Proliferation kinetics and prognosis in GC after resection. Eur J Cancer. 2001;37:1635–41.
39. Nesi G, Bruno L, Saieva C, et al. DNA ploidy and S-phase fraction as prognostic factors in surgically resected gastric carcinoma: a 7-year prospective study. Anticancer Res. 2007;27:4435–41.
40. Suzuki K, Ohnami S, Tanabe C, et al. The genomic damage estimated by arbitrarily primed PCR DNA fingerprinting is useful for the prognosis of GC. Gastroenterology. 2003;125:1330–40.
41. Kimura Y, Noguchi T, Kawahara K, et al. Genetic alterations in 102 primary GCs by comparative genomic hybridization: gain of 20q and loss of 18q are associated with tumor progression. Mod Pathol. 2004;17:1328–37.
42. Panani AD. Cytogenetic and molecular aspects of GC: clinical implications. Cancer Lett. 2008;266:99–115.
43. Leung WK, To KF, Chu ES, et al. Potential diagnostic and prognostic values of detecting promoter hypermethylation in the serum of patients with GC. Br J Cancer. 2005;92:2190–4.
44. Ksiaa F, Ziadi S, Amara K, et al. Biological significance of promoter hypermethylation of tumor-related genes in patients with gastric carcinoma. Clin Chim Acta. 2009;404:128–33.
45. Buffart TE, Overmeer RM, Steenbergen RD, et al. MAL promoter hypermethylation as a novel prognostic marker in GC. Br J Cancer. 2008;99:1802–7.
46. de Maat MF, van de Velde CJ, Umetani N, et al. Epigenetic silencing of cyclooxygenase-2 affects clinical outcome in GC. J Clin Oncol. 2007;25:4887–94.
47. Toyota M, Ahuja N, Suzuki H, et al. Aberrant methylation in GC associated with the CpG island methylator phenotype. Cancer Res. 1999;59:5438–42.
48. An C, Choi IS, Yao JC, et al. Prognostic significance of CpG island methylator phenotype and microsatellite instability in gastric carcinoma. Clin Cancer Res. 2005;11:656–63.
49. Etoh T, Kanai Y, Ushijima S, et al. Increased DNA methyltransferase 1 (DNMT1) protein expression correlates significantly with poorer tumor differentiation and frequent DNA hypermethylation of multiple CpG islands in GCs. Am J Pathol. 2004;164:689–99.
50. Kusano M, Toyota M, Suzuki H, et al. Genetic, epigenetic, and clinicopathologic features of gastric carcinomas with the CpG island methylator phenotype and an association with Epstein-Barr virus. Cancer. 2006;106:1467–79.
51. Lin W, Kao HW, Robinson D, et al. Tyrosine kinases and GC. Oncogene. 2000;19:5680–9.
52. Tsugawa K, Yonemura Y, Hirono Y, et al. Amplification of the c-met, c-erbB-2 and epidermal growth factor receptor gene in human GCs: correlation to clinical features. Oncology. 1998;55:475–81.
53. Nakajima M, Sawada H, Yamada Y, et al. The prognostic significance of amplification and overexpression of c-met and c-erb B-2 in human gastric carcinomas. Cancer. 1999;85:1894–902.
54. Huang TJ, Wang JY, Lin SR, et al. Overexpression of the c-met protooncogene in human gastric carcinoma–correlation to clinical features. Acta Oncol. 2001;40:638–43.
55. Drebber U, Baldus SE, Nolden B, et al. The overexpression of c-met as a prognostic indicator for gastric carcinoma compared to p53 and p21 nuclear accumulation. Oncol Rep. 2008;19:1477–83.
56. Ushijima T, Sasako M. Focus on GC. Cancer Cell. 2004;5:121–5.
57. Oue N, Mitani Y, Motoshita J, et al. Accumulation of DNA methylation is associated with tumor stage in GC. Cancer. 2006;106:1250–9.
58. Graziano F, Arduini F, Ruzzo A, et al. Prognostic analysis of E-cadherin gene promoter hypermethylation in patients with surgically resected, node-positive, diffuse GC. Clin Cancer Res. 2004;10:2784–9.
59. Yu J, Tao Q, Cheng YY, et al. Promoter methylation of the Wnt/beta-catenin signaling antagonist Dkk-3 is associated with poor survival in GC. Cancer. 2009;115:49–60.
60. Kang YH, Lee HS, Kim WH. Promoter methylation and silencing of PTEN in gastric carcinoma. Lab Invest. 2002;82:285–91.

61. Park TJ, Han SU, Cho YK. Methylation of O(6)-methylguanine-DNA methyltransferase gene is associated significantly with K-ras mutation, lymph node invasion, tumor staging, and disease free survival in patients with gastric carcinoma. Cancer. 2001;92:2760–8.
62. Wanajo A, Sasaki A, Nagasaki H, et al. Methylation of the calcium channel-related gene, CACNA2D3, is frequent and a poor prognostic factor in GC. Gastroenterology. 2008;135:580–90.
63. Yu J, Cheng YY, Tao Q. Methylation of protocadherin 10, a novel tumor suppressor, is associated with poor prognosis in patients with GC. Gastroenterology. 2009;136:640–51.
64. Otsubo T, Akiyama Y, Yanagihara K, et al. SOX2 is frequently downregulated in GCs and inhibits cell growth through cell-cycle arrest and apoptosis. Br J Cancer. 2008;98:824–31.
65. Buffart TE, van Grieken NC, Tijssen M, et al. High resolution analysis of DNA copy-number aberrations of chromosomes 8, 13, and 20 in GCs. Virchows Arch. 2009;455:213–23.
66. Ren J, Singh BN, Huang Q, Li Z, Gao Y, Mishra P, et al. DNA hypermethylation as a chemotherapy target. Cell Signal. 2011;23(7):1082–93.
67. Ooi CH, Ivanova T, Wu J, et al. Oncogenic pathway combinations predict clinical prognosis in gastric cancer. PLoS Genet. 2009;5(10):e1000676.
68. Hanahan D, Weinberg RA. The hallmarks of cancer. Cell. 2000;100(1):57–70.
69. Gravalos C, Jimeno A. HER2 in GC: a new prognostic factor and a novel therapeutic target. Ann Oncol. 2008;19:1523–9.
70. Matsubara J, Yamada Y, Hirashima Y, et al. Impact of insulin-like growth factor type 1 receptor, epidermal growth factor receptor, and HER2 expressions on outcomes of patients with GC. Clin Cancer Res. 2008;14:3022–9.
71. Allgayer H, Babic R, Gruetzner KU, et al. c-erbB-2 is of independent prognostic relevance in GC and is associated with the expression of tumor-associated protease systems. J Clin Oncol. 2000;18:2201–9.
72. Zhang XL, Yang YS, Xu DP, et al. Comparative study on overexpression of HER2/neu and HER3 in GC. World J Surg. 2009;33:2112–8.
73. Yonemura Y, Ninomiya I, Yamaguchi A, et al. Evaluation of immunoreactivity for erbB-2 protein as a marker of poor short term prognosis in GC. Cancer Res. 1991;51:1034–8.
74. Sasano H, Date F, Imatani A, et al. Double immunostaining for c-erbB-2 and p53 in human stomach cancer cells. Hum Pathol. 1993;24:584–9.
75. Tanner M, Hollmen M, Junttila TT, et al. Amplification of HER-2 in gastric carcinoma: association with Topoisomerase IIalpha gene amplification, intestinal type, poor prognosis and sensitivity to trastuzumab. Ann Oncol. 2005;16:273–8.
76. Qian X, LeVea CM, Freeman JK, Dougall WC, Greene MI. Heterodimerization of epidermal growth factor receptor and wild-type or kinase-deficient Neu: a mechanism of interreceptor kinase activation and transphosphorylation. Proc Natl Acad Sci USA. 1994;91:1500–4.
77. Hu T, Li C. Convergence between Wnt-b-catenin and EGFR signaling in cancer. Mol Cancer. 2010;9:236.
78. Lemoine NR, Jain S, Silvestre F, et al. Amplification and overexpression of the EGF receptor and c-erbB-2 protooncogenes in human stomach cancer. Br J Cancer. 1991;64(1):79–83.
79. Kimura M, Tsuda H, Morita D, et al. Usefulness and limitation of multiple endoscopic biopsy sampling for epidermal growth factor receptor and c-erbB-2 testing in patients with gastric adenocarcinoma. Jpn J Clin Oncol. 2005;35:324–31.
80. Kimura M, Tsuda H, Morita D, et al. A proposal for diagnostically meaningful criteria to classify increate epidermal growth factor receptor and c-erbB-2 gene copy numbers in gastric carcinoma, based on correlation of fluorescence in situ hybridization and immunohistochemical measurements. Virchows Arch. 2004;45:255–62.
81. Kim MA, Lee HS, Lee HE, et al. EGFR in gastric carcinomas: prognostic significance of protein overexpression and high gene copy number. Histopathology. 2008;52:738–46.
82. Takehana T, Kunitomo K, Suzuki S, et al. Expression of epidermal growth factor receptor in gastric carcinomas. Clin Gastroenterol Hepatol. 2003;1:438–45.
83. Moutinho C, Mateus AR, Milanezi F, et al. Epidermal growth factor receptor structural alterations in gastric cancer. BMC Cancer. 2008;16:8–10.
84. Okines A, Cunningham D, Chau I. Targeting the human EGFR family in esophagogastric cancer. Nat Rev Clin Oncol. 2011. doi:10.1038/nrclinonc.2011.45.
85. Corso G, Velho S, Paredes J, et al. Oncogenic mutations in gastric cancer with microsatellite instability. Eur J Cancer. 2011;47(3):443–51.
86. Toyokawa T, Yashiro M, Hirakawa K. Co-expression of keratinocyte growth factor and K-sam is an independent prognostic factor in gastric carcinoma. Oncol Rep. 2009;21:875–80.
87. Uchino S, Noguchi M, Ochiai A, et al. p53 mutation in GC: a genetic model for carcinogenesis is common to gastric and colorectal cancer. Int J Cancer. 1993;54:759–64.
88. Rugge M, Shiao YH, Busatto G, et al. The p53 gene in patients under the age of 40 with GC: mutation rates are low but are associated with a cardiac location. Mol Pathol. 2000;53:207–10.

89. Shiao YH, Palli D, Caporaso NE, et al. Genetic and immunohistochemical analyses of p53 independently predict regional metastasis of GCs. Cancer Epidemiol Biomarkers Prev. 2000;9:631–3.
90. Liu XP, Tsushimi K, Tsushimi M, et al. Expression of p53 protein as a prognostic indicator of reduced survival time in diffuse-type gastric carcinoma. Pathol Int. 2001;51:440–4.
91. Kouraklis G, Katsoulis IE, Theocharis S, et al. Does the expression of cyclin E, pRb, and p21 correlate with prognosis in gastric adenocarcinoma? Dig Dis Sci. 2009;54:1015–20.
92. Seo YH, Joo YE, Choi SK, et al. Prognostic significance of p21 and p53 expression in GC. Korean J Intern Med. 2003;18:98–103.
93. Okuyama T, Maehara Y, Kabashima A, et al. Combined evaluation of expressions of p53 and p21 proteins as prognostic factors for patients with gastric carcinoma. Oncology. 2002;63:353–61.
94. Sgambato A, Migaldi M, Leocata P, et al. Loss of p27Kip1 expression is a strong independent prognostic factor of reduced survival in N0 gastric carcinomas. Cancer. 2000;89:2247–57.
95. Nitti D, Belluco C, Mammano E, et al. Low level of p27(Kip1) protein expression in gastric adenocarcinoma is associated with disease progression and poor outcome. J Surg Oncol. 2002;81:167–75.
96. Wiksten JP, Lundin J, Nordling S, et al. The prognostic value of p27 in GC. Oncology. 2002;63:180–4.
97. Ninomiya I, Yonemura Y, Matsumoto H, et al. Expression of c-myc gene product in gastric carcinoma. Oncology. 1991;8:149–53.
98. Han S, Kim HY, Park K, et al. c-myc expression is related with cell proliferation and associated with poor clinical outcome in human GC. J Korean Med Sci. 1999;14:526–30.
99. Junttila MR, Westermarck J. Mechanisms of MYC stabilization in human malignancies. Cell Cycle. 2008;7:592–6.
100. Khanna A, Böckelman C, Hemmes A, et al. MYC-dependent regulation and prognostic role of CIP2A in GC. J Natl Cancer Inst. 2009;101:793–805.
101. Feakins RM, Nickols CD, Bidd H, et al. Abnormal expression of pRb, p16, and cyclin D1 in gastric adenocarcinoma and its lymph node metastases: relationship with pathological features and survival. Hum Pathol. 2003;34:1276–82.
102. Lee HK, Lee HS, Yang HK, et al. Prognostic significance of Bcl-2 and p53 expression in GC. Int J Colorectal Dis. 2003;18:518–25.
103. Anagnostopoulos GK, Stefanou D, Arkoumani E, et al. Expression of Bax protein in gastric carcinomas. A clinicopathological and immunohistochemical study. Acta Gastroenterol Belg. 2007;70:285–9.
104. Kolligs FT, Bommer G, Göke B. Wnt/beta-catenin/tcf signaling: a critical pathway in gastrointestinal tumorigenesis. Digestion. 2002;66(3):131–44.
105. Nishimura T. Total number of genome alterations in sporadic gastrointestinal cancer inferred from pooled analyses in the literature. Tumour Biol. 2008;29(6):343–50.
106. Koppert LB, van der Velden AW, van de Wetering M, et al. Frequent loss of the AXIN1 locus but absence of AXIN1 gene mutations in adenocarcinomas of the gastro-oesophageal junction with nuclear beta-catenin expression. Br J Cancer. 2004;90(4):892–9.
107. Lee SH, Kang HJ, Shin DH, et al. Expression of beta-catenin and its mechanism of delocalization in intestinal-type early gastric cancer based on mucin expression. Histol Histopathol. 2009;24(7):831–8.
108. Carneiro F, Oliveira C, Seruca R. Pathology and genetics of familial gastric cancer. Int J Surg Pathol. 2010;18(3 Suppl):33S–6.
109. Becker KF, Keller G, Hoefler H. The use of molecular biology in diagnosis and prognosis of GC. Surg Oncol. 2000;9:5–11.
110. Ohno T, Aihara R, Kamiyama Y, et al. Prognostic significance of combined expression of MUC1 and adhesion molecules in advanced GC. Eur J Cancer. 2006;42:256–63.
111. Chan AO, Lam SK, Chu KM, et al. Soluble E-cadherin is a valid prognostic marker in gastric carcinoma. Gut. 2001;48:808–11.
112. Chan AO, Chu KM, Lam SK, et al. Soluble E-cadherin is an independent pretherapeutic factor for long-term survival in gastric cancer. J Clin Oncol. 2003;21:2288–93.
113. Zhang HK, Zhang QM, Zhao TH, Li YY, Yi YF. Expression of mucins and E-cadherin in gastric carcinoma and their clinical significance. World J Gastroenterol. 2004;10:3044–7.
114. Kondo K, Kohno N, Yokoyama A, Hiwada K. Decreased MUC1 expression induces E-cadherin-mediated cell adhesion of breast cancer cell lines. Cancer Res. 1998;58:2014–9.
115. Nekarda H, Schmitt M, Ulm K, et al. Prognostic impact of urokinase-type plasminogen activator and its inhibitor PAI-1 in completely resected gastric cancer. Cancer Res. 1994;54:2900–7.
116. Beyer BC, Heiss MM, Simon EH, et al. Urokinase system expression in gastric carcinoma: prognostic impact in an independent patient series and first evidence of predictive value in preoperative biopsy and intestinal metaplasia specimens. Cancer. 2006;106:1026–35.
117. Wu CY, Wu MS, Chen YJ, et al. Clinicopathological significance of urokinase-type plasminogen activator genotypes in gastric cancer. Hepatogastroenterology. 2008;55:1890–4.

118. Tang H, Wang J, Bai F, et al. Positive correlation of osteopontin, cyclooxygenase-2 and vascular endothelial growth factor in gastric cancer. Cancer Invest. 2008;26:60–7.
119. Duff SE, Li C, Jeziorska M, et al. Vascular endothelial growth factors C and D and lymphangiogenesis in gastrointestinal tract malignancy. Br J Cancer. 2003;89:426–30.
120. Vidal O, Metges JP, Elizalde I, et al. High preoperative serum vascular endothelial growth factor levels predict poor clinical outcome after curative resection of gastric cancer. Br J Surg. 2009;96:1443–51.
121. Hsu PI, Hsieh HL, Lee J, et al. Loss of RUNX3 expression correlates with differentiation, nodal metastasis, and poor prognosis of gastric cancer. Ann Surg Oncol. 2009;16:1686–94.
122. Ogasawara N, Tsukamoto T. Mizoshita et al. RUNX3 expression correlates with chief cell differentiation in human gastric cancers. Histol Histopathol. 2009;24:31–40.
123. Wei D, Gong W, Oh SC, et al. Loss of RUNX3 expression significantly affects the clinical outcome of gastric cancer patients and its restoration causes drastic suppression of tumor growth and metastasis. Cancer Res. 2005;65:4809–16.
124. Sumiyoshi Y, Kakeji Y, Egashira A, et al. Overexpression of hypoxia-inducible factor 1alpha and p53 is a marker for an unfavorable prognosis in gastric cancer. Clin Cancer Res. 2006;12:5112–7.
125. Griffiths EA, Pritchard SA, Valentine HR, et al. Hypoxia-inducible factor-1alpha expression in the gastric carcinogenesis sequence and its prognostic role in gastric and gastro-oesophageal adenocarcinomas. Br J Cancer. 2007;96:95–103.
126. Ma J, Zhang L, Ru GQ, Zhao ZS, Xu WJ. Upregulation of hypoxia inducible factor 1alpha mRNA is associated with elevated vascular endothelial growth factor expression and excessive angiogenesis and predicts a poor prognosis in gastric carcinoma. World J Gastroenterol. 2007;13:1680–6.
127. Chen Z, Han ZC. STAT3: a critical transcription activator in angiogenesis. Med Res Rev. 2008;28(2):185–200.
128. Jackson CB, Giraud AS. STAT3 as a prognostic marker in human gastric cancer. J Gastroenterol Hepatol. 2009;24(4):505–7.
129. Aggarwal BB, Sethi G, Ahn KS, Sandur SK, Pandey MK, Kunnumakkara AB, et al. Targeting signal-transducer-and-activator-of-transcription-3 for prevention and therapy of cancer: modern target but ancient solution. Ann NY Acad Sci. 2006;1091:151–6.
130. Balch C, Fang F, Matei DE, Huang TH, Nephew KP. Minireview: epigenetic changes in ovarian cancer. Endocrinology. 2009;150:4003–11.
131. Faber C, Kirchner T, Hlubek F. The impact of microRNAs on colorectal cancer. Virchows Arch. 2009;454:359–67.
132. Mott JL. MicroRNAs involved in tumor suppressor and oncogene pathways: implications for hepatobiliary neoplasia. Hepatology. 2009;50:630–7.
133. Wang QZ, Xu W, Habib N, Xu R. Potential uses of microRNA in lung cancer diagnosis, prognosis, and therapy. Curr Cancer Drug Targets. 2009;9:572–94.
134. Rachagani S, Kumar S, Batra SK. MicroRNA in pancreatic cancer: pathological, diagnostic and therapeutic implications. Cancer Lett. 2010;292:8–16.
135. Ueda T, Volinia S, Okumura H, Shimizu M, Taccioli C, Rossi S, et al. Relation between microRNA expression and progression and prognosis of GC: a microRNA expression analysis. Lancet Oncol. 2010;11:136–46.
136. Carthew RW, Sontheimer EJ. Origins and Mechanisms of miRNAs and siRNAs. Cell. 2009;136:642–55.
137. Petrocca F, Visone R, Onelli MR, Shah MH, Nicoloso MS, de Martino I, et al. E2F1-regulated microRNAs impair TGFbeta-dependent cell-cycle arrest and apoptosis in GC. Cancer Cell. 2008;13:272–86.
138. Rodriguez A, Griffiths-Jones S, Ashurst JL, Bradley A. Identification of mammalian microRNA host genes and transcription units. Genome Res. 2004;14:1902–10.
139. Kim MS, Oh JE, Kim YR, Park SW, Kang MR, Kim SS, et al. Somatic mutations and losses of expression of microRNA regulation- related genes AGO2 and TNRC6A in gastric and colorectal cancers. J Pathol. 2010;221:139–46.
140. Kimura Y, Noguchi T, Kawahara K, Kashima K, Daa T, Yokoyama S. Genetic alterations in 102 primary GCs by comparative genomic hybridization: gain of 20q and loss of 18q are associated with tumor progression. Mod Pathol. 2004;17:1328–37.
141. Calin GA, Croce CM. Chromosomal rearrangements and microRNAs: a new cancer link with clinical implications. J Clin Invest. 2007;117:2059–66.
142. Saito Y, Suzuki H, Tsugawa H, Nakagawa I, Matsuzaki J, Kanai Y, et al. Chromatin remodeling at Alu repeats by epigenetic treatment activates silenced microRNA-512-5p with down- regulation of Mcl-1 in human GC cells. Oncogene. 2009;28:2738–44.
143. Mirnezami AH, Pickard K, Zhang L, Primrose JN, Packham G. MicroRNAs: key players in carcinogenesis and novel therapeutic targets. Eur J Surg Oncol. 2009;35:339–47.
144. Ruan K, Fang X, Ouyang G. MicroRNAs: novel regulators in the hallmarks of human cancer. Cancer Lett. 2009;285:116–26.
145. Inui M, Martello G, Piccolo S. MicroRNA control of signal transduction. Nat Rev Mol Cell Biol. 2010;11:252–63.

146. Wu WK, Lee CW, Cho CH, Fan D, Wu K, Yu J, Sung JJ. MicroRNA dysregulation in GC: a new player enters the game. Oncogene 2010;29:5761–5771. Supplementary Information accompanies the paper on the Oncogene website (http://www.nature.com/onc).
147. Wang J, Wang Q, Liu H, Hu B, Zhou W, Cheng Y. MicroRNA expression and its implication for the diagnosis and therapeutic strategies of GC. Cancer Lett. 2010;297:137–43.
148. Yoo YD, Choi JY, Lee SJ, Kim JS, Min BR, Lee YI, et al. TGF-beta-induced cell-cycle arrest through the p21(WAF1/CIP1)- G1 cyclin/Cdks-p130 pathway in gastric-carcinoma cells. Int J Cancer. 1999;83:512–7.
149. Kim YK, Yu J, Han TS, Park SY, Namkoong B, Kim DH, et al. Functional links between clustered microRNAs: suppression of cell-cycle inhibitors by microRNA clusters in GC. Nucleic Acids Res. 2009;37:1672–81.
150. Shen WW, Wu J, Cai L, Liu BY, Gao Y, Chen GQ, et al. Expression of anion exchanger 1 sequestrates p16 in the cytoplasm in gastric and colonic adenocarcinoma. Neoplasia. 2007;9:812–9.
151. Wu Q, Jin H, Yang Z, Luo G, Lu Y, Li K, et al. MiR-150 promotes GC proliferation by negatively regulating the proapoptotic gene EGR2. Biochem Biophys Res Commun. 2010;392:340–5.
152. Vogiatzi P, De Falco G, Claudio PP, Giordano A. How does the human RUNX3 gene induce apoptosis in GC? Latest data, reflections and reactions. Cancer Biol Ther. 2006;5:371–4.
153. Yamamura Y, Lee WL, Inoue K, Ida H, Ito Y. RUNX3 cooperates with FoxO3a to induce apoptosis in GC cells. J Biol Chem. 2006;281:5267–76.
154. Lai KW, Koh KX, Loh M, Tada K, Subramaniam MM, Lim XY, et al. MicroRNA-130b regulates the tumour suppressor RUNX3 in GC. Eur J Cancer. 2010;46:1456–63.
155. Xia L, Zhang D, Du R, Pan Y, Zhao L, Sun S, et al. miR-15b and miR-16 modulate multidrug resistance by targeting BCL2 in human GC cells. Int J Cancer. 2008;123:372–9.
156. Zhu W, Shan X, Wang T, Shu Y, Liu P. miR-181b modulates multidrug resistance by targeting BCL2 in human cancer cell lines. Int J Cancer. 2010;127:2520–9.
157. He L, He X, Lim LP, de Stanchina E, Xuan Z, Liang Y, et al. A microRNA component of the p53 tumour suppressor network. Nature. 2007;447:1130–4.
158. Ji Q, Hao X, Meng Y, Zhang M, Desano J, Fan D, et al. Restoration of tumor suppressor miR-34 inhibits human p53- mutant GC tumorspheres. BMC Cancer. 2008;8:266.
159. He XX, Yang J, Ding YW, Liu W, Shen QY, Xia HH. Increased epithelial and serum expression of macrophage migration inhibitory factor (MIF) in GC: potential role of MIF in gastric carcinogenesis. Gut. 2006;55:797–802.
160. Beswick EJ, Pinchuk IV, Suarez G, Sierra JC, Reyes VE. Helicobacter pylori CagA-dependent macrophage migration inhibitory factor produced by gastric epithelial cells binds to CD74 and stimulates procarcinogenic events. J Immunol. 2006;176:6794–801.
161. Beswick EJ, Reyes VE. Macrophage migration inhibitory factor and interleukin-8 produced by gastric epithelial cells during Helicobacter pylori exposure induce expression and activation of the epidermal growth factor receptor. Infect Immun. 2008;76:3233–40.
162. Li GQ, Xie J, Lei XY, Zhang L. Macrophage migration inhibitory factor regulates proliferation of GC cells via the PI3K/Akt pathway. World J Gastroenterol. 2009;15:5541–8.
163. Bandres E, Bitarte N, Arias F, Agorreta J, Fortes P, Agirre X, et al. microRNA-451 regulates macrophage migration inhibitory factor production and proliferation of gastrointestinal cancer cells. Clin Cancer Res. 2009;15:2281–90.
164. Tsukamoto Y, Nakada C, Noguchi T, Tanigawa M, Nguyen LT, Uchida T, et al. MicroRNA-375 is downregulated in gastric carcinomas and regulates cell survival by targeting PDK1 and 14-3- 3zeta. Cancer Res. 2010;70:2339–49.
165. Shen HM, Tergaonkar V. NFkappaB signaling in carcinogenesis and as a potential molecular target for cancer therapy. Apoptosis. 2009;14:348–63.
166. Sasaki N, Morisaki T, Hashizume K, Yao T, Tsuneyoshi M, Noshiro H, et al. Nuclear factor-kappaB p65 (RelA) transcription factor is constitutively activated in human gastric carcinoma tissue. Clin Cancer Res. 2001;7:4136–42.
167. Gao C, Zhang Z, Liu W, Xiao S, Gu W, Lu H. Reduced microRNA-218 expression is associated with high nuclear factor kappa B activation in GC. Cancer. 2010;116:41–9.
168. Wan HY, Guo LM, Liu T, Liu M, Li X, Tang H. Regulation of the transcription factor NF-kappaB1 by microRNA-9 in human gastric adenocarcinoma. Mol Cancer. 2010;9:16.
169. Zhang Z, Li Z, Gao C, Chen P, Chen J, Liu W, et al. miR-21 plays a pivotal role in GC pathogenesis and progression. Lab Invest. 2008;88:1358–66.
170. Motoyama K, Inoue H, Mimori K, Tanaka F, Kojima K, Uetake H, et al. Clinicopathological and prognostic significance of PDCD4 and microRNA-21 in human GC. Int J Oncol. 2010;36:1089–95.
171. Xiao B, Guo J, Miao Y, Jiang Z, Huan R, Zhang Y, et al. Detection of miR-106a in gastric carcinoma and its clinical significance. Clin Chim Acta. 2009;400:97–102.

172. Tie J, Pan Y, Zhao L, Wu K, Liu J, Sun S, et al. MiR-218 inhibits invasion and metastasis of GC by targeting the Robo1 receptor. PLoS Genet. 2010;6:e1000879.
173. Pfannkuche K, Summer H, Li O, Hescheler J, Droge P. The high mobility group protein HMGA2: a co-regulator of chromatin structure and pluripotency in stem cells? Stem Cell Rev. 2009;5:224–30.
174. Motoyama K, Inoue H, Nakamura Y, Uetake H, Sugihara K, Mori M. Clinical significance of high mobility group A2 in human GC and its relationship to let-7 microRNA family. Clin Cancer Res. 2008;14:2334–40.
175. Liu T, Tang H, Lang Y, Liu M, Li X. MicroRNA-27a functions as an oncogene in gastric adenocarcinoma by targeting prohibitin. Cancer Lett. 2009;273:233–42.
176. Fusaro G, Dasgupta P, Rastogi S, Joshi B, Chellappan S. Prohibitin induces the transcriptional activity of p53 and is exported from the nucleus upon apoptotic signaling. J Biol Chem. 2003;278:47853–61.
177. Rastogi S, Joshi B, Dasgupta P, Morris M, Wright K, Chellappan S. Prohibitin facilitates cellular senescence by recruiting specific corepressors to inhibit E2F target genes. Mol Cell Biol. 2006;26:4161–71.
178. Theiss AL, Jenkins AK, Okoro NI, Klapproth JM, Merlin D, Sitaraman SV. Prohibition inhibits tumor necrosis factor alpha-induced nuclear factor-kappa B nuclear translocation via the novel mechanism of decreasing importin alpha3 expression. Mol Biol Cell. 2009;20:4412–23.
179. Wada R, Akiyama Y, Hashimoto Y, Fukamachi H, Yuasa Y. miR-212 is downregulated and suppresses methyl-CpG-binding protein MeCP2 in human GC. Int J Cancer. 2009;127:1106–14.
180. Arisawa T, Tahara T, Shibata T, Nagasaka M, Nakamura M, Kamiya Y, et al. A polymorphism of microRNA 27a genome region is associated with the development of gastric mucosal atrophy in Japanese male subjects. Dig Dis Sci. 2007;52:1691–7.
181. Peng S, Kuang Z, Sheng C, Zhang Y, Xu H, Cheng Q. Association of MicroRNA-196a-2 gene polymorphism with GC risk in a Chinese population. Dig Dis Sci. 2010;55:2288–93.
182. Tsujiura M, Ichikawa D, Komatsu S, Shiozaki A, Takeshita H, Kosuga T, et al. Circulating microRNAs in plasma of patients with GCs. Br J Cancer. 2010;102:1174–9.
183. Zhou H, Guo JM, Lou YR, Zhang XJ, Zhong FD, Jiang Z, et al. Detection of circulating tumor cells in peripheral blood from patients with GC using microRNA as a marker. J Mol Med. 2010;88:709–17.
184. Catalano V, Labianca R, Beretta GD, Gatta G, de Braud F, Van Cutsem E. Gastric cancer. Crit Rev Oncol Hematol. 2009;71:127–64.
185. Li X, Zhang Y, Ding J, Wu K, Fan D. Survival prediction of GC by a seven-microRNA signature. Gut. 2010;59:579–85.
186. Katada T, Ishiguro H, Kuwabara Y, Kimura M, Mitui A, Mori Y, et al. microRNA expression profile in undifferentiated GC. Int J Oncol. 2009;34:537–42.
187. Weiler J, Hunziker J, Hall J. Anti-miRNA oligonucleotides (AMOs): ammunition to target miRNAs implicated in human disease? Gene Ther. 2006;13:496–502.
188. Wang S, Nath N, Adlam M, Chellappan S. Prohibitin, a potential tumor suppressor, interacts with RB and regulates E2F function. Oncogene. 1999;18:3501–10.
189. Kidd M, Modlin IM, Mane SM, et al. Utility of molecular genetic signatures in the delineation of gastric neoplasia. Cancer. 2006;106(7):1480–8.
190. Yap YL, Zhang XW, Smith D, Soong R, Hill J. Molecular gene expression signature patterns for gastric cancer diagnosis. Comput Biol Chem. 2007;31(4):275–87.

Chapter 8
Gene Signatures in Colorectal Cancer

Alessandro Lugli and Inti Zlobec

Introduction

The Tumor Node Metastasis (TNM) staging system from the American Joint Committee on Cancer and the "Union Internationale Contre le Cancer" (AJCC/UICC) remains the most reliable prognostic indicator for patients with colorectal cancer [1]. Overall 5-year survival rates are reported at 65% and correspond closely to disease progression; patients with stage I disease have more favorable prognoses with 5-year survival rates exceeding 80–90%. In contrast, patients with stage II, III and IV disease experience progressively worse outcomes with varying 5-year survival rates of 70–85%, 44–80% and <10%, respectively [2]. The role of accurate staging has in recent years been expanded to include the selection of patients for more individualized treatment regimens. In particular, patients with lymph node positive, nonmetastatic (TNM stage III, AJCC 6th edition) colorectal cancers are considered for adjuvant therapies, patients with pT3-4 or lymph node (TNM stage IIB or stage III, AJCC 6th edition) positive rectal cancers may be suitable for preoperative neoadjuvant therapy while patients with TNM stage IV (AJCC 6th edition), and thus metastatic diseases are currently offered anti-EGFR therapies based on molecular analysis of *K-RAS* gene mutations. According to changes in treatment and improvements in surgery, therapy, overall and disease-specific survival, newly outlined and updated tumor staging for patients with colorectal cancer is warranted. In November 2009, the AJCC published the 7th edition of the Cancer Staging Manual where several important changes in colorectal cancer staging have been made; for example, T classification now includes T4a (tumor penetrates the surface of the peritoneum) and T4b (tumor directly invades or is histologically adherent to other organs or structures) [3]. This change has consequently led to modifications in stage II, III and IV with the aim of improving prognostication and tailoring therapy.

Lessons from previous TNM staging systems have demonstrated that despite having identically staged tumors, not all patients within the same TNM stage neither experience similar clinical outcomes nor rates of recurrence or responses to therapy. For this reason, highly relevant and promising prognostic and predictive features are currently being explored including a range of tumor- and host-related histomorphological, protein and molecular factors.

Although many studies have proposed novel prognostic and predictive markers in colorectal cancer, their implementation into daily diagnostic work has until now not been performed for

A. Lugli (✉) • I. Zlobec
Institute of Pathology, University of Bern, Bern, Switzerland
e-mail: alessandro.lugli@pathology.unibe.ch

A. Russo et al. (eds.), *Diagnostic, Prognostic and Therapeutic Value of Gene Signatures*, Current Clinical Pathology, DOI 10.1007/978-1-61779-358-5_8, © Springer Science+Business Media, LLC 2012

several reasons, such as lack of standardized scoring systems, missing inter- and intraobserver and laboratory variability studies and concise validation in prospective patient cohorts. The recently published REMARK guidelines aim towards standardization of the approach used to investigate novel prognostic and predictive biomarkers and increase the likelihood of their inclusion as part of pre- or postoperative management of colorectal cancer patients [4].

In the last years a considerable progress has been made in the field of gene signatures proposing prognostic and predictive biomarkers based on Affymetrix, Oligonucleotide, cDNA and miRNA arrays. The aim of this chapter is to give the reader a general overview on prognostic and predictive factors in colorectal cancer on various cellular/molecular levels including DNA, RNA, miRNA and protein.

Prognostic and Predictive Role of Histomorphological, Protein and Molecular Markers in Colorectal Cancer: Current Perspectives

Tumor-Related Histomorphological Markers

According to the UICC, the most pertinent and prognostically relevant histomorphological factors after TNM staging recommended for routine pathological diagnosis include venous and lymphatic invasion (Fig. 8.1) [5]. Tumor grade, histological subtype, perineural invasion, tumor border configuration and tumor budding are considered "additional" prognostic factors. Indeed, tumor budding defined as single cells or clusters of up to five cells at the invasive front, has consistently been shown to have adverse effects on clinical outcome and recurrence, with strong associations with lymph node positivity and the presence of distant metastatic spread [6–12]. Problems with standardization in the reporting of tumor budding constitute the most important limiting factor for this feature and therefore studies focusing on a simple and reproducible tumor budding score are still missing. In contrast, the tumor border in colorectal cancer is clearly defined and includes the infiltrating, pushing and mixed pattern [13]. In a recent study, the addition of tumor border configuration to TNM stage (AJCC 6th edition) was found to improve the prognostic classification of colorectal cancer patients by almost 20% [14].

Host-Related Histomorphological Markers

The presence of conspicuous peritumoral lymphocytic (PTL) inflammation, viewed as a distinctive "encapsulating" connective tissue mantle cap at the invasive front of colorectal cancer, is inversely correlated with the presence of tumor budding and positively associated with improved survival. Jass demonstrated that PTL infiltration in rectal cancer decreased with more advanced Dukes' stage to 53%, 28% and 13% with Dukes' A, B and C cases respectively [15]. In addition, the significantly worsened prognosis in patients lacking PTL inflammation at the tumor border was highlighted, while patients with moderate or pronounced infiltration performed significantly better independently of disease stage. The results have also been confirmed by other study groups. However, the presence of PTL inflammation at the invasive front does not appear to be an independent prognostic factor in patients with this disease [14]. Nonetheless, PTL inflammation seems to be intimately

a

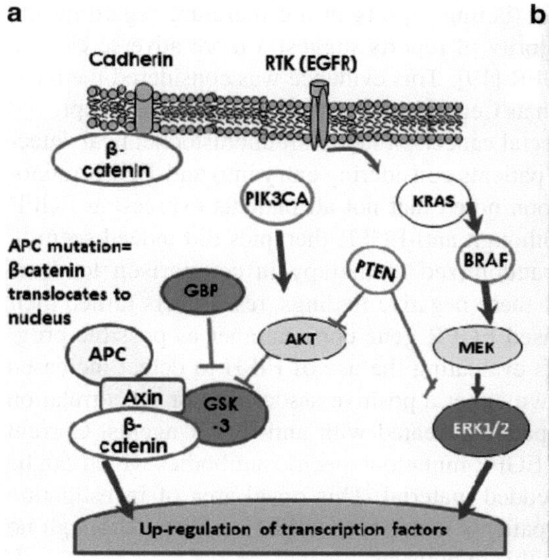

b

Feature	Prognostic	Predictive
TNM stage	+++	
Tumor border configuration	++	
Tumor budding	++	
Vacular invasion	++	
Lymphatic invasion	++	
Perineural invasion	++	
Tumor grade	+	
Tumor infiltrating lymphocytes	++	
EGFR gene copy	+	+
KRAS gene	+	+++
BRAF gene	+	+
CpG Island Methylator Phenotype	+	+
Microsatellite Instability	++	++

Fig. 8.1 The most important prognostic and predictive factors in colorectal cancer. (**a**) Two major pathways involved in colorectal cancer progression. Up-regulation of WNT pathway signaling, as a consequence of APC mutation, leads to translocation of beta-catenin to the nucleus. RAS/MAPK signaling is up-regulated as a consequence of EGFR over-expression, *K-RAS* mutation and *B-RAF* mutation. (**b**) Strength of evidence supporting the most important features as prognostic and predictive factors

linked with abundant CD8+ tumor infiltrating T-lymphocytes, further implicating tumor immunity in the defense against colorectal cancer.

Tumor-Related Molecular and Protein Markers

In 2007, a molecular classification of colorectal cancer including the *K-RAS*, *B-RAF*, microsatellite instability (MSI), methylation of O^6-methylguanine-DNA methyltransferase (*MGMT*) and the CpG Island Methylator Phenotype (CIMP) status was proposed [16].

K-RAS and B-RAF Status

The pathogenesis and oncogenic behavior of colorectal cancer is in large part regulated by the RAS/MAP kinase signaling pathway [17]. In fact, several key components of this pathway, including epidermal growth factor receptor (EGFR), *K-RAS* and *B-RAF* are being intensively investigated for their potential predictive value in patients with metastatic disease treated with anti-EGFR monoclonal antibodies [18]. The EGFR is a tyrosine kinase receptor, which upon ligand binding, dimerizes, is autophosphorylated and activates downstream signaling molecules. Initial efforts to uncover the prognostic value of EGFR in both unselected and metastatic patients focused on the immunohistochemical detection of the protein in colorectal cancers. The use of over ten different definitions to characterize EGFR over-expression has led to conflicting reports in the literature regarding the prognostic value of this protein, although the majority of reports suggest a more adverse clinical outcome in patients with "over-expression" of EGFR [19]. This evidence was considered the basis for the development of monoclonal antibodies such as Cetuximab and Panitumumab, now approved for the treatment of patients with metastatic colorectal cancer. In fact, immunohistochemical detection of EGFR was used as a selection criterion for patients considering entry into anti-EGFR monoclonal antibody clinical trials. However, it was soon noted that not all patients expressing EGFR experienced an objective response to treatment although anti-EGFR therapies did indeed seem to improve the overall survival rates of patients randomized to therapy in comparison to those receiving best supportive care [20]. In response to these negative findings, researchers turned their focus toward EGFR gene amplification or increased EGFR gene copy number as possible prognostic or predictive markers [21]. The few reports evaluating the use of FISH to detect increased gene copy numbers in colorectal cancer have shown either a positive association or no correlation with response or clinical outcome in metastatic patients treated with anti-EGFR agents. Current technical advances have led to the development of EGFR mutation-specific antibodies which can be used for immunohistochemistry on paraffin-embedded material. This novel area of investigation may also have an impact on the identification of patients with poor clinical outcome, although no studies to date have been performed to investigate this hypothesis.

The lack of predictive value of EGFR for identifying potential metastatic colorectal cancer patients responsive to anti-EGFR therapy has recently promoted the investigation of downstream RAS/MAPK signaling molecules and their potential predictive value. Mutations of *K-RAS* occur in approximately 30–40% of all colorectal cancer patients. Although the majority of studies indicate that mutations in *K-RAS* have a negative effect on prognosis, these findings have not been unanimously described. The largest meta-analysis to date performed by Andreyev and colleagues in 1998 and again in 2001 seems to suggest that the overall increase in the relative risk of death in patients with mutation of *K-RAS* is only 1.3 times greater than those patients without *K-RAS* mutation [22, 23]. A possible explanation for these relatively weak findings may be due to the possible differential prognostic effects of specific point mutations in the *K-RAS* gene. In particular, three mutations (codon 12 Gly→Asp (G12D); codon 12 Gly→Val (G12V) and codon 13 Gly→Asp (G13D)) occur with a relative increased frequency compared to all other mutations, namely in 14%, 8.5% and 7.3% of cases, respectively [24]. Works by Finkelstein and colleagues have demonstrated the significantly different clinical behaviors of tumors bearing each of these three mutations in addition to patients with wild-type tumors [25, 26]. The most aggressive form of *K-RAS* appears to be mutation in G12D followed by tumors with wild-type *K-RAS*. More indolent than the wild-type tumors were those with G12V mutation and finally the least aggressive of all were tumors with mutations in G13D. Therefore, the proportion of patients with specific point mutations may to some extent influence the overall relative risks of death attributed to *K-RAS* mutation. Despite this controversy, a clear association of *K-RAS* gene mutation and lack of response to anti-EGFR therapies has been consistently described, to the point where now *K-RAS* mutational investigations are routinely performed in molecular pathology laboratories [27]. However, not all patients with *K-RAS* wild-type tumors respond to monoclonal antibodies targeting EGFR and conversely not all responders are those with *K-RAS* wild-type tumors.

Downstream of *K-RAS* in RAS/MAP kinase signaling lays *B-RAF*, a gene that is mutated in approximately 10–15% of CRCs [17]. Interestingly, *K-RAS* and *B-RAF* mutations appear to occur as mutually exclusive events. On a clinico-pathological level, *B-RAF* mutations have been described as occurring more frequently in colon versus rectal cancers and often found in proximal compared to distal tumors. Moreover, mutation in *B-RAF* seems to be specific for sporadic disease, with mutation status suggested as an exclusion criterion for suspected Lynch syndrome-associated/hereditary nonpolyposis CRC (HNPCC) [28]. Molecularly, *B-RAF* mutations have been linked to high levels of MSI (MSI-H), MLH1 hypermethylation as well as CIMP-high status [29]. On a protein level, down-regulation of *B-RAF* in cell lines has been shown to significantly decrease ERK1/2 phosphorylation and Cyclin D1 expression as well as increase expression of p27, results which have been confirmed using tissue from CRC patients [30–33]. Additional differences between *B-RAF* mutated and wild-type CRCs seem to include decreased expression of CDX2, loss of p16, positivity for DNA methyltransferase-3B, a marker of de novo CpG island methylation or SIRT1 histone deacetylase expression [31, 34, 35].

Only recently have a hand-full of large studies confirmed the poorer outcome in patients with *B-RAF* mutation. Ogino et al. report a significant adverse effect of *B-RAF* mutation in a large cohort of more than 600 CRC patients [29]. French et al. investigated combinations of *B-RAF* mutation and MSI status and found that those with MSI cancers and *B-RAF* wild-type gene status had a significantly improved outcome [36]. In MSS cases, Kakar and colleagues report a significantly poorer outcome with *B-RAF* mutation and a significant association with chromosomal instability [37]. The recent results from the PETACC-3 study also show significantly worse overall survival with *B-RAF* mutation in patients with MSI-low or MSS CRCs [38]. Richman and colleagues reporting the findings of MRC FOCUS Trial find as well a poorer prognostic effect in patients with *B-RAF* mutation compared to those with wild-type gene status [39]. In the metastatic setting, patients with *B-RAF* mutation are found to be nonresponsive to anti-EGFR therapies. However, the role of *B-RAF* mutation as a general prognostic or specifically as a predictive factor in metastatic patients requires further assessment in large randomized trials.

Microsatellite Instability Status

Microsatellite instability is a term used to characterize changes occurring in the number of microsatellite loci, a phenomenon observed in certain tumor types such as colorectal cancer, endometrial and gastric carcinomas. The importance of MSI status in colorectal cancer was first recognized through investigations on HNPCC, or Lynch syndrome, a familial form of colorectal cancer affecting approximately 3–5% of patients with this disease. HNPCC cancers exhibit a high degree of MSI (MSI-H) manifested by germline mutations in one or more of the genes associated with DNA mismatch repair (MMR) [40]. Colorectal cancers arising in the setting of HNPCC are distinct from sporadic tumors in that patients are generally younger, seem to have an improved survival and have a decreased risk of metachronous tumor development compared to their sporadic counterparts. Importantly, MSI-H status is found in approximately 15% of sporadic colorectal cancers caused by biallelic hypermethylation of the promoter region of MLH1 leading to inactivation of MMR genes [41]. MSI-H status can be established by analyzing mutations in a panel of five MSI markers which include mono- and dinucleotide markers, namely BAT25, BAT26, D2S123, D5S346 and D17S250 which can also be supplemented by the more complex repeat MYCL1 [42]. Sporadic MSI-H tumors appear to originate from the serrated pathway (i.e., arising through serrated adenomas), tend to occur more often in female patients and are located on the right side of the colon [43].

Histomorphologically, they appear frequently as poorly differentiated tumors, with pushing/ expanding tumor borders with little tumor budding, and with a significant PTL infiltration (or Crohn's-like reaction) at the invasive tumor front [44, 45]. Additionally, these MSI-H tumors tend to have abundant CD8+ tumour infiltrating lymphocytes (TILs) suggesting an inherent immunogenicity not seen in tumors with no instability at any of the five MSI loci. Sporadic MSI-H colorectal cancers seem to have a favorable prognosis despite being less responsive to chemotherapeutic agents such as 5-FU [41] but these findings have recently been challenged [46]. The remaining 85% of colorectal cancer cases are thought to follow to a large extent the classical adenoma–carcinoma sequence first proposed by Volgelstein and Fearon and are microsatellite stable (MSS). The adenoma–carcinoma sequence proposes that colorectal cancers may arise through a linear accumulation of genetic mutations occurring at various stages of adenoma and carcinoma progression including mutations in APC, *K-RAS* and p53 [47]. Mutations in the p53 gene, particularly in patients with rectal cancer seem to have a considerable negative effect on clinical outcome and local recurrence rates [48]. Although mutation in p53 may occur in up to 50% of cases, only 10% of all colorectal cancers seem to harbor mutations in APC, *K-RAS* and p53 simultaneously. Loss of heterozygosity (LOH) at 18q is also being proposed as a prognostic factor for patients with MSS, or chromosomal instable tumors, but no consensus has been reached [49, 50]. Interestingly, an intermediate subset of colorectal cancers has been hypothesized based on the number of low-level MSI, or MSI-L colorectal cancers, which are defined, although ambiguously, as instable in <30–40% of microsatellite loci [51–53]. Although anecdotal evidence suggests that MSI-L cancers lead to a highly adverse outcome, the definition of MSI-L status is equivocal since the larger the number of MSI markers evaluated, the larger is the probability of finding instability in at least one of them. Several groups have evaluated either 5-panel markers, according to the recommended Bethesda guidelines, while others assess 10-panel markers or have evaluated only the more complex MYCL1 in order to establish MSI-L status. For this reason, the frequency of MSI-L in the literature is difficult to ascertain, ranging from 5 to 35% in certain reports. MSI-L colorectal cancers, if they truly exist, have yet to be characterized.

CIMP Status

MSI-H is linked to high-level CIMP. CpG islands are short sequences rich in the CpG dinucleotide and can be found in the 5′ region of about half of all human genes [54]. Methylation of cytosine within 5′ CpG islands is linked to transcriptional gene silencing, and occurs in approximately 40% of colorectal cancers. This process described as a nonrandom event, is usually associated with absence of coding region mutations suggesting that CpG island methylation is a highly relevant mechanism for gene inactivation. In fact, CIMP status plays a fundamental role in the current model of colorectal tumorigenesis and is currently thought to predispose tumors to MSI. CIMP-High/CIMP-positive (CIMP-H/CIMP+) colorectal cancers are often right-sided, more frequent in older female patients, are commonly of high tumor grade, and mucinous histology [55]. At the molecular level, CIMP-H cancers are often associated with *B-RAF* mutation, are less common in *K-RAS* mutated cases and are most often, but not exclusively, MSI-H. Despite the more "favorable" prognosis typically associated with MSI-H tumors, outcome in patients with CIMP-H colorectal cancers may be dependent on MSI status and/or *B-RAF* mutation [29, 55–57]. Older female individuals compose a recognized group of patients poorly responsive to 5-FU-based therapies and are reportedly at greater risk of developing CIMP-H colorectal cancers as suggested by their higher rates of hypermethylation in normal adjacent colonic mucosa compared to males and younger individuals. In 2003, Van Rijnsoever and colleagues investigated two groups of 103 matched stage III patients receiving either surgery or surgery plus 5-FU/leucovorin and found that CIMP+patients had a significant survival benefit from 5-FU compared to CIMP-negative patients [58]. Jover and colleagues have recently shown the opposite [59].

MGMT Status

CIMP has itself been linked with methylation of *MGMT* [60]. *MGMT* acts as a DNA repair enzyme, which works to removes alkyl groups that have been added to guanine residues, often occurring as a result of chemotherapy. Over-expression of *MGMT* has been shown, in patients with glioblastoma, to lead to tumor cell resistance [61]. In contrast, methylation of *MGMT* leads to down-regulation of the gene and decreased expression has been linked to improved therapeutic success. In fact, methylation of *MGMT* is currently incorporated into diagnostic molecular pathology laboratories to help with clinical decision making for patients with gliomas. In colorectal cancer, it is currently being proposed that methylation of *MGMT* in the precancerous state (termed field cancerization) produces a selective pressure for the subsequent development of MSI. If *MGMT* is impaired, the MMR system is itself unable to repair lesions caused by the addition of methyl adducts thus leading to double-strand breaks and apoptosis, while MMR defects would lead to "methylation tolerance" [62]. Only a handful of studies have investigated the clinical impact of *MGMT* and its relationship with molecular features. Loss of *MGMT* expression may be linked to Temozolomide response in patients with metastatic colorectal cancer [63] while Ogino and colleagues have shown no effect of *MGMT* methylation status on prognosis [60]. *MGMT* methylation has never been evaluated in the context of tumor budding.

Molecular and Protein Biomarkers of the PIK3CA/PTEN Signaling Pathway

The PIK3CA gene appears to be mutated in 20% of colorectal cancers. PIK3CA mutations occurring in the "hotspots" located in exon 9 (E542K, E545K) and exon 20 (H1047R) are oncogenic in colorectal cancer cellular models. The PIK3CA gene encodes for a lipid kinase that regulates, alongside with *K-RAS*, signaling pathways downstream of EGFR. Moreover, the p110a subunit of PI3K, encoded by PIK3CA, can be activated by interaction with RAS proteins. PI3K-initiated signaling is normally inhibited by phosphatase and tensin homologue deleted on chromosome ten (PTEN). Recent evidence suggests that both mutation of PIK3CA and negative immunohistochemical expression of PTEN leads to a lack of response in metastatic colorectal cancer patients treated with anti-EGFR therapies [64, 65].

Host-Related Molecular and Protein Markers

Most studies to date confirm that a high rate of TILs, in particular those located intraepithelially characterized by CD4+ and CD8+ tumor-associated antigens are beneficial for patient outcome. An abundant TIL count appears to be linked to earlier Dukes' stage, decreased local recurrence rate following curative surgery and improved overall and disease-free survival time both in nonmetastatic and metastatic patients undergoing hepatic resection. Galon et al. evaluated by gene expression profiling and immunohistochemistry, the type, density and location (whether at the invasive margin or the tumor center) of TILs in a large number of cases. They evaluated CD3, CD8, granzyme B and memory CD45RO T cells and demonstrated a significant independent and positive effect of TILs on both recurrence and survival [66]. Pages et al. performed a comprehensive analysis of TILs focusing on early metastatic invasion [67]. They found, by RT-PCR on 75 cases that mRNA levels of CD8, granzyme B and granulysin were significantly greater in patients without vascular emboli, lymphatic and perineural invasion (collectively known as VELIPI) compared to those with these features and that CD45RO+ cells had independent prognostic value [67]. New evidence suggest that TIA-1 may be a marker for an activated cytotoxic phenotype and helps to further stratify the prognostic effect associated with high levels of CD8+ T-cells [67]. Increased numbers of interepithelial CD3, CD4, CD8, CD20 and CD16 have been reported to significantly improve overall outcome [68, 69]. Moreover, regulatory T-cells expressing FOXP3+ and IL-17-expressing cells have been shown to correlate with improved outcome independently of TNM stage [70, 71]. However, evidence suggests that the prognostic value of inflammatory and immunological biomarkers is surely confounded by MSI status.

Gene Signatures of Relapse and Prognosis

An estimated 102,900 and 39,670 new cases of colon and rectal cancer, respectively, were diagnosed in the United States in 2010 [72]. Colorectal cancer remains the third most important cancer-related killer in Western countries with patient prognosis closely linked to TNM stage of disease. Five-year survival rates are stage I: more than 90%, stage IIA and IIB: 60–85%, stage IIIA: 55–60%, stage IIIB: 35–42%, stage IIIC: 25–27% and finally stage IV: 5–7% [2, 73]. In addition to TNM stage, other prognostic factors have been linked to poor survival including fewer than 12 lymph nodes examined, poor differentiated histology, lymphovascular invasion, perineural invasion, bowel obstruction, or perforation and tumor involvement of surgical resection margins [73]. At diagnosis, more than 75% of patients undergo curative resection, but up to 40% of patients present at first diagnosis with distant metastases. Postoperative adjuvant therapy has had a major positive impact on the treatment of patients with colorectal cancer, which is primarily limited to stage III disease. Randomized trials support 6 months of postoperative fluorouracil, leucovorin and oxaliplatin and may be beneficial to both younger and older patients [74]. Interestingly, patients with stage II disease represent a subgroup of colorectal cancer patients that may differ widely in terms of prognosis. In fact, one of the most interesting areas of colorectal cancer research today is the study of prognostic or predictive biomarkers in stage II patients that could identify those who may derive a benefit from adjuvant chemotherapy. Additional histomorphological prognostic factors, such as the presence of tumor budding and infiltrating tumor border configuration with absence of PTL inflammation and low numbers of CD8+ TILs and possibly other immunological cell types have all been shown, mostly in retrospective studies to stratify stage II patients into high- and low-risk groups [75–78]. The potential for gene signatures to help complement TNM stage and identify patients with lymph node-negative disease who may be at risk of local recurrence and distant metastasis would lead to the identification of high-risk patients who may benefit from additional adjuvant, or even neo-adjuvant, therapies (Table 8.1).

Using cDNA or oligonucleotide arrays, Wang and colleagues describe a 23-gene signature to predict cancer recurrence in Dukes' B (node-negative) patients, which led to a 78% accuracy in a validation group of 36 independent patients [98]. Moreover, recurrence-free survival time differences between high- and low-risk groups had independent prognostic value when adjusting for age, T stage, grade, and tumor size in multivariate analysis. Of the 23 genes included in their signature, several are involved in cell proliferation, cell signaling and immune response. Bandres and colleagues, using two different statistical methods, found eight genes, five of which were validated by real-time PCR, to predict high-risk of recurrence [80]. They show that down-regulation of *CHD2* (Chromodomain helicase DNA binding protein 2), *RPS5* (Ribosomoal protein S5), *ZNF148* (Zinc-finger protein 148), *BRI3* (Brain protein I3) and MGC23401 (a hypothetical protein with unknown function) discriminate between patients with and without relapse. Arango and colleagues made use of a unique subset of Dukes' C patients only treated by surgery to investigate high- and low-risk groups within this stage of disease [79]. They performed gene expression analysis first on fresh-frozen material from these adjuvantly nontreated patients and identified, among the 218 distinct genes differentially expressed between tumors leading to "good" and "bad" prognosis, the gene *RHOA*, a Ras homologue. Using an independent set of more than 100 patients, they performed immunohistochemistry for RHOA and survival analysis, which seemed to confirm that down-regulation of this gene had a considerable negative prognostic effect. Eschrich and colleagues identify a 43-gene signature to discriminate between prognostic groups of patients with Dukes B and C cancers. Of these genes, two, namely *osteopontin* (a ligand for CD44) and *neuregulin* (a ligand for ERBB) were singled out as having possibly important biological significance [88]. Barrier and colleagues identified a 30-gene prognostic predictor in a training set of stage II patients, which yielded an 80% prognostic accuracy in the validation set. These 30 genes included many

Table 8.1 Gene signatures of colorectal cancer prognosis

First author [Reference]	Year	Platform/assay	Samples	Aim/design	Results	Comment
Arango [79]	2005	Affymetrix	Frozen ($n=25$)	Prognosis after surgery alone	5-Gene profile outperformed previously established markers *RHOA*, *CLTC*, *MTMR1*, *ARCN1* and *IDH3G* (like TP53 and *K-RAS*, LOH at 18q)	137 Dukes' C independent samples to validated RHOA (Ras homolog) by IHC and TMA
Bandres [80]	2007	Oligo.arrays	Frozen ($n=16$)	Predicting high risk of recurrence	Two different statistical techniques used, eight overlapping genes in both	27 Dukes B and C Independent samples, five genes validated by real-time PCR
Barrier [81]	2007	Affymetrix	Frozen ($n=24$)	Prognosis in stage II	70-Gene signature	
Barrier [82]	2005	Affymetrix	Frozen ($n=18$)	Prognosis in stage III	30 Tumor gene signature and 70 normal mucosa based predictor	
Barrier [83]	2005	Affymetrix	Frozen ($n=12$)	Prognosis prediction	47-Gene signature	
Bertucci [84]	2004	cDNA	($n=50$)	Prognostic prediction	194 Discriminating gene signature	382 Samples of tumor and normal, IHC using TMA
Carvalho [85]	2011	cDNA	($n=52$)	Validation of previously established prognostic gene signature	Although differences between normal tissue and cancer confirmed the 17-gene signature, only 1 had a survival effect: PTTG1	
Cavalieri [86]	2007	cDNA	($n=19$)	Prognosis stage III and IV	8-gene signature	Validation: RT-PCR
D'Arrigo [87]	2005	cDNA	Frozen ($n=10$)	Prediction of metastatic potential	29-Gene signature	Validation: RT-PCR
Eschrich [88]	2005	cDNA	Frozen ($n=75$)	Prognosis in stage II and III	43-Gene signature	
Hao [89]	2010	Affymetrix	In vivo mouse model	Prediction of metastatic potential	5-Gene signature to predict metastasis (LYN, SDCBP, MAP4K4, MID1, DKK1)	181 Clinical samples additional
Jorissen [90]	2009	Affymetrix	Frozen ($n=293$)	Predict outcome in Dukes B and C patients	128-Gene signature predicts prognosis in stage B and C	33 Dukes' B and C from independent set
Kalady [91]	2010	cDNA	Frozen ($n=100$)	Predict recurrence in stage I–II colorectal cancer patients	36-Gene signature predicts recurrence in early stage disease	
Kwon [92]	2004	cDNA	Frozen ($n=12$)	Difference in normal and cancer	112 Discriminating genes	Validation: RT-PCR
Matsuyama [93]	2010	Oligo. array	Froezn ($n=100$)	Predict distant metastasis	MUC12 expression predicts poor survival	
Peng [94]	2010	cDNA	Frozen ($n=95$)	Predict 3-year disease-free survival	8-Gene signature predicts relapse-free survival, outperforms TNM 6th	

(continued)

Table 8.1 (continued)

First author [Reference]	Year	Platform/assay	Samples	Aim/design	Results	Comment
Pillaire [95]	2010	cDNA	Frozen ($n=74$)	Test prognostic effect of 47 DNA replication gene signature on 74 patients	Overexpression of MCM7 (helicase) or POLQ (polymerase) was significantly related to poor patients survival	IHC and TMA
Salazar [96]	2011	Oligo. array	Frozen ($n=188$)	Predict disease relapse	18-Gene signature identified	ColoPrint®
Smith [97]	2010	Affymetrix	In vitro, in vivo mouse model and Frozen ($n=55$)	Predict high-risk stage II and low-risk stage III	34-Gene classifier associated with high-risk of metastasis	177 Patients from a second institution
Wang [98]	2004	Affymetrix	Frozen ($n=74$)	Predict relapse	Identification of 23-gene signature predicting recurrence	36 Independent patients. Performance accuracy 78%
Watanabe [99]	2010	cDNA	Frozen ($n=189$)	Predict liver metastasis	10-Gene signature to predict liver metastasis (including epiregulin, amphiregulin, COX2)	29 Independent patients as validation RT-PCR
Webber [100]	2010	cDNA	FFPE	Predict prognosis in stage II	7 Prognostic gene signature for stage II colorectal cancer	Onco*type* Dx®
Yamasaki [101]	2007	cDNA	Frozen ($n=58$ primary; 34 liver metastases)	Prediction of metastatic potential	119-Gene signatures	

FFPE Formalin fixed paraffin-embedded, *Oligo.* oligonucleotide ribosomal proteins, calreticulin, and insulin-like growth factor binding protein 1, among others [82]. In a second study, the same group investigated a prognostic predictor for stage II patients using nonneoplastic mucosa gene expression profiles and identify a 70-gene signature leading to 81.8% accuracy [81].

These studies highlight the fact that despite investigating colorectal cancers from a relatively "homogeneous" histological background, the number of genes included in particular signatures, the biological functions of such genes and the prognostic accuracies of gene signatures vary widely.

Prognostic Gene Signatures Available on the Market

There are several commercial prognostic gene signatures either already available on the market or in the last phases of development. The ChipDx® Colon Cancer module offers a 163-gene prognostic gene expression signature to predict risk of recurrence in patients with stage II or III colon cancer [102]. The Onco*type* Dx® Colon Cancer Assay is a quantitative multigene real-time PCR assay, which offers a 7-gene relapse-free prognostic gene signature to yield a prognostic recurrence score for patients with stage II disease [100]. The genes included in the assay are: *BGN* (biglycan), *MYC*, *FAP* (fibroblast activating protein), *GADD45B* (growth arrest and DNA-damage-inducible, beta), *INHBA* (inhibin, beta A), *MK167* (antigen Ki67) and *MYBL2* (v-myb myeloblastosis viral oncogene homolog (avian)-like 2) and five reference normalization genes. These are associated with activated stroma (*BGN*, *INHBA* and *FAP*), represent cell cycle (*MK167*, *MYBL2*, and *MYC*) and genotoxic stress pathways (*GADD45B*).

ColoPrint® is a prognostic gene expression profile to predict the risk of recurrence in stage II and III colon cancer patients [96]. It is based on an 18-gene signature, which when classifying patients into low- and high-risk groups seems to have independent prognostic impact as shown by multivariate analysis for relapse free survival. Genes included in the ColoPrint® assay are *MCTP1* (Multiple C2 domains, transmembrane 1), *LAMA3* (Laminin, alpha 3), *CTSC* (Cathepsin C), *PYROXD1* (Pyridine nucleotide-disulphide oxidoreductase domain 1; unknown function), *EDEM1* (ER degradation enhancer, mannosidase alpha-like 1), *IL2RB* (Interleukin2 receptor beta), *ZNF697* (Zinc finger 697; unknown function), *SLC6A11* (Solute carrier family 6 (neurotransmitter transporter, GABA), member 11), *IL2RA* (Interleukin 2 receptor alpha), *CYFIP2* (Cytoplasmic FMR1 interacting protein 2), *PIM3* (Pim-3 oncogene), *LIF* (Leukemia inhibitory factor) (cholinergic differentiation factor), *PLIN3* (Mannose-6-phosphate receptor binding protein 1), *HSD3B1* (Hydroxy-delta-5-steroid dehydrogenase, 3 beta- and steroid delta-isomerase 1), *ZBED4* (Zinc finger, BED-type containing 4), *PPARA* (Peroxisome proliferator-activated receptor alpha), *THNSL2* (Threonine synthase-like 2 [*Saccharomyces cerevisiae*]), and *CA438802* (function unknown). Interestingly, no genes proposed by either of these three commercial assays overlap.

Gene Signatures of Treatment Response

Over the last decade, considerable efforts have been initiated to identify biomarkers of treatment response to therapies with the hopes of identifying gene signatures leading to individualized and more tailored treatment regimens for patients with colorectal cancer. The focus has been on three areas: metastatic colorectal cancer patients and response to anti-EGFR-based therapies, stage II or III colorectal cancer patients and response to adjuvant chemotherapies and finally rectal cancer patients and response to preoperative radio/chemotherapies. Some examples are found in Table 8.2.

Table 8.2 Gene signatures of colorectal cancer treatment response

First author [Reference]	Year	Platform	Samples	Aim/design	Results	Comment
Allen [103]	2008	cDNA	In vitro	Predict response to irinotecan-based therapy	16-Gene signature with 75% overall accuracy (81.8% sensitivity and 66.6% specificity)	Validation by RT-PCR
Arango [104]	2004	cDNA	In vitro	Predict response to oxaliplatin	80-Gene signature	
Baker [105]	2011	Affymetrix	FFPE ($n=144$)	Predict response to anti-EGFR therapy in *K-RAS* wild-type patients	EREG and AREG identified	–
Del Rio [106]	2007	Affymetrix	Frozen ($n=21$)	Predict response to FOLFIRI	14-Gene signature with 100% specificity and 92% sensitivity	–
Eschrich [107]	2009	Affymetrix	In vitro and frozen ($n=14$)	Predict response to chemo-radiotherapy	Ten-gene (AR, c-Jun, PKC, RelA, c-Abl, SUMO-1, HDAC1, CDK1, IRF1) radiosensitivity index predicted response	
Ghadimi [108]	2005	cDNA	Fresh ($n=30$)	Predict response to chemo-radiotherapy	54-Gene signature	
Huang [109]	2011	Oligo. array	In vitro	Predict response to radiation	15-Gene signature	Validation: RT-PCR ($n=110$)
Khambata-Ford [110]	2007	Affymetrix	($n=80$)	Predict response to anti-EGFR therapy	EREG and AREG identified	
Kim [111]	2007	Affymetrix	Frozen ($n=31$)	Predict response to chemo-radiotherapy	95-Gene signature with 84% accuracy	Validation on 15 samples
Koukourakis [112]	2009	Illumina gene array	Fresh ($n=22$)	Pre- and post-bevacizumab	Down-regulation of 30 genes in involved in DNA repair, proliferation	
Liersch [113]	2009	cDNA and Oligo. array	Fresh ($n=23$)	Test the clinical/prognostic value of a 54-gene signature previously established to discriminate responders and nonresponders	20-Gene signature predicts recurrence, 7 of these part of the 54-gene signature for response	Seven patients as validation
Mariadason [114]	2003	cDNA	In vitro	Predict response to 5-FU and camptothecin	Set of 50 different predictive genes	–
Martinez-Cardus [115]	2009	Oligo. array	In vitro	Predict response to oxaliplatin	6 candidate genes identified to predict resistance	Validation: qt-PCR
Maxwell [116]	2003	cDNA	In vitro	Prediction of resistance to 5-FU	Increased expression of SSAT, annexin II, thymosin beta-10, cheperonin-10, MAT-8	
Mencia [117]	2010	Oligo. array	In vitro	Differential expression patterns between methotrexate-resistant and -sensitive cell lines	S100A4 over-expression in 5/7 resistant cell lines. Beta-catenin may have a role in S100A4 up-regulation	Validation: real-time PCR, Western blot

Meynard [118]	2007	Atlas Plastic Human 8 K Microarray	In vitro	Cytotoxicities to oxaliplatin and cisplatin	394-Gene signature correlating with oxalilpatin; 40-gene signature correlating with cisplatin	Validation: RT-PCR
Ojima [119]	2007	Oligo. array	In vitro	Predict response to pre-operative radiotherapy	159 Differentially expressed genes	Validation: RT-PCR
Oliveras-Ferraros [120]	2011	Agilent whole human array	In vitro	Predict response to anti-EGFR therapy among *K-RAS* wild-type patients		Validation: qt real-time PCR, IHC, flow cytometry
Petty [121]	2009	Affymetrix	Fresh ($n=8$)	Predicting resistance to 5-FU (specifically marker APRIL/TNSF13)	APRIL up-regulated with 5-FU	IHC using TMA for 234 independent samples
Pitts [122]	2010	Affymetrix	In vitro	Predict response to IGF1R tyrosine kinase inhibitor (OSI-906)	Establishment of k-top scoring pair classifier with 100% accuracy	Eight human CRC explants in vivo
Rimkus [123]	2008	cDNA	Frozen ($n=43$)	Predict response to chemo-radiotherapy	42-Gene signature. Sensitivity 71%, specificity 86%	
Shimizu [124]	2005	cDNA	In vitro	Prediction to -5FU	81-gene signature	
Shin [125]	2009	miRNA microarray	In vitro	miRNAs, p53 and radiation response	12 miRNAs affected by p53, radiation or a combination of both	
Snipstad [126]	2010	cDNA	Frozen ($n=21$) matched tumor and normal	Molecular targets for chemo/radiotherapy	1,327 differentially expressed genes between normal and irradiated tumors, significant enrichment wof genes involved in adhesion an leucocyte migration	Validation: real-time PCR, IHC
Spitzner [127]	2010	cDNA	In vitro	Identification of chemo/radiosensitive genes	Insulin- and WNT-related genes including STAT3, RASSF1, DOK3 and ERBB2	Validation: real-time PCR
Tentler [128]	2010	Affymetrix	In vitro	Predictive markers of response to MEK1/2 inhibitor in *K-RAS* mutated patients	WNT-related genes over-expressed in resistant cell lines. 71% accuracy	
Wallin [129]	2010	Oligo. array	In vitro	Predict response to SN-38 (irinotecan)	3,974 differentially expressed genes involved in receptor and kinase activity, signal transduction, apoptosis	
Zhou [130]	2010	miRNA microarray	In vitro	Effect of 5-FU and oxaliplatin on miRNAs	56 up-regulated and 50 down-regulated miRNAs in treated cell lines compared to controls	Validation: real-time RT-PCR

FFPE Formalin fixed paraffin-embedded. *Oligo.* oligonucleotide

Treatment Response to Anti-EGFR Therapies

In 2004 and 2006, two monoclonal antibodies targeting EGFR received FDA approval for the treatment of patients with EGFR-expressing, metastatic colorectal cancer with disease progression on or following fluoropyrimidine-, oxaliplatin-, and irinotecan-containing chemotherapy regimens: Cetuximab and Panitumumab. Cetuximab, in addition, has also been approved for use as a single agent in patients who are intolerant to irinotecan-based chemotherapy. It is now generally accepted that patients with *K-RAS* mutated cancers respond little to anti-EGFR-based therapies [18]. As a consequence, the ASCO recommends that metastatic patients who may be candidates for Cetuximab or Panitumumab be tested for *K-RAS* codon 12 and 13 mutations, and this test is now available in many diagnostic molecular pathology laboratories [131]. Interesting is, however, the relatively low clinical response rates, which range from 13 to 61% even in the *K-RAS* wild-type setting [27]. Although some of the lack of response in these patients may be explained by additional mutations in *B-RAF* (and possibly mutations in other downstream effectors of the RAS/MAPK signaling pathway), gene signatures that predict clinical responsiveness in the *K-RAS* wild-type setting are sought after. The work by Oliveras-Ferraros and colleagues is of interest since it attempts to answer the question of response prediction among *K-RAS* wild-type patients by evaluating different anti-EGFR therapies in vitro using Agilent whole genome arrays [120]. They identify down-regulation of markers related to epithelial mesenchymal transition and up-regulation of E-cadherin as involved in Cetuximab response. Baker and colleagues and Khambata-Ford and colleagues could perform gene expression analysis using formalin-fixed paraffin-embedded material from metastatic colorectal cancer patients. They identify from an original probe set of 110 genes, a 4-gene classifier capable or predicting response and progression-free survival. Included in their classifier are epiregulin and amphiregulin, two ligands for EGFR [105, 110]. In fact, expression of these two genes has been previously found to have an effect on patients receiving either monotherapy or cetuximab in combination with chemotherapy [110, 132].

Although *K-RAS* mutation has a negative effect on clinical response, several studies have nonetheless reported up to 30% response rates. Several questions have arisen. Although high rates of concordance between the primary colorectal cancer and the distant site of metastasis have been reported by several groups, should primary tumor or metastatic site be molecularly investigated? What is the effect of possible intratumoral heterogeneity of *K-RAS* (and *B-RAF*) mutations? How many blocks or tissue samples should be evaluated? Gene expression studies and the identification of gene signatures predicting clinical response in metastatic colorectal cancer, particularly in the *K-RAS*-wild type setting could have significant implications for patient management in the future.

Treatment Response to Preoperative Radiochemotherapy

About 1/3 of colorectal cancers are indeed located in the rectum. The clinical management of rectal cancers differs considerably from those located in the colon. Patients with clinically staged cT3-4 or cN+ disease are considered for neoadjuvant radio/chemotherapy and relatively high rates of complete pathologic response are reported with various treatment protocols [133]. The ability to predict from the pretreatment rectal cancer biopsy which patients are more or less likely to respond to various regimens would have an important impact on clinical decision making. To date, dozens of potential immunohistochemical biomarkers have been investigated for their predictive value, including Ki67, EGFR, p53, thymidylate synthase, p21, bax/bcl2 with mixed results [134]. Predictive models based on immunohistochemistry would have many advantages: practicality, ease and cost-effectiveness. They are, however, not without their problems: fixation, staining protocols and evaluation of "positive" staining are recognized as sources of variability making it difficult to standardize protein markers, despite their considerable potential as prognostic and predictive biomarkers.

Molecular investigations on rectal cancer biopsy material are significantly fewer and only a handful of research groups have performed gene expression profiling to identify gene signatures predictive of tumor response. Ghadimi and colleagues using cDNA microarrays analyzed 23 pre-treatment rectal tumor biopsies and found 54 differentially expressed genes between responders and nonresponders [108]. Using their algorithm, 78% of responders and 86% of nonresponders could be correctly predicted. A validation set confirmed differential expression of 39 out of the 54 genes and response for 6 out of 7 tumors was correctly predicted. Rimkus and colleagues analyzed 43 rectal cancer patients [123]. With their 42-gene signature, 71% of responders and 86% nonresponders were correctly identified and led to a similar accuracy in a validation set of 5 patients. Interestingly, despite the nearly identical rates of discrimination in both studies, there was no overlap between differentially expressed genes.

Liersch and colleagues tested the prognostic value of a 54-gene signature previously established to discriminate between responsive and nonresponsive rectal cancer patients [113]. Using two different platforms (cDNA and oligonucleotide arrays) and biopsies from 23 rectal cancer patients, they identify a 20-gene signature predicting tumor recurrence, with seven genes previously included in the original 54-gene signature. These included: *ZFP106*, *KTN1 (kinectin 1)*, *RBM25 S164 protein*, *AP3D1*, *PAK1 p21/Cdc42/Rac1-activated kinase 1*, *MLL myeloid/lymphoid or mixed-lineage leukemia*, *PPP1R10 protein phosphatase 1, regulatory subunit 10* predict recurrence. Eschrich and coworkers used 48 different cancer cell lines to establish a predictive gene signature for rectal and esophageal cancer which could then be subsequently tested on human rectal cancer tissues. They describe a ten-gene profile including *androgen receptor*, *c-Jun*, *PKC*, *RelA*, *c-Abl*, *SUMO-1*, *HDAC1*, *CDK1*, *IRF1* [88].

These studies distinctly show the wide variability in genes differentially expressed between different study groups. This problem is compounded with the issue of defining "tumor response" since at least five different methods are frequently used to assign patients into responsive/partially responsive/nonresponsive subgroups [135]. Even in cases where gene signatures have been established, accuracy of prediction is relatively low. The feasibility and clinical value of gene expression arrays on preoperative biopsy material from patients with rectal cancer remains to be proven. Nonetheless, this technology may be useful to identify candidate genes which should then be tested and validated on larger patient cohorts.

Treatment Response to Adjuvant Chemotherapy

Gene signatures identifying potentially responsive stage II and III patients to different adjuvant treatment regimens would be a significant stride forward toward more tailored treatment of these patients. Cancer cell lines grown in culture or mouse xenografts have been the major resources used to test drug sensitivity. In vitro studies have the advantage of being able to analyze a wide range of different chemotherapeutic agents and combine therapies prior to testing gene signatures on data from real patients. The group of McLeod and colleagues have investigated the changes in normal and tumor tissue from Dukes' C patients treated with 5-FU and, in a second study, with irinotecan in order to identify molecular differences in relevant genes which may contribute to the variability in response [136, 137]. Although their data was not related to clinical outcome, the evaluation of genes specifically targeting the metabolism of chemotherapeutic agents could provide important directions in which to focus future studies.

MicroRNA and Methylation Profiling of Colorectal Cancer

MicroRNA Profiling

miRNAs are posttranscriptional regulators, only 18–25 nucleotides long, that bind to complementary sequences on target mRNA transcripts, usually resulting in translational repression and gene silencing. Several miRNAs have been linked to colorectal cancer and are found to be significantly overexpressed in tumor compared to normal tissue including miR-20a, miR-21, miR-106a, miR-181b, and miR-203 [138]. miR-200 is proposed as an integral part of a regulatory feedback loop with Zinc-finger enhancer binding transcription factors (ZEB1 and ZEB2) to regulate EM [139]. Investigations into miRNAs have provided researchers with insight into the fundamental differences between colon and rectal cancers and the possible interactions between miRNA and DNA methylation [140]. Grady and colleagues found that transfection of miR-342 into colon cancer cell lines may lead to apoptosis which can be inhibited by methylation [141, 142]. Others have found that miR-124a, let-7a-3, miR-10a and miR-342 may be silenced by DNA hypomethylation [143]. miR-155 and miR-146 have been associated with NFkB, suppressor of cytokine signaling 1 (SOCS1), interleukin 6 (IL6) and tumor necrosis factor-alpha (TNF-a). miR-21 may be important for IL6, signal transducer and activator of transcription 3 (STAT3), phosphatase and tensin homolog (PTEN), and Interleukin 12 (IL12) and others have been linked to SMAD4, COX2, PTGS2, VEGF and CCND1.

Upregulation of miR-21 has been related to shorter disease-free survival in patients with stage II colon cancers [142]. miRNA profiling may also be useful to identify new targets for therapy. Ragusa and colleagues investigated 667 miRNAs in two human colorectal cancer cell lines, one resistant and the other sensitive to anti-EGFR monoclonal antibody Cetuximab using real-time PCR [144]. They identify that upregulation of MiR-17 is a marker of Cetuximab resistance. Global network functional analysis (based on miRNA targets) showed a significant overrepresentation of cancer-related biological processes and networks centered on critical nodes involved in EGFR internalization and ubiquitin-mediated degradation.

Methylation Profiling

As mentioned previously, the current model of colorectal tumorigenesis encompasses to a significant degree epigenetic aberrations in genes such as *MLH1*, *MGMT* and high-level CIMP. CpG islands are short sequences rich in the CpG dinucleotide and can be found in the 5¢ region of about half of all human genes [54]. Methylation of cytosine within 5¢ CpG islands is linked to transcriptional gene silencing, and occurs in approximately 40% of colorectal cancers. Not only could methylation profiling help to identify novel tumor suppressor genes, but the identification of commonly and aberrantly methylated genes could have significant diagnostic and therapeutic implications. Kim and colleagues [145] analyzed the methylation profile of 27,578 CpG sites spanning more than 14,000 genes in CRC and in the adjacent normal mucosa with bead-chip array-based technology. They found that chromosomes 18 and 5 had the most frequently hypermethylated genes whereas chromosomes 22, 17 and 15 contained the most hypomethylated genes. Among the genes validated by pyrosequencing were ten hypermethylated genes *ADHFE1* (alcohol dehydrogenase, iron containing, 1), *BOLL*, *SLC6A15* (solute carrier family 6 (neutral amino acid transporter), member 15), *ADAMTS5* (ADAM metallopeptidase with thrombospondin type 1 motif, 5), *TFPI2* (tissue factor pathway inhibitor 2), *EYA4* (yes absent homolog 4), *NPY* (neuropeptide *Y*), *TWIST1* (twist homolog 1), *LAMA1* (laminin, alpha 1), and *GAS7* (growth arrest-specific 7). Interestingly, significant overlap between methylated genes has been reported between different studies [146, 147]. Ang and colleagues used the same technique to evaluated 1,505 CpG sites in 807 cancer-related genes using both normal colon and 91 colorectal cancer samples. Their methylation profile which included 202 CpG sites differentially methylated between tumor and normal tissue showed distinct clinic-pathological and molecular features [56]. Xu and coworkers performed methylation-specific PCR on 31 genes in colon cancer, normal mucosa and colorectal adenoma and identified changes in methylation in the early phases of colon cacrcinogenesis only [148]. Intriguing is the reported lack of correlation between hypermethylation and changes in protein expression.

Summary

Our understanding of colorectal cancer progression is continuously changing. Although in 2007, a new molecular classification of colorectal cancer was proposed based on the *K-RAS*, *B-RAF*, *MGMT*, CIMP and the mismatch-repair status, the growing number of genes and proteins, which seems to be intimately involved in tumor progression will likely lead to further fine-tuning of the current proposal. Recent technological advances have opened many novel avenues of investigation, notably the extraction of good quality RNA from paraffin-embedded tissues, and are another step into the cellular micro-cosmos of colorectal cancer pathogenesis. Nonetheless, the goal of personalized medicine is far from being reached since examples of molecular analysis helping to guide clinical decision-making strategies are at this time limited to the metastatic setting.

References

1. Sobin LH, Gospodarowicz MK, Wittekind C. TNM classification of malignant tumours. 7th ed. New York: Wiley; 2010.
2. O'Connell JB, Maggard MA, Ko CY. Colon cancer survival rates with the new American Joint Committee on Cancer sixth edition staging. J Natl Cancer Inst. 2004;96:1420–5.
3. AJCC. AJCC cancer staging manual. 7th ed. Chicago, IL: American Joint Committee on Cancer; 2010.
4. McShane LM, Altman DG, Sauerbrei W, et al. Reporting recommendations for tumor marker prognostic studies (REMARK). J Natl Cancer Inst. 2005;97:1180–4.
5. Compton CC. Prognostic factors in cancer. New York: Wiley; 2005.
6. Hase K, Shatney C, Johnson D, et al. Prognostic value of tumor "budding" in patients with colorectal cancer. Dis Colon Rectum. 1993;36:627–35.
7. Nakamura T, Mitomi H, Kanazawa H, et al. Tumor budding as an index to identify high-risk patients with stage II colon cancer. Dis Colon Rectum. 2008;51:568–72.
8. Park KJ, Choi HJ, Roh MS, et al. Intensity of tumor budding and its prognostic implications in invasive colon carcinoma. Dis Colon Rectum. 2005;48:1597–602.
9. Prall F, Nizze H, Barten M. Tumour budding as prognostic factor in stage I/II colorectal carcinoma. Histopathology. 2005;47:17–24.
10. Ueno H, Mochizuki H, Hashiguchi Y, et al. Predictors of extrahepatic recurrence after resection of colorectal liver metastases. Br J Surg. 2004;91:327–33.
11. Ueno H, Murphy J, Jass JR, et al. Tumour 'budding' as an index to estimate the potential of aggressiveness in rectal cancer. Histopathology. 2002;40:127–32.
12. Wang HS, Liang WY, Lin TC, et al. Curative resection of T1 colorectal carcinoma: risk of lymph node metastasis and long-term prognosis. Dis Colon Rectum. 2005;48:1182–92.
13. Compton C, Fenoglio-Preiser CM, Pettigrew N, et al. American Joint Committee on Cancer prognostic factors consensus conference: Colorectal Working Group. Cancer. 2000;88:1739–57.
14. Zlobec I, Baker K, Minoo P, et al. Tumor border configuration added to TNM staging better stratifies stage II colorectal cancer patients into prognostic subgroups. Cancer. 2009;115:4021–9.
15. Jass JR, Atkin WS, Cuzick J, et al. The grading of rectal cancer: historical perspectives and a multivariate analysis of 447 cases. Histopathology. 1986;10:437–59.

16. Jass JR. Classification of colorectal cancer based on correlation of clinical, morphological and molecular features. Histopathology. 2007;50:113–30.
17. Fang JY, Richardson BC. The MAPK signalling pathways and colorectal cancer. Lancet Oncol. 2005;6:322–7.
18. Siena S, Sartore-Bianchi A, Di Nicolantonio F, et al. Biomarkers predicting clinical outcome of epidermal growth factor receptor-targeted therapy in metastatic colorectal cancer. J Natl Cancer Inst. 2009;101:1308–24.
19. Zlobec I, Lugli A. Prognostic and predictive factors in colorectal cancer. J Clin Pathol. 2008;61:561–9.
20. Cunningham D, Humblet Y, Siena S, et al. Cetuximab monotherapy and cetuximab plus irinotecan in irinotecan-refractory metastatic colorectal cancer. N Engl J Med. 2004;351:337–45.
21. Martin V, Mazzucchelli L, Frattini M. An overview of the epidermal growth factor receptor fluorescence in situ hybridisation challenge in tumour pathology. J Clin Pathol. 2009;62:314–24.
22. Andreyev HJ, Norman AR, Cunningham D, et al. Kirsten ras mutations in patients with colorectal cancer: the 'RASCAL II' study. Br J Cancer. 2001;85:692–6.
23. Andreyev HJ, Norman AR, Cunningham D, et al. Kirsten ras mutations in patients with colorectal cancer: the multicenter "RASCAL" study. J Natl Cancer Inst. 1998;90:675–84.
24. Neumann J, Zeindl-Eberhart E, Kirchner T, et al. Frequency and type of K-RAS mutations in routine diagnostic analysis of metastatic colorectal cancer. Pathol Res Pract. 2009;205:858–62.
25. Finkelstein SD, Sayegh R, Bakker A, et al. Determination of tumor aggressiveness in colorectal cancer by K-RAS-2 analysis. Arch Surg 1993;128:526–531; discussion 531–522.
26. Finkelstein SD, Sayegh R, Christensen S, et al. Genotypic classification of colorectal adenocarcinoma. Biologic behavior correlates with K-RAS-2 mutation type. Cancer. 1993;71:3827–38.
27. Plesec TP, Hunt JL. K-RAS mutation testing in colorectal cancer. Adv Anat Pathol. 2009;16:196–203.
28. Seruca R, Velho S, Oliveira C, et al. Unmasking the role of K-RAS and B-RAF pathways in MSI colorectal tumors. Expert Rev Gastroenterol Hepatol. 2009;3:5–9.
29. Ogino S, Nosho K, Kirkner GJ, et al. CpG island methylator phenotype, microsatellite instability, B-RAF mutation and clinical outcome in colon cancer. Gut. 2009;58:90–6.
30. Baba Y, Nosho K, Shima K, et al. Prognostic significance of AMP-activated protein kinase expression and modifying effect of MAPK3/1 in colorectal cancer. Br J Cancer. 2010;103:1025–33.
31. Ogino S, Nosho K, Irahara N, et al. A cohort study of cyclin D1 expression and prognosis in 602 colon cancer cases. Clin Cancer Res. 2009;15:4431–8.
32. Ogino S, Shima K, Nosho K, et al. A cohort study of p27 localization in colon cancer, body mass index, and patient survival. Cancer Epidemiol Biomarkers Prev. 2009;18:1849–58.
33. Zlobec I, Kovac M, Erzberger P, et al. Combined analysis of specific K-RAS mutation, B-RAF and microsatellite instability identifies prognostic subgroups of sporadic and hereditary colorectal cancer. Int J Cancer. 2010;127:2569–75.
34. Ogino S, Nosho K, Baba Y, et al. A cohort study of STMN1 expression in colorectal cancer: body mass index and prognosis. Am J Gastroenterol. 2009;104:2047–56.
35. Zlobec I, Bihl MP, Schwarb H, et al. Clinicopathological and protein characterization of B-RAF- and K-RAS-mutated colorectal cancer and implications for prognosis. Int J Cancer. 2010;127:367–80.
36. French AJ, Sargent DJ, Burgart LJ, et al. Prognostic significance of defective mismatch repair and B-RAF V600E in patients with colon cancer. Clin Cancer Res. 2008;14:3408–15.
37. Kakar S, Deng G, Sahai V, et al. Clinicopathologic characteristics, CpG island methylator phenotype, and B-RAF mutations in microsatellite-stable colorectal cancers without chromosomal instability. Arch Pathol Lab Med. 2008;132:958–64.
38. Roth AD, Tejpar S, Delorenzi M, et al. Prognostic role of K-RAS and B-RAF in stage II and III resected colon cancer: results of the translational study on the PETACC-3, EORTC 40993, SAKK 60–00 trial. J Clin Oncol. 2010;28:466–74.
39. Richman SD, Seymour MT, Chambers P, et al. K-RAS and B-RAF mutations in advanced colorectal cancer are associated with poor prognosis but do not preclude benefit from oxaliplatin or irinotecan: results from the MRC FOCUS trial. J Clin Oncol. 2009;27:5931–7.
40. Jass JR, Young J, Leggett BA. Evolution of colorectal cancer: change of pace and change of direction. J Gastroenterol Hepatol. 2002;17:17–26.
41. Boland CR, Goel A. Microsatellite instability in colorectal cancer. Gastroenterology. 2010;138:2073–87.
42. Umar A, Boland CR, Terdiman JP, et al. Revised Bethesda Guidelines for hereditary nonpolyposis colorectal cancer (Lynch syndrome) and microsatellite instability. J Natl Cancer Inst. 2004;96:261–8.
43. Leggett B, Whitehall V. Role of the serrated pathway in colorectal cancer pathogenesis. Gastroenterology. 2010;138:2088–100.
44. Jass JR, Barker M, Fraser L, et al. APC mutation and tumour budding in colorectal cancer. J Clin Pathol. 2003;56:69–73.
45. Jenkins MA, Hayashi S, O'Shea AM, et al. Pathology features in Bethesda guidelines predict colorectal cancer microsatellite instability: a population-based study. Gastroenterology. 2007;133:48–56.

46. Hutchins G, Southward K, Handley K, et al. Value of mismatch repair, K-RAS, and B-RAF mutations in predicting recurrence and benefits from chemotherapy in colorectal cancer. J Clin Oncol. 2011;29:1261–70.
47. Vogelstein B, Fearon ER, Hamilton SR, et al. Genetic alterations during colorectal-tumor development. N Engl J Med. 1988;319:525–32.
48. Munro AJ, Lain S, Lane DP. P53 abnormalities and outcomes in colorectal cancer: a systematic review. Br J Cancer. 2005;92:434–44.
49. Ogino S, Nosho K, Irahara N, et al. Prognostic significance and molecular associations of 18q loss of heterozygosity: a cohort study of microsatellite stable colorectal cancers. J Clin Oncol. 2009;27:4591–8.
50. Popat S, Houlston RS. A systematic review and meta-analysis of the relationship between chromosome 18q genotype, DCC status and colorectal cancer prognosis. Eur J Cancer. 2005;41:2060–70.
51. Graham T, Halford S, Page KM, et al. Most low-level microsatellite instability in colorectal cancers can be explained without an elevated slippage rate. J Pathol. 2008;215:204–10.
52. Halford SE, Sawyer EJ, Lambros MB, et al. MSI-low, a real phenomenon which varies in frequency among cancer types. J Pathol. 2003;201:389–94.
53. Tomlinson I, Halford S, Aaltonen L, et al. Does MSI-low exist? J Pathol. 2002;197:6–13.
54. Issa JP. CpG island methylator phenotype in cancer. Nat Rev Cancer. 2004;4:988–93.
55. Weisenberger DJ, Siegmund KD, Campan M, et al. CpG island methylator phenotype underlies sporadic microsatellite instability and is tightly associated with B-RAF mutation in colorectal cancer. Nat Genet. 2006;38:787–93.
56. Ang PW, Loh M, Liem N, et al. Comprehensive profiling of DNA methylation in colorectal cancer reveals subgroups with distinct clinicopathological and molecular features. BMC Cancer. 2010;10:227.
57. Dahlin AM, Palmqvist R, Henriksson ML, et al. The role of the CpG island methylator phenotype in colorectal cancer prognosis depends on microsatellite instability screening status. Clin Cancer Res. 2010;16:1845–55.
58. Van Rijnsoever M, Elsaleh H, Joseph D, et al. CpG island methylator phenotype is an independent predictor of survival benefit from 5-fluorouracil in stage III colorectal cancer. Clin Cancer Res. 2003;9:2898–903.
59. Jover R, Nguyen T-P, Pérez-Carbonell L, et al. 5-Fluorouracil adjuvant chemotherapy does not increase survival in patients with CpG island methylator phenotype colorectal cancer. Gastroenterology. 2011;140(4):1174–81.
60. Ogino S, Kawasaki T, Kirkner GJ, et al. Molecular correlates with MGMT promoter methylation and silencing support CpG island methylator phenotype-low (CIMP-low) in colorectal cancer. Gut. 2007;56:1564–71.
61. Hegi ME, Diserens AC, Gorlia T, et al. MGMT gene silencing and benefit from temozolomide in glioblastoma. N Engl J Med. 2005;352:997–1003.
62. Svrcek M, Buhard O, Colas C, et al. Methylation tolerance due to an O6-methylguanine DNA methyltransferase (MGMT) field defect in the colonic mucosa: an initiating step in the development of mismatch repair-deficient colorectal cancers. Gut. 2010;59:1516–26.
63. Shacham-Shmueli E, Beny A, Geva R, et al. Response to temozolomide in patients with metastatic colorectal cancer with loss of MGMT expression: a new approach in the era of personalized medicine? J Clin Oncol. 2011;29:e262–5.
64. Di Nicolantonio F, Arena S, Tabernero J, et al. Deregulation of the PI3K and K-RAS signaling pathways in human cancer cells determines their response to everolimus. J Clin Invest. 2010;120:2858–66.
65. Sartore-Bianchi A, Martini M, Molinari F, et al. PIK3CA mutations in colorectal cancer are associated with clinical resistance to EGFR-targeted monoclonal antibodies. Cancer Res. 2009;69:1851–7.
66. Galon J, Costes A, Sanchez-Cabo F, et al. Type, density, and location of immune cells within human colorectal tumors predict clinical outcome. Science. 2006;313:1960–4.
67. Pages F, Berger A, Camus M, et al. Effector memory T cells, early metastasis, and survival in colorectal cancer. N Engl J Med. 2005;353:2654–66.
68. Zlobec I, Karamitopoulou E, Terracciano L, et al. TIA-1 cytotoxic granule-associated RNA binding protein improves the prognostic performance of CD8 in mismatch repair-proficient colorectal cancer. PLoS One. 2010;5:e14282.
69. Milasiene V, Stratilatovas E, Norkiene V. The importance of T-lymphocyte subsets on overall survival of colorectal and gastric cancer patients. Medicina (Kaunas). 2007;43:548–54.
70. Sconocchia G, Zlobec I, Lugli A, et al. Tumor infiltration by FcgammaRIII (CD16) + myeloid cells is associated with improved survival in patients with colorectal carcinoma. Int J Cancer. 2011;128:2663–72.
71. Salama P, Phillips M, Grieu F, et al. Tumor-infiltrating FOXP3+ T regulatory cells show strong prognostic significance in colorectal cancer. J Clin Oncol. 2009;27:186–92.
72. Tosolini M, Kirilovsky A, Mlecnik B, et al. Clinical impact of different classes of infiltrating T cytotoxic and helper cells (Th1, th2, treg, th17) in patients with colorectal cancer. Cancer Res. 2011;71:1263–71.
73. Jemal A, Siegel R, Xu J, et al. Cancer statistics, 2010. CA Cancer J Clin. 2010;60:277–300.
74. Wolpin BM, Meyerhardt JA, Mamon HJ, et al. Adjuvant treatment of colorectal cancer. CA Cancer J Clin. 2007;57:168–85.
75. Lombardi L, Morelli F, Cinieri S, et al. Adjuvant colon cancer chemotherapy: where we are and where we'll go. Cancer Treat Rev. 2010;36 Suppl 3:S34–41.

76. Compton CC. Pathologic prognostic factors in the recurrence of rectal cancer. Clin Colorectal Cancer. 2002;2:149–60.
77. Compton CC. Optimal pathologic staging: defining stage II disease. Clin Cancer Res. 2007;13:6862s–70.
78. Lugli A, Karamitopoulou E, Panayiotides I, et al. CD8+ lymphocytes/ tumour-budding index: an independent prognostic factor representing a 'pro-/anti-tumour' approach to tumour host interaction in colorectal cancer. Br J Cancer. 2009;101:1382–92.
79. Arango D, Laiho P, Kokko A, et al. Gene-expression profiling predicts recurrence in Dukes' C colorectal cancer. Gastroenterology. 2005;129:874–84.
80. Bandres E, Malumbres R, Cubedo E, et al. A gene signature of 8 genes could identify the risk of recurrence and progression in Dukes' B colon cancer patients. Oncol Rep. 2007;17:1089–94.
81. Barrier A, Roser F, Boelle PY, et al. Prognosis of stage II colon cancer by non-neoplastic mucosa gene expression profiling. Oncogene. 2007;26:2642–8.
82. Barrier A, Lemoine A, Boelle PY, et al. Colon cancer prognosis prediction by gene expression profiling. Oncogene. 2005;24:6155–64.
83. Barrier A, Boelle PY, Lemoine A, et al. Gene expression profiling of nonneoplastic mucosa may predict clinical outcome of colon cancer patients. Dis Colon Rectum. 2005;48:2238–48.
84. Bertucci F, Salas S, Eysteries S, et al. Gene expression profiling of colon cancer by DNA microarrays and correlation with histoclinical parameters. Oncogene. 2004;23:1377–91.
85. Carvalho L, Yu J, Schwartsmann G, et al. RNA expression of the molecular signature genes for metastasis in colorectal cancer. Oncol Rep. 2011;25:1321–7.
86. Cavalieri D, Dolara P, Mini E, et al. Analysis of gene expression profiles reveals novel correlations with the clinical course of colorectal cancer. Oncol Res. 2007;16:535–48.
87. D'Arrigo A, Belluco C, Ambrosi A, et al. Metastatic transcriptional pattern revealed by gene expression profiling in primary colorectal carcinoma. Int J Cancer. 2005;115:256–62.
88. Eschrich S, Yang I, Bloom G, et al. Molecular staging for survival prediction of colorectal cancer patients. J Clin Oncol. 2005;23:3526–35.
89. Hao JM, Chen JZ, Sui HM, et al. A five-gene signature as a potential predictor of metastasis and survival in colorectal cancer. J Pathol. 2010;220:475–89.
90. Jorissen RN, Gibbs P, Christie M, et al. Metastasis-associated gene expression changes predict poor outcomes in patients with Dukes Stage B and C colorectal cancer. Clin Cancer Res. 2009;15:7642–51.
91. Kalady MF, Dejulius K, Church JM, et al. Gene signature is associated with early stage rectal cancer recurrence. J Am Coll Surg. 2010;211:187–95.
92. Kwon HC, Kim SH, Roh MS, et al. Gene expression profiling in lymph node-positive and lymph node-negative colorectal cancer. Dis Colon Rectum. 2004;47:141–52.
93. Matsuyama T, Ishikawa T, Mogushi K, et al. MUC12 mRNA expression is an independent marker of prognosis in stage II and stage III colorectal cancer. Int J Cancer. 2010;127:2292–9.
94. Peng J, Wang Z, Chen W, et al. Integration of genetic signature and TNM staging system for predicting the relapse of locally advanced colorectal cancer. Int J Colorectal Dis. 2010;25:1277–85.
95. Pillaire MJ, Selves J, Gordien K, et al. A 'DNA replication' signature of progression and negative outcome in colorectal cancer. Oncogene. 2010;29:876–87.
96. Salazar R, Roepman P, Capella G, et al. Gene expression signature to improve prognosis prediction of stage II and III colorectal cancer. J Clin Oncol. 2011;29:17–24.
97. Smith JJ, Deane NG, Wu F, et al. Experimentally derived metastasis gene expression profile predicts recurrence and death in patients with colon cancer. Gastroenterology. 2010;138:958–68.
98. Wang Y, Jatkoe T, Zhang Y, et al. Gene expression profiles and molecular markers to predict recurrence of Dukes' B colon cancer. J Clin Oncol. 2004;22:1564–71.
99. Watanabe T, Kobunai T, Yamamoto Y, et al. Prediction of liver metastasis after colorectal cancer using reverse transcription-polymerase chain reaction analysis of 10 genes. Eur J Cancer. 2010;46:2119–26.
100. Webber EM, Lin JS, Evelyn PW (2010) Oncotype DX tumor gene expression profiling in stage II colon cancer. Application: prognostic, risk prediction. PLoS Curr 2010;2:RRN1177.
101. Yamasaki M, Takemasa I, Komori T, et al. The gene expression profile represents the molecular nature of liver metastasis in colorectal cancer. Int J Oncol. 2007;30:129–38.
102. Van Laar RK. An online gene expression assay for determining adjuvant therapy eligibility in patients with stage 2 or 3 colon cancer. Br J Cancer. 2010;103:1852–7.
103. Allen WL, Coyle VM, Jithesh PV, et al. Clinical determinants of response to irinotecan-based therapy derived from cell line models. Clin Cancer Res. 2008;14:6647–55.
104. Arango D, Wilson AJ, Shi Q, et al. Molecular mechanisms of action and prediction of response to oxaliplatin in colorectal cancer cells. Br J Cancer. 2004;91:1931–46.
105. Baker JB, Dutta D, Watson D, et al. Tumour gene expression predicts response to cetuximab in patients with K-RAS wild-type metastatic colorectal cancer. Br J Cancer. 2011;104:488–95.

106. Del Rio M, Molina F, Bascoul-Mollevi C, et al. Gene expression signature in advanced colorectal cancer patients select drugs and response for the use of leucovorin, fluorouracil, and irinotecan. J Clin Oncol. 2007;25:773–80.
107. Eschrich SA, Pramana J, Zhang H, et al. A gene expression model of intrinsic tumor radiosensitivity: prediction of response and prognosis after chemoradiation. Int J Radiat Oncol Biol Phys. 2009;75:489–96.
108. Ghadimi BM, Grade M, Difilippantonio MJ, et al. Effectiveness of gene expression profiling for response prediction of rectal adenocarcinomas to preoperative chemoradiotherapy. J Clin Oncol. 2005;23:1826–38.
109. Huang MY, Wang JY, Chang HJ, et al. CDC25A, VAV1, TP73, BRCA1 and ZAP70 gene overexpression correlates with radiation response in colorectal cancer. Oncol Rep. 2011;25:1297–309.
110. Khambata-Ford S, Garrett CR, Meropol NJ, et al. Expression of epiregulin and amphiregulin and K-RAS mutation status predict disease control in metastatic colorectal cancer patients treated with cetuximab. J Clin Oncol. 2007;25:3230–7.
111. Kim IJ, Lim SB, Kang HC, et al. Microarray gene expression profiling for predicting complete response to preoperative chemoradiotherapy in patients with advanced rectal cancer. Dis Colon Rectum. 2007;50:1342–53.
112. Koukourakis MI, Giatromanolaki A, Sheldon H, et al. Phase I/II trial of bevacizumab and radiotherapy for locally advanced inoperable colorectal cancer: vasculature-independent radiosensitizing effect of bevacizumab. Clin Cancer Res. 2009;15:7069–76.
113. Liersch T, Grade M, Gaedcke J, et al. Preoperative chemoradiotherapy in locally advanced rectal cancer: correlation of a gene expression-based response signature with recurrence. Cancer Genet Cytogenet. 2009;190:57–65.
114. Mariadason JM, Arango D, Shi Q, et al. Gene expression profiling-based prediction of response of colon carcinoma cells to 5-fluorouracil and camptothecin. Cancer Res. 2003;63:8791–812.
115. Martinez-Cardus A, Martinez-Balibrea E, Bandres E, et al. Pharmacogenomic approach for the identification of novel determinants of acquired resistance to oxaliplatin in colorectal cancer. Mol Cancer Ther. 2009;8:194–202.
116. Maxwell PJ, Longley DB, Latif T, et al. Identification of 5-fluorouracil-inducible target genes using cDNA microarray profiling. Cancer Res. 2003;63:4602–6.
117. Mencia N, Selga E, Rico I, et al. Overexpression of S100A4 in human cancer cell lines resistant to methotrexate. BMC Cancer. 2010;10:250.
118. Meynard D, Le Morvan V, Bonnet J, et al. Functional analysis of the gene expression profiles of colorectal cancer cell lines in relation to oxaliplatin and cisplatin cytotoxicity. Oncol Rep. 2007;17:1213–21.
119. Ojima E, Inoue Y, Miki C, et al. Effectiveness of gene expression profiling for response prediction of rectal cancer to preoperative radiotherapy. J Gastroenterol. 2007;42:730–6.
120. Oliveras-Ferraros C, Vazquez-Martin A, Cufi S, et al. Stem cell property epithelial-to-mesenchymal transition is a core transcriptional network for predicting cetuximab (Erbitux) efficacy in K-RAS wild-type tumor cells. J Cell Biochem. 2011;112:10–29.
121. Petty RD, Samuel LM, Murray GI, et al. APRIL is a novel clinical chemo-resistance biomarker in colorectal adenocarcinoma identified by gene expression profiling. BMC Cancer. 2009;9:434.
122. Pitts TM, Tan AC, Kulikowski GN, et al. Development of an integrated genomic classifier for a novel agent in colorectal cancer: approach to individualized therapy in early development. Clin Cancer Res. 2010;16:3193–204.
123. Rimkus C, Friederichs J, Boulesteix AL, et al. Microarray-based prediction of tumor response to neoadjuvant radiochemotherapy of patients with locally advanced rectal cancer. Clin Gastroenterol Hepatol. 2008;6:53–61.
124. Shimizu D, Ishikawa T, Ichikawa Y, et al. Prediction of chemosensitivity of colorectal cancer to 5-fluorouracil by gene expression profiling with cDNA microarrays. Int J Oncol. 2005;27:371–6.
125. Shin S, Cha HJ, Lee EM, et al. MicroRNAs are significantly influenced by p53 and radiation in HCT116 human colon carcinoma cells. Int J Oncol. 2009;34:1645–52.
126. Snipstad K, Fenton CG, Kjaeve J, et al. New specific molecular targets for radio-chemotherapy of rectal cancer. Mol Oncol. 2010;4:52–64.
127. Spitzner M, Emons G, Kramer F, et al. A gene expression signature for chemoradiosensitivity of colorectal cancer cells. Int J Radiat Oncol Biol Phys. 2010;78:1184–92.
128. Tentler JJ, Nallapareddy S, Tan AC, et al. Identification of predictive markers of response to the MEK1/2 inhibitor selumetinib (AZD6244) in K-RAS-mutated colorectal cancer. Mol Cancer Ther. 2010;9:3351–62.
129. Wallin A, Francis P, Nilbert M, et al. Gene expression profile of colon cancer cell lines treated with SN-38. Chemotherapy. 2010;56:17–25.
130. Zhou J, Zhou Y, Yin B, et al. 5-Fluorouracil and oxaliplatin modify the expression profiles of microRNAs in human colon cancer cells in vitro. Oncol Rep. 2010;23:121–8.
131. Allegra CJ, Jessup JM, Somerfield MR, et al. American Society of Clinical Oncology provisional clinical opinion: testing for K-RAS gene mutations in patients with metastatic colorectal carcinoma to predict response to anti-epidermal growth factor receptor monoclonal antibody therapy. J Clin Oncol. 2009;27:2091–6.

132. Jacobs B, De Roock W, Piessevaux H, et al. Amphiregulin and epiregulin mRNA expression in primary tumors predicts outcome in metastatic colorectal cancer treated with cetuximab. J Clin Oncol. 2009;27:5068–74.
133. Glimelius B, Pahlman L, Cervantes A. Rectal cancer: ESMO clinical practice guidelines for diagnosis, treatment and follow-up. Ann Oncol. 2010;21 Suppl 5:v82–6.
134. Kuremsky JG, Tepper JE, McLeod HL. Biomarkers for response to neoadjuvant chemoradiation for rectal cancer. Int J Radiat Oncol Biol Phys. 2009;74:673–88.
135. Zlobec I. Assessing downgrading of locally advanced rectal cancer after chemo-radiotherapy. Eur J Cancer. 2011;47(8):1138–45.
136. Kidd EA, Yu J, Li X, et al. Variance in the expression of 5-Fluorouracil pathway genes in colorectal cancer. Clin Cancer Res. 2005;11:2612–9.
137. Yu J, Shannon WD, Watson MA, et al. Gene expression profiling of the irinotecan pathway in colorectal cancer. Clin Cancer Res. 2005;11:2053–62.
138. Schetter AJ, Leung SY, Sohn JJ, et al. MicroRNA expression profiles associated with prognosis and therapeutic outcome in colon adenocarcinoma. JAMA. 2008;299:425–36.
139. Brabletz S, Brabletz T. The ZEB/miR-200 feedback loop–a motor of cellular plasticity in development and cancer? EMBO Rep. 2010;11:670–7.
140. Slattery ML, Wolff E, Hoffman MD, et al. MicroRNAs and colon and rectal cancer: differential expression by tumor location and subtype. Genes Chromosomes Cancer. 2011;50:196–206.
141. Grady WM, Parkin RK, Mitchell PS, et al. Epigenetic silencing of the intronic microRNA hsa-miR-342 and its host gene EVL in colorectal cancer. Oncogene. 2008;27:3880–8.
142. Nielsen BS, Jorgensen S, Fog JU, et al. High levels of microRNA-21 in the stroma of colorectal cancers predict short disease-free survival in stage II colon cancer patients. Clin Exp Metastasis. 2011;28:27–38.
143. Lujambio A, Ropero S, Ballestar E, et al. Genetic unmasking of an epigenetically silenced microRNA in human cancer cells. Cancer Res. 2007;67(4):1424–9.
144. Ragusa M, Majorana A, Statello L, et al. Specific alterations of microRNA transcriptome and global network structure in colorectal carcinoma after cetuximab treatment. Mol Cancer Ther. 2010;9:3396–409.
145. Kim YH, Lee HC, Kim SY, et al. Epigenomic Analysis of Aberrantly Methylated Genes in Colorectal Cancer Identifies Genes Commonly Affected by Epigenetic Alterations. Ann Surg Oncol. 2011;18(8):2338–47.
146. Estecio MR, Yan PS, Ibrahim AE, et al. High-throughput methylation profiling by MCA coupled to CpG island microarray. Genome Res. 2007;17:1529–36.
147. Schuebel KE, Chen W, Cope L, et al. Comparing the DNA hypermethylome with gene mutations in human colorectal cancer. PLoS Genet. 2007;3:1709–23.
148. Xu XL, Yu J, Zhang HY, et al. Methylation profile of the promoter CpG islands of 31 genes that may contribute to colorectal carcinogenesis. World J Gastroenterol. 2004;10:3441–54.

Chapter 9
The Role of Epigenetics in Cancer: From Molecular Function to High-Throughput Assays

Aleksandra Pekowska, Joaquin Zacarias-Cabeza, Jinsong Jia, Pierre Ferrier, and Salvatore Spicuglia

Why Epigenetics?

Malignant cells are characterised by global cancer-specific changes in gene expression patterns. cDNA abundance, analysed by microarray, has been successfully used to determine different types and subtypes of cancer cells [1]. However, these approaches suffer from important limitations when aimed to distinguish groups among a same tumour type as the analysed specimens may display considerable heterogeneity. Conversely, even very similar gene expression pattern of groups of samples may be the outcome of fundamentally different underlying molecular mechanisms, which can be, in turn, cancer specific and associated with distinct clinical prognosis [2]. These obstacles can, however, be overcome when combining transcriptomic approaches with epigenetic analyses of cancer specimens [3]. A comprehensive analysis framework can lead, therefore, to a more sensitive and discriminative cancer classification, and ultimately, to a better comprehension of fundamental mechanisms acting to establish malignant cell phenotype. Currently, much effort is being put in the dissection of the role of epigenetics in cancer development and maintenance [4]. Aberrant epigenetic profiles of various loci are analysed in search of markers for cancer aggressiveness. Finally, the reversal of cancer phenotypes by means of a controlled epigenome modification(s) is envisioned in practise and the first clinical trials using the so-called "epidrugs" open new perspectives in modern cancer treatment [5]. In this chapter, we introduce fundamental notions in epigenetics and its association with cancer and discuss progress in the unbiased use of epigenomics as an additional tool in cancer therapy.

Fundamental Concepts in Epigenetics

Epigenetics can be viewed as "the structural adaptation of chromosomal regions so as to register, signal or perpetuate altered activity states" [6]. At a molecular level, it is reflected by specific modifications of the chromatin, which ensure proper function and activity of the genome and may act as

---

A. Pekowska • J. Zacarias-Cabeza • P. Ferrier • S. Spicuglia (✉)
Centre d'Immunologie de Marseille-Luminy, Université Aix-Marseille, France; CNRS, UMR6102, Marseille, France; Inserm, U631, Marseille, France
e-mail: spicuglia@ciml.univ-mrs.fr

J. Jia
Centre d'Immunologie de Marseille-Luminy, Université Aix-Marseille, France; CNRS, UMR6102, Marseille, France; Inserm, U631, Marseille, France; Institute of Hematology, People's Hospital, Beijing University, Beijing, People's Republic of China

A. Russo et al. (eds.), *Diagnostic, Prognostic and Therapeutic Value of Gene Signatures*, Current Clinical Pathology, DOI 10.1007/978-1-61779-358-5_9,

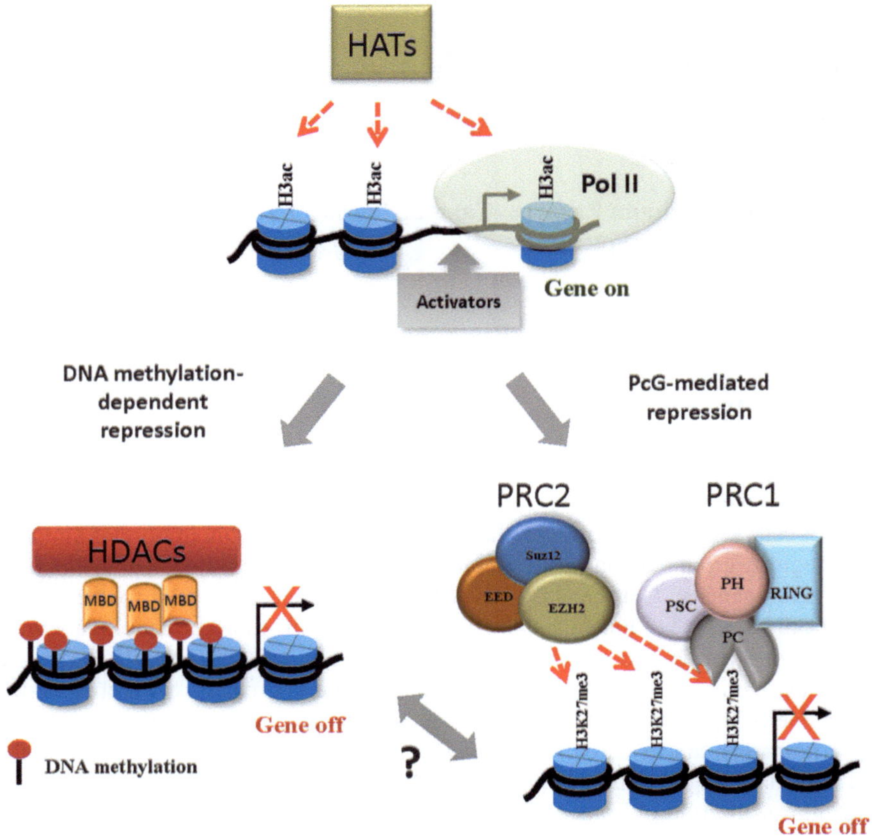

Fig. 9.1 Schematic view of the regulation of gene expression by epigenetic mechanisms. Acetylated chromatin (H3ac) at gene promoters is generally permissive to binding of transcriptional activators and recruitment of the RNA Polymerase II (Pol II). This conformation is favourable to transcription. Gene silencing can be mediated by either DNA methylation-dependent or polycomb Group (PcG)-mediated mechanisms. Hypermethylation of DNA inhibits the binding of transcriptional activators and is associated with methyl binding (MBD) proteins, which in turn recruit histone deacetylases (HDACs). Alternatively, PcG (formed by PRC1 and PRC2 complexes) methylates H3K27 via the PRC2 complex, which in turn allow the binding of PRC1 and establishment of a more compact chromatin, repressive towards transcription

heritable determinants of cell-type-specific phenotypes [7]. Various modifications of the chromatin fibre have been described, including post-translational modifications of histone proteins, DNA methylation and nucleosome displacement (Fig. 9.1) [8]. They are crucial to the regulation of nuclear processes, including gene and microRNA expression, DNA repair as well as replication, silencing of transposable elements, X-chromosome inactivation and gene imprinting [7]. The fundamental role of epigenetics in normal development and physiology of the cell is highlighted by the observation that many diseases, including cancer, develop when chromatin is altered [9].

DNA Methylation

DNA methylation is probably the most extensively characterised epigenetic modification [10, 11]. In mammals, the vast majority of DNA methylation occurs at position 5 of the cytosine pyrimidine ring (5mC) in the case of CpG dinucleotides. CpGs are underrepresented in the mammalian genome in comparison to other dinucleotide combinations [10]. However, stretches of DNA with unexpectedly high content of CpGs are also found on the human genome and form the so-called CpG islands [12, 13]. CpG islands are found in about 60% of gene promoters and are usually unmethylated in normal cells, although some CpG islands associated to tissue-specific genes might become methylated during cell differentiation [10]. The majority of methylated DNA is thus found in regions displaying low CpG content.

DNA methylation is catalysed by specialised DNA methyl transferases (DNMTs), namely DNMT1, DNMT3a and DNMT3b [10]. DNMT1, the so-called maintenance methyltransferase, is able to methylate DNA during replication based on the DNA methylation pattern of the mother strand. DNMT3a and DNMT3b insure de novo methylation of DNA. Whether a specific DNA demethylase exists in mammalian cell is still an open question; the removal of this mark is proposed to be mediated by DNA repair machineries [14]. The recent discovery that the TET family of proteins could catalyse conversion of 5mC to 5-hydroximethylcytosine (5hmC), raises the possibility that DNA demethylation might be achieved via TET-mediated hydroxylation [15, 16]. Interestingly, TET1 is a frequent fusion partner of MLL in AML patients [17].

Initial studies using 5¢ aza-deoxycitidine, a drug which inhibits the enzymatic activity of DNMTs, showed a positive effect on gene expression, thus establishing CpG methylation as an epigenetic modification involved in gene silencing [10]. Subsequently, DNA methylation has been linked to the regulation of tissue-specific gene expression throughout development [11]. However, recent pieces of evidence point to a more complex interplay between DNA methylation and transcription; while DNA methylation in promoter regions anti-correlates with transcription, its presence in gene-bodies displays a positive correlation with gene expression level [18, 19]. The repressive role of DNA methylation was furthermore associated with gene imprinting, which consists in parent-of-origin specific silencing of one of two alleles resulting in heterozygous gene expression [20].

DNA methylation at promoters can inhibit gene expression through various mechanisms [11], including (1) recruitment of methyl-CpG binding domains (MBDs) proteins, which in turn recruit histone modifying and chromatin-remodelling complexes to methylated DNA and induce a compacted chromatin structure (Fig. 9.1); or (2) the inhibition of the binding of specific transcription factors (TFs) to their target sites. On the contrary, the absence of DNA methylation at promoter-associated CpG islands generates a chromatin structure which favours gene expression [21].

Histone Post-translational Modifications

A nucleosome, the basic unit of chromatin, consists of an octameric histone core surrounded by genomic DNA [22]. Post-translational modifications of histones occur mainly on the unstructured N-terminal tail of histone proteins protruding from the nucleosome surface. These modifications include: methylation, acetylation, phosphorylation, ADP ribosylation and sumoylation [7]. The deposition of specific histone modifications are catalysed by the following dedicated enzymes: histone acetyl-transferases and deacetylases (HATs and HDACs, respectively), histone methyl transferases and demethylases (KMT and KDMs, respectively) [8]. Plethora of possible combinatorial patterns of histone modifications at a given genomic position, along with the existence of dedicated protein domains recognising distinct modifications, has led to the "histone code hypothesis" [23]. According to this concept, distinct combinations of epigenetic modifications encode specific information. Indeed, histone modifications can serve as docking platforms for various factors, including components of the general transcription machinery and DNA repair enzymes, thus modulating underlying nuclear processes, such as transcription and DNA repair. Protein domains able to recognise histone marks include, among others: bromodomains recognising acetylated histones, and PHD as well as chromodomains binding methylated histones [8].

Distinct patterns of histone modifications positively or negatively correlate with gene expression. Transcriptionally active genes are enriched for histone 3 acetylation (H3ac) as well as trimethyl form of histone 3 lysine 4 (H3K4me3) established by the MLL family of HMTs, whereas the lack of histone acetylation along with H3K9me3 and H3K27me3 deposited by Suv39h and the polycomb group (PcG) complex, respectively, is primarily associated with transcriptional repression [7].

The presence of specific chromatin modifications modulates the affinity of TFs to underlying DNA sequences, which results in a locus-specific transcriptional programme (Fig. 9.1). The marks H3K4me3 and H3ac stabilise the binding of the basic transcription machinery, thus promoting gene expression. In addition, these histone modifications favour nucleosome displacement resulting in a more "open" chromatin structure prone to TF binding, hence perpetuating active gene states. At inactive promoters, the lack of these modifications together with the deposition of H3K27me3 and/or H3K9me2 and DNA methylation [8], induce a more condensed chromatin structure, less permissive to transcription.

Role of Epigenetic Alterations in Cancer: The Epimutation Concept

As mentioned earlier, epigenetic information can be transmitted mitotically [6]. This leaves the possibility that an aberrant epigenetic pattern can, in turn, also be perpetuated through cell division. This assumption lays the basis of the epimutation concept [24], in which perpetuated alterations in the epigenetic state may cause heritable aberrant transcription pattern(s) observed in various human diseases, such as cancer (Fig. 9.2). As epimutations may occur at a much higher frequency than genetic mutations, they are likely to have a greater impact on a selection of subpopulations of cells during tumour progression or acquisition of resistance to anticancer drugs [25]. Indeed, in addition to classic genetic mutations, cancer cells display a profoundly altered epigenetic landscape, characterised by global changes in DNA methylation, histone modifications and expression of chromatin modifying enzymes [9]. What could be at the origin of epigenetic aberrations found in cancer? Environmental factors (asbestos, tobacco, etc.) in conjunction with ageing have been shown to be significantly associated with epigenomic alterations (Fig. 9.2). Their already recognised oncogenic potential further points to a fundamental role of epimutations in common carcinogen-driven oncogenesis [26, 27]. Additional mechanisms include aberrant expression and/or function of chromatin-modifying enzymes [9]. As discussed in detail below, various chromatin-modifying enzymes display altered levels of expression in cancer, and the corresponding genes are often implicated in chromosomal translocations.

Methods for Assessing the Epigenetic Modifications at the Genome-Wide Scale

During the past two decades, considerable advances have been made in mapping and analysing epigenetic modifications [28]. The completion of the human genome project, the advent and versatility of array technologies and, more recently, next generation sequencing (NGS) technologies have led to a growing number of genome-wide studies assessing DNA methylation and histone post-translational modifications in normal and cancer cells [4, 29, 30].

Various assays have been developed to analyse DNA methylation at a global scale, which mainly differ by the means used for isolation and detection of 5mC (Fig. 9.3). In restriction enzyme-based approaches, methylation sensitive and insensitive isoschizomers are used to generate restriction maps of the total genomic DNA, which directly reflects the genome-wide distribution of methylated

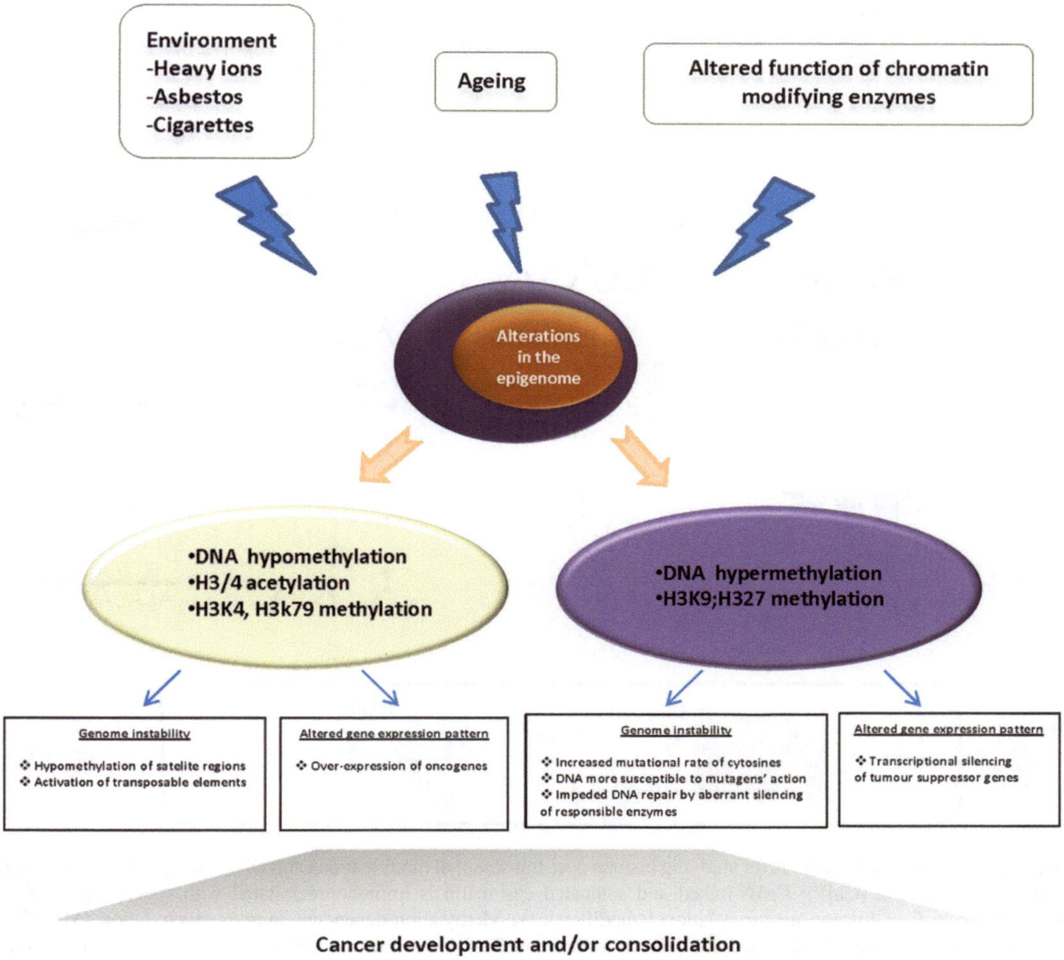

Fig. 9.2 Possible causes and consequences of epigenetic alterations in the development of cancer. Various environmental clues may influence the epigenome of the cell. Alterations in the epigenome are reflected by aberrant changes in the level of DNA methylation and key histone post-translational modifications, which in turn induce genome instability and/or altered gene expression patterns. These aberrant epigenetic landscapes, along with other genetic events, might eventually lead to cancer

cytosines. The methylation pattern is further analysed by direct NGS or hybridisation to dedicated microarrays covering either the whole genome or chosen areas, such as gene promoters or CpG islands [31, 32]. This technique enables the detection of hyper or hypomethylated regions with a moderate resolution depending on the restriction enzyme in use. Another approach consists in the affinity-based enrichment of methylated DNA by a 5mC-specific antibody (MeDIP) [33] or a recombinant MBD protein (MethylCap) [34]. MeDIP was originally combined with CpG or promoter microarrays [33, 35, 36] and subsequently adapted to NGS [37, 38]. Limitations of these techniques include decreased affinity towards CpG-poor regions and potential bias against CG-rich regions [35, 36, 39]. However, to improve the quantification efficiency, fully in vitro methylated DNA might be used as a control [2, 39, 40]. Finally, in bisulfite-based methods, DNA is first treated with sodium bisulfite. This treatment converts all unmethylated cytosines to uracyl while methylated cytosines remain unchanged. A bisulfite-based method has been adapted to microarrays in the

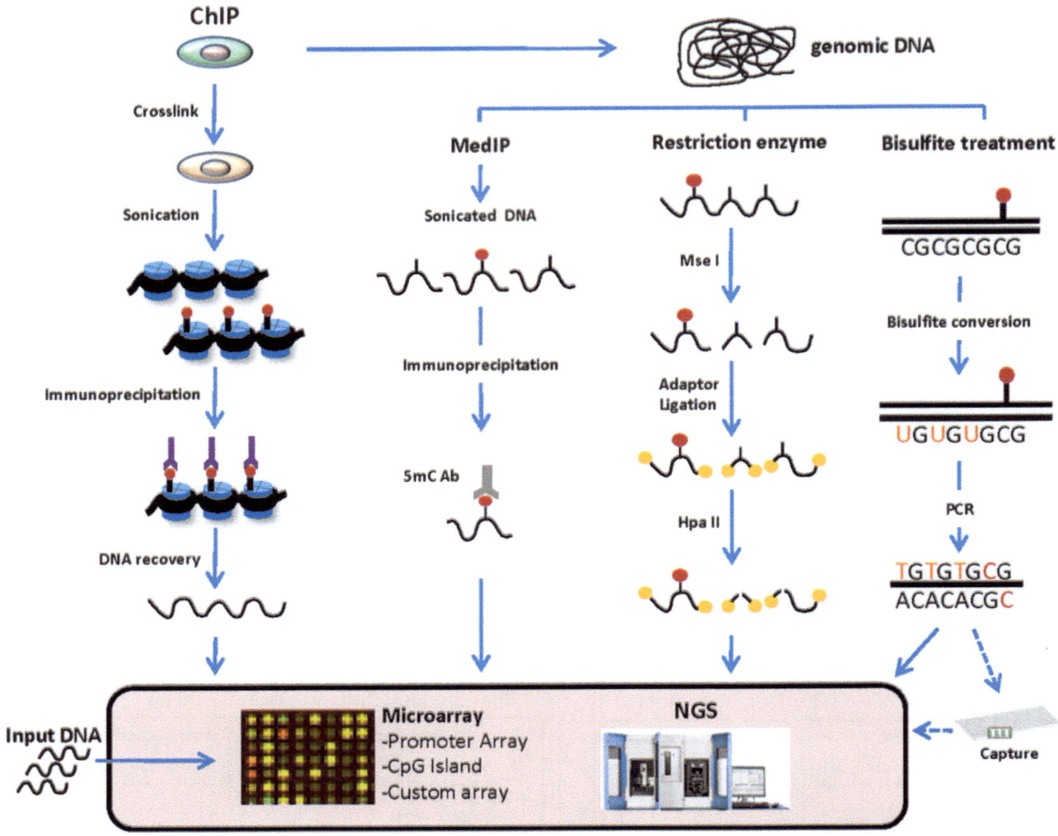

Fig. 9.3 Genome-wide methods for studying histone post-translational modifications and DNA methylation. Chromatin immunoprecipitation (ChIP): Cross-linked and sonicated chromatin is immunoprecipitated with specific antibodies recognising distinct histone post-translational modifications. Methylation-dependent immunoprecipitation (MeDIP): Genomic DNA is first sonicated and an antibody specific to 5mC is used to recover methylated regions. Restriction enzyme: DNA methylation can be identified by a combination of restriction enzymes that differentially recognise methylated and unmethylated CpGs (Hpa II is shown as an example). Bisulfite treatment: the treatment of DNA with bisulfite changes all unmethylated cytosines into uracils, leaving methylated cytosines unchanged. Any of these methods can be combined with either microarray hybridisation or next generation sequencing (NGS). Alternatively, bisulfite-treated samples can be enriched for regions of interest by using a capture approach before NGS

Infinium HumanMethylation27 assay developed by Illumina, Inc (California, USA) [34]. Bisulfite-treated samples can also be used directly for high-throughput sequencing (BS-Seq) to precisely map the positions where 5mC is present in a genome-wide manner [41]. Although to date, this technique enables the most accurate and unbiased localisation of methylated DNA, it still remains expensive, as it requires full genome sequencing [30]. More affordable bisulfite-based sequencing methods have been developed, including "reduced representation bisulfite sequencing" (RRBS) [42–44] and bisulfite capture [45]. A recent study provided a quantitative comparison of genome-wide DNA methylation mapping approaches applied to clinical samples [46]. This study concluded that all tested methods allow for an efficient detection of differentially methylated regions between samples, but differ in terms of sensitivity and genomic coverage.

Histone post-translational modifications are commonly assessed by chromatin immunoprecipitation (ChIP), which consists in immuno-enrichment of cross-linked and randomly sheared chromatin fragments using highly specific antibodies against a given chromatin modification (Fig. 9.3) [29]. To obtain a genome-wide view of histone modification patterns, ChIP samples are either hybridised onto tiling microarrays (ChIP-on-chip) or directly used for high-throughput sequencing (ChIP-Seq). This latter method provides genome-wide coverage and a less-biased analysis outcome, but generally requires a higher number of cells. Although ChIP-on-chip and ChIP-Seq procedures are clearly more technically demanding than DNA methylation approaches, it is possible to adapt it to large-scale studies on primary blasts (see below).

General Alterations of DNA Methylation in Cancer

Aberrant DNA methylation events are believed to play a fundamental role in cancer development and maintenance [47]. Global diminution of the level of DNA methylation in cancer measured by HPLC was the very first link between alteration in this epigenetic modification and cancer [48, 49]. Global diminution of methylated DNA was further found to be molecularly associated with decreased genome stability via the activation of transposable elements and repetitive sequences [50, 51]. The determining role of DNA hypomethylation in oncogenic processes was formally demonstrated later on by genetic disruption of the *DNMT1* gene in mouse, which resulted in global loss of DNA methylation and correlated with the occurrence of aggressive tumours and genomic instability [52–54].

In addition to global loss of DNA methylation, cancer-specific DNA methylomes can also be characterised by site-specific hypo- and hypermethylation of CpG residues [55]. Initial evidences of tumour type-specific epigenetic alterations have emerged from gene-by-gene analyses of promoter DNA methylation. On the one hand, hypomethylation at specific promoters can activate the aberrant expression of oncogenes, such as *RAS* [56], *SERPINB5* [57] and *S100P* [58]. *L*oss *O*f gene *I*mprinting (LOI) is a prominent example of DNA hypomethylation induced expression of growth promoting genes commonly observed in cancer [59, 60], such as *IGF2* [61]. Conversely, aberrant transcriptional down-regulation of various tumour suppressor genes, including *Rb*, *p53*, *MGMT* or *p16INK4a*, was shown to involve abnormally frequent DNA methylation of their promoter regions; and was generally associated to a less favourable clinical prognosis [55, 62–65]. DNA hypermethylation is also tightly linked to the so-called *L*oss *O*f *H*eterozygosity (LOH) phenomenon, which consists in the complete silencing of imprinted loci and is commonly observed in cancer [66, 67].

Initial studies performing genome-wide profiling of DNA methylation of various cell lines, tumour samples and normal tissues, revealed both aberrantly methylated CpGs shared by distinct tumour types and tumour-specific signatures [35, 68–70]. Furthermore, in colon cancer and T-cell acute lymphoblastic leukaemia (T-ALL), some patients were shown to be more susceptible to aberrant DNA methylation of CpG-rich promoters of cancer suppressor genes [71–74]. This led to the "CpG Island Methylator Phenotype" (CIMP) hypothesis, whereby increased DNA methylation frequency and pronounced transcriptional down-regulation of surrounding genes are linked, together with a less favourable prognosis [75]. Finally, even though most studies have focused on CpG islands located near promoters, recent findings revealed that most of the hypermethylated regions in cancer do not reside in CpG islands themselves but rather in CpG-rich flanking sequences termed "the CpG island shores" [58, 76].

DNA Methylation Signatures

The above described reports evidenced severe genome-wide alterations of DNA methylation profiles in cancer versus normal tissues. This point to the existence of a cancer-specific epigenome potentially associated with distinct clinical outcomes [77]. Several labs have performed promoter-wide screens of DNA methylation profiles in collections of leukaemias [74, 78–83] and solid tumours [69, 80, 84–87]. Collectively, these studies suggest that DNA methylation distributes into specific patterns in cancer cells, and that these methylation profiles reflect critical biological differences with practical clinical and therapeutic implications. Furthermore, the study of epigenetic profiles of a cohort of two groups of acute myeloid leukaemia (AML) patients who shared a common gene expression profile, but differed in the way the *CEBPA* locus was inactivated, revealed that these two groups could be readily segregated in an unsupervised fashion, based on their DNA methylation profiles alone [81]. This latter study highlights the interest of performing integrative analyses combining genomic, transcriptomic and epigenomic approaches.

However, in parallel with these outstanding discoveries, a substantial variability of DNA methylation profiles has been observed, pointing to an increased biological noise in epigenetic profiles, which eventually hinders the intended analysis. Likely, epigenomic heterogeneity among distinct group of cancers is the rule and reflects their various ontogenic histories. For example, clustering of follicular lymphoma (FL) samples based on DNA methylation profiles distinguished well between FL and normal lymph node samples, but showed minor differences between subtypes of FL, suggesting that the global epigenome is highly conserved among different FL subtypes [79, 80]. This obstacle can probably be circumvented by using larger collection of samples.

To date, the most comprehensive study of the interplay between DNA methylome and clinical outcome was performed by Figueroa et al. [82]. The DNA methylation signatures of a large cohort of 344 AML patients were analysed with methylation sensitive restriction enzyme recovery of methylated DNA, coupled with microarray detection [82]. This study revealed the existence of distinct DNA methylation patterns in AML and identified novel, biologically and clinically relevant defined AML subgroups. Moreover, these authors described a "15 gene DNA methylation classifier" capable of predicting overall survival in an independent cohort of patients. Thus, large-scale DNA methylation profiling demonstrates the potential benefit of using epigenetic markers, even with patients for whom clinical biomarkers are not currently available.

General Alteration of Histone Modifications in Cancer

Despite the fact that a plethora of histone-modifying enzymes has been shown to undergo mutation in cancer, the association between aberrant histone modifications and cancer is still not completely understood [9, 88]. Pioneering quantification of global levels of histone modifications revealed a general loss of H3K16 acetylation and H4K20me3 modifications in various tumours, which were associated with diminished level of these modifications in repetitive sequences of the genome [89]. In a parallel study, global levels of five histone post-translational modifications assessed by tissue microarray technology were also associated with clinical outcome of prostate cancer patients [90]. Since then, several reports pointed to important alterations in the global level of various histone modifications, such as H3K27me3, and demonstrated their association with cancer prognosis (Table 9.1) [88]. Likewise, low levels of the histone variants macroH2A1.1 and macroH2A2 are poor-prognosis factors in prostate cancer [100]. Although the genetic causes of global alteration in histone modifications are not clearly defined, they are likely due to aberrant expression of chromatin-modifying enzymes. Indeed, histone modification profiles have been extensively studied in cancer models, where chromatin-modifying enzymes are directly implicated in the oncogenic process, due to defined genetic alterations involving notably RARa, MLL and NUP98 as fusion partners (Fig. 9.4).

Expression of the PML–RARa oncofusion protein is characteristic of a subset of acute promyelocytic leukaemia (APL). PML–RARa has been shown to recruit HDACs, PcG complexes and HMTs and to aberrantly silence genes in APL [101]. Yet, recent genome-wide ChIP-Seq analyses

Table 9.1 Alterations in global histone modification levels associated with cancer

Modification	Cancer	Alteration	First author [Reference]
H3K4me1	Prostate	Decreased	Ellinger et al. [91]
H3K4me2	kidney, breast, pancreatic adenocarcinoma	Decreased	Manuyakorn et al. [92] Elsheikh et al. [93]
	Prostate cancer	Increased	Ellinger et al. [94] Seligson et al. [90]
	Renal carcinoma	Increased	Bianco-Miotto et al. [95] Ellinger et al. [91]
H3K4me3	Prostate	Increased	Ellinger et al. [94]
H3K27me3	Breast, ovarian, pancreatic	Decreased	Wei et al. [96]
H3K9me2	Pancreatic adenocarcinoma, prostate, kidney	Decreased	Manuyakorn et al. [92] Seligson et al. [90]
H3K9me3	Gastric adenocarcinomas	Increased	Ellinger et al. [94]
	Prostate	Decreased	Ellinger et al. [94]
H4K20me3	Various tumour cell lines, primary lymphomas and colorectal adenocarcinomas	Decreased	Fraga et al. [89]
K3K79me3	Leukaemia	Decreased	Lin et al. [124]
H2AK119ubq	Prostate cancer	Increased	Berezovska et al. [97]
H3ac	Prostate	Decreased	Ellinger et al. [94]
H4ac	Renal carcinoma	Decreased	Moshasvilli et al. [98]
	Prostate		Ellinger et al. [94]
H4K16ac	Various tumour cell lines, primary lymphomas and colorectal adenocarcinomas, breast cancer	Decreased	Fraga et al. [89] Elsheikh et al. [93]
H3K9ac	Renal carcinoma	Decreased	Moshasvilli et al. [98]
H3K18ac	Renal carcinoma, pancreas	Decreased	Moshasvilli et al. [98]
	Prostate	Increased	Manuyakorn et al. [92] Bianco-Miotto et al. [95]
H3K9ac H3K18ac H4K12ac H3K9me2	Esophagal squamous cell carcinoma	Distinct combinatorial pattern of modifications	I H et al. [99]

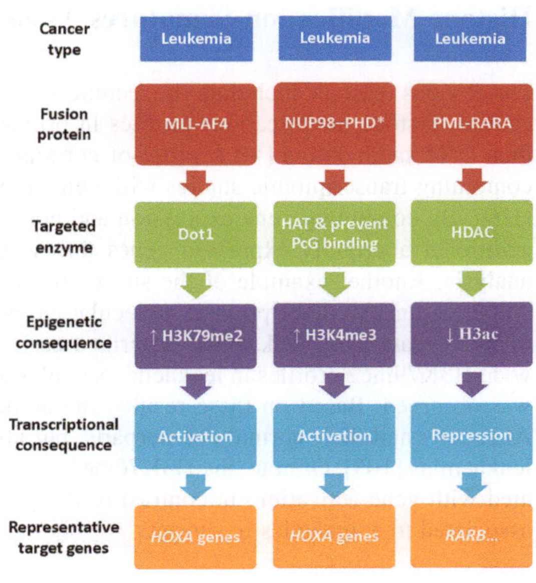

Fig. 9.4 Oncogenic processes directly involving chromatin-modifying enzymes and altered epigenetic profiles

revealed that alteration in histone acetylation is the essential epigenetic outcome associated with PML–RARa oncogenic function(s) (Fig. 9.4) [38]. Interestingly, PLZF–RARa, another oncofusion protein associated with APL, was shown to recruit the PcG complex and to induce H3K27 methylation of the silenced RAR target *RARb* [102]. It is therefore possible that depending on the RARa fusion partner, distinct epigenetic mechanisms may be involved in the leukaemogenic process.

Translocations of MLL histone methyltransferase are found within 10% of human leukaemias and involve some 50 different fusion partners [103]. The resulting product is often devoid of the SET domain of MLL, and therefore loses its H3K4 methylation function. Interestingly, MLL often fuses to other chromatin-modifying enzymes, such as the HATs CBP and p300. Several MLL oncofusion partners, including AF4, AF9, AF10 and ENL TFs, have been reported to associates with the HMT DOT1L [3]. DOT1L mediates H3K79 methylation, a histone mark associated with transcriptional elongation. Ectopic recruitment of DOT1L to natural MLL targets, such as the *HOXA9* gene, resulted in increased level of H3K79me2 and gene activation. The best-characterised example of the interplay between MLL and DOT1L chromatin modifiers in leukaemia is provided by the oncofusion protein MLL–AF4 (Fig. 9.4) [3]. Genome-wide mapping of MLL–AF4 targets revealed that regions bound by the fusion protein undergo widespread chromatin remodelling associated with enrichment in the H3K79me2 histone mark [104]. In a parallel study, Krivtsov et al. also demonstrated massive increases of H3K79me2 across MLL–AF4 targets in both murine and human cells [105]. Whether other MLL oncofusion proteins that associate with DOT1L also induce a similar epigenotype still needs to be investigated.

Finally, translocation of the nucleoporin-98 (*NUP98*) gene in AML often fuses NUP98 to either NSD1, which is an H3K36 histone methyltransferase, or to the PHD domain-containing proteins PHF23 and JARID1A [106]. Conversely, NUP98–NSD1 binds genomic elements adjacent to *HoxA7* and *HoxA9*, maintains H3K36 methylation and histone acetylation, and prevents EZH2-mediated transcriptional repression of the *Hox-A* locus [107]. NUP98–PHD fusions protein, however, binds to H3K4me2/3 via the PHD domain, thus preventing the repression of loci encoding lineage-specific TFs, such as the *HOX* genes (Fig. 9.4) [108]. Hence, NUP98 fusions result in similar behaviour regarding gene expression but, depending on the fusion partner, differ in the target histone modification and hence in the intrinsic mechanism that may be in cause.

Histone Modification Signatures Associated with Cancer

For obvious reasons, including the technical complexity and variety of post-translational modifications, histone modifications profiles in primary cancer cells have thus far been less explored than DNA methylation [4]. A proof-of-concept study investigated several cases of leukaemia by combining transcriptomic studies with genome-wide analysis of H3K9ac and DNA methylation [109]. By combining gene expression and epigenetic analyses, the authors were able to identify additional aberrantly expressed genes that would have been missed in a less comprehensive analysis. Another example of the successful use of ChIP-on-chip technology in cancer patient classification and description of molecular bases of oncogene function was provided by the study of MLL-rearranged leukaemia, described earlier in this chapter [105]. By comparing genome-wide H3K79me2 profiles in leukaemia samples and normal B-cell precursors, a striking signature was observed. Based on these results, the authors were not only able to discriminate between ALL cells and their normal counterparts, but also between MLL-rearranged and not rearranged leukaemias [105]. Of note, the H3K76me2 signature in MLL-rearranged ALL was mostly associated with gene activation, in contrast to the gene repression behaviour generally believed to be associated to epigenetics in cancer.

Much attention has been put into the description of the role of the mark H3K27me3 in cancer cells as the overexpression of Ezh2 is associated with poor clinical outcome [110]. A large-scale study addressing the question of H3K27me3 distribution in primary cancer cells has been conducted on gastric cancer specimens [111]. The authors observed substantial differences in H3K27me3 signal. Further insights into the relationship between distribution of this modification and oncogenesis have been gained by comparison of H3K27me3 ChIP-on-chip profiling between prostate cancer cell lines and normal prostate epithelial cells [112–115]. These studies demonstrated a strong fingerprint of PcG-mediated transcriptional repression in metastatic prostate cancer. Interestingly, significant subsets of these genes are also targets of PcG in embryonic stem cells, highlighting the link between stem cell self-renewal and cancer [110]. In addition, in prostate cancer, the mark H3K27me3 was associated with silencing of genes harbouring CpG-rich promoters, independently of DNA methylation, thus pointing to an alternative (DNA methylation independent) epigenetic mechanism of transcriptional down-regulation [114]. However, other labs have reported that H3K27me3 targets loci for the novo DNA methylation in cancer cells [116–118]. Large-scale promoter profiling of H3K27me3 and methylated DNA marks in normal versus prostate cancer cell lines revealed frequent replacement of one mark by the other, without change in the expression level of the corresponding genes [2]. This switch mechanism might serve as a modulator of epigenetic plasticity potentially associated with the growth advantage of cancer cells. Thus, even though the link between aberrant DNA methylation and PcG/H3K27me3 chromatin signature in cancer is not clear yet (Fig. 9.1), it may provide important clues for the design of new therapeutic strategies targeting both DNA methylases and PcG components.

Summary and Future Directions

As discussed in this chapter, cancer cells can be described by altered profiles of epigenetic marks, including both DNA methylation and histone post-translational modifications [9]. The alterations in these epigenetic modifications can thus be descriptive of a cancer-cell stage, as they occur gradually starting from the onset of the earliest malignant cells. In addition, they can constitute an additional source of information on the molecular determinants of the observed phenotype. Extensive mapping of epigenetic alterations in normal versus malignant cells can highlight molecular mechanisms guiding cancer formation and maintenance and, ultimately, can lead to the isolation of potential markers of cancer aggressiveness. In particular, the elevated stability of DNA molecules, in comparison to RNA molecules, and the development of highly sensitive DNA methylation assays render possible the epigenotyping of various tumour samples and early detection of cancer [119]. The continuous decrease in the cost of genome-wide approaches will allow large-scale mapping of epigenetic profiles with growing number of cancer samples, including paraffin-embedded patient samples [120]. Also the development of new sequencing technologies, such as single DNA molecule sequencing, will allow, for instance, performing unbiased BS-Seq approaches in a cost-effective way [30, 121]. Moreover, the reactivation of expression of tumour suppressor genes can be obtained by the use of drugs modulating the epigenetic profile of the cell. Such epidrugs are currently either already employed in cancer therapy or being tested in clinical trials [5]. The successful development of epigenetic inhibitors for use in cancer therapy depends, however, on the precise understanding of epigenetic mechanisms involved. Examples of such knowledge-based approaches have recently been published [122, 123]. In conclusion, it is the combination of comprehensive genotyping, gene expression pattern analyses, along with detailed epigenomic profiling of cancer cells that will eventually lead to a more profound understanding of cancer as a disease and ultimately arm clinicians with more reliable and robust means of cancer classification and treatment.

Acknowledgements Work in the PF laboratory is supported by institutional grants from Inserm and the CNRS, and by specific grants from the "Fondation Princesse Grace de Monaco", the "Association pour la Recherche sur le Cancer" (ARC), the "Agence Nationale de la Recherche" (ANR), the "Institut National du Cancer" (INCa) and the Commission of the European Communities. We extend our acknowledgements to Pierre Cauchy for critical proofreading of the manuscript. AP was supported by a Marie Curie research training fellowship (MRTN-CT-2006-035733) and is now supported by the "Fondation pour la Recherche Medicale" (FRM). JJ was supported by the "Fondation Franco-Chinoise pour la Science et ses Applications", the China Scholarship Council and Marseille-Nice Genopole.

References

1. Golub TR. Molecular classification of cancer: class discovery and class prediction by gene expression monitoring. Science. 1999;286(5439):531–7.
2. Gal-Yam EN, Egger G, Iniguez L, et al. Frequent switching of Polycomb repressive marks and DNA hypermethylation in the PC3 prostate cancer cell line. Proc Natl Acad Sci USA. 2008;105(35):12979–84.
3. Neff T, Armstrong SA. Chromatin maps, histone modifications and leukemia. Leukemia. 2009;23(7):1243–51.
4. Esteller M. Cancer epigenomics: DNA methylomes and histone-modification maps. Nat Genet. 2007;8(4): 286–98.
5. Kaiser J. Epigenetic drugs take on cancer. Science. 2010;330(6004):576–8.
6. Bird A. Perceptions of epigenetics. Nature. 2007;447(7143):396–8.
7. Meissner A. Epigenetic modifications in pluripotent and differentiated cells. Nat Biotechnol. 2010;28(10):1079–88.
8. Kouzarides T. Chromatin modifications and their function. Cell. 2007;128(4):693–705.
9. Portela A, Esteller M. Epigenetic modifications and human disease. Nat Biotechnol. 2010;28(10):1057–68.
10. Klose RJ, Bird AP. Genomic DNA methylation: the mark and its mediators. Trends Biochem Sci. 2006;31(2): 89–97.
11. Cedar H, Bergman Y. Linking DNA methylation and histone modification: patterns and paradigms. Nat Rev Genet. 2009;10(5):295–304.
12. Bock C, Walter J, Paulsen M, Lengauer T. CpG island mapping by epigenome prediction. PLoS Comput Biol. 2007;3(6):e110.
13. Takai D, Jones P. Comprehensive analysis of CpG islands in human chromosomes 21 and 22. Proc Natl Acad Sci USA. 2002;99(6):3740–5.
14. Ooi SKT, Bestor TH. The colorful history of active DNA demethylation. Cell. 2008;133(7):1145–8.
15. Tahiliani M, Koh KP, Shen Y, et al. Conversion of 5-methylcytosine to 5-hydroxymethylcytosine in mammalian DNA by MLL partner TET1. Science. 2009;324(5929):930–5.
16. Ito T, Ando H, Suzuki T, et al. Identification of a primary target of thalidomide teratogenicity. Science. 2010;327(5971):1345–50.
17. Langemeijer SMC, Aslanyan MG, Jansen JH. TET proteins in malignant hematopoiesis. Cell Cycle. 2009;8(24): 4044–8.
18. Rakyan VK, Down T, Thorne NP, et al. An integrated resource for genome-wide identification and analysis of human tissue-specific differentially methylated regions (tDMRs). Genome Res. 2008;18(9):1518–29.
19. Ball MP, Li JB, Gao Y, et al. Targeted and genome-scale strategies reveal gene-body methylation signatures in human cells. Nat Biotechnol. 2009;27(4):361–8.
20. Kacem S, Feil R. Chromatin mechanisms in genomic imprinting. Mamm Genome. 2009;20(9–10):544–56.
21. Thomson JP, Skene PJ, Selfridge J, et al. CpG islands influence chromatin structure via the CpG-binding protein Cfp1. Nature. 2010;464(7291):1082–6.
22. Kornberg RD, Lorch Y. Twenty-five years of the nucleosome, fundamental particle of the eukaryote chromosome. Cell. 1999;98(3):285–94.
23. Jenuwein T, Allis CD. Translating the histone code. Science. 2001;293(5532):1074–80.
24. Holliday. Epigenetic defects, assisted reproductive technology, and clinical practice: a call for clinicians and genetic counselors. Clin Genet. 1987;22(5):133–482.
25. Glasspool RM, Teodoridis JM, Brown R. Epigenetics as a mechanism driving polygenic clinical drug resistance. Br J Cancer. 2006;94(8):1087–92.
26. Christensen BC, Houseman EA, Marsit CJ, et al. Aging and environmental exposures alter tissue-specific DNA methylation dependent upon CpG island context. PLoS Genet. 2009;5(8):e1000602.
27. Marsit CJ. Promoter hypermethylation is associated with current smoking, age, gender and survival in bladder cancer. Carcinogen. 2007;28(8):1745–51.

28. Bernstein BE, Meissner A, Lander ES. The mammalian epigenome. Cell. 2007;128(4):669–81.
29. Schones DE, Zhao K. Genome-wide approaches to studying chromatin modifications. Nat Genet. 2008;9(3):179–91.
30. Lister R, Ecker JR. Finding the fifth base: Genome-wide sequencing of cytosine methylation. Genome Res. 2009;19(6):959–66.
31. Brunner AL, Johnson DS, Kim SW, et al. Distinct DNA methylation patterns characterize differentiated human embryonic stem cells and developing human fetal liver. Genome Res. 2009;19(6):1044–56.
32. Irizarry R, Ladd-Acosta C, Carvalho B, et al. Comprehensive high-throughput arrays for relative methylation (CHARM). Genome Res. 2008;18(5):780–90.
33. Weber M, Davies JJ, Wittig D, et al. Chromosome-wide and promoter-specific analyses identify sites of differential DNA methylation in normal and transformed human cells. Nat Genet. 2005;37(8):853–62.
34. Brinkman AB, Simmer F, Ma K, et al. Whole-genome DNA methylation profiling using MethylCap-seq. Methods. 2010;52(3):232–6.
35. Keshet I, Schlesinger Y, Farkash S, et al. Evidence for an instructive mechanism of de novo methylation in cancer cells. Nat Genet. 2006;38(2):149–53.
36. Weber M, Hellmann I, Stadler MB, et al. Distribution, silencing potential and evolutionary impact of promoter DNA methylation in the human genome. Nat Genet. 2007;39(4):457–66.
37. Ruike Y, Imanaka Y, Sato F, Shimizu K, Tsujimoto G. Genome-wide analysis of aberrant methylation in human breast cancer cells using methyl-DNA immunoprecipitation combined with high-throughput sequencing. BMC Genomics. 2010;11:137.
38. Martens JH et al. PML-RARalpha/RXR alters the epigenetic landscape in acute promyelocytic lekumia. Cancer Cell. 2010;17(2):173–85.
39. Jia J, Pekowska A, Jaeger S, et al. Assessing the efficiency and significance of methylated DNA immunoprecipitation (MeDIP) assays in using in vitro methylated genomic DNA. BMC Res Notes. 2010;3:240.
40. Pelizzola M, Koga Y, Urban AE, et al. MEDME: an experimental and analytical methodology for the estimation of DNA methylation levels based on microarray derived MeDIP-enrichment. Genome Res. 2008;18(10):1652–9.
41. Laurent L, Wong E, Li G, et al. Dynamic changes in the human methylome during differentiation. Genome Res. 2010;20(3):320–31.
42. Meissner A, Mikkelsen TS, Gu H, et al. Genome-scale DNA methylation maps of pluripotent and differentiated cells. Nature. 2008;454(7205):766–70.
43. Lister R, Pelizzola M, Dowen RH, et al. Human DNA methylomes at base resolution show widespread epigenomic differences. Nature. 2009;462(7271):315–22.
44. Gu H, Bock C, Mikkelsen TS, et al. Genome-scale DNA methylation mapping of clinical samples at single-nucleotide resolution. Nat Genet. 2010;7(2):133–6.
45. Hodges E, Smith AD, Kendall J, et al. High definition profiling of mammalian DNA methylation by array capture and single molecule bisulfite sequencing. Genome Res. 2009;19(9):1593–605.
46. Bock C, Tomazou EM, Brinkman AB, et al. Quantitative comparison of genome-wide DNA methylation mapping technologies. Nat Biotechnol. 2010;28:1106–14.
47. Jones P, Baylin SB. The fundamental role of epigenetic events in cancer. Nat Genet. 2002;3(6):415–28.
48. Goelz SE, Vogelstein B, Hamilton SR, Feinberg AP. Hypomethylation of DNA from benign and malignant human colon neoplasms. Science. 1985;228(4696):187–90.
49. Feinberg AP, Kuo KC. Reduced genomic 5-Methylcytosine content in human colonic neoplasi and content as afraction of total. Cancer Res. 1988;48(5):1159–61.
50. Yoder J, Walsh CP, Bestor TH. Cytosine methylation and the ecology of intragenomic parasites. Trends Genet. 1997;13(8):335–40.
51. Xu GL, Bestor TH, Bourc'his D, et al. Chromosome instability and immunodeficiency syndrome caused by mutations in a DNA methyltransferase gene. Nature. 1999;402(6758):187–91.
52. Gaudet F, Hodgson JG, Eden A, et al. Induction of tumors in mice by genomic hypomethylation. Science. 2003;300(5618):489–92.
53. Eden A, Gaudet F, Waghmare A, Jaenisch R. Chromosomal instability and tumors promoted by DNA hypomethylation. Science. 2003;300(5618):455.
54. Dodge JE, Okano M, Dick F, et al. Inactivation of Dnmt3b in mouse embryonic fibroblasts results in DNA hypomethylation, chromosomal instability, and spontaneous immortalization. J Biol Chem. 2005;280(18):17986–91.
55. Esteller M, Corn PG, Baylin SB, Herman JG. Perspectives in Cancer Res a gene hypermethylation profile of human cancer 1. Cancer Res. 2001;61:3225–9.
56. Wilson AS, Power BE, Molloy PL. DNA hypomethylation and human diseases. Biochem Biophys Acta. 2007;1775(1):138–62.
57. Futscher BW, Oshiro MM, Wozniak RJ, et al. Role for DNA methylation in the control of cell type–specific maspin expression. Nat Genet. 2002;31(2):175–9.

58. Irizarry R, Ladd-Acosta C, Wen B, et al. The human colon cancer methylome shows similar hypo- and hypermethylation at conserved tissue-specific CpG island shores. Nat Genet. 2009;41(2):178–86.
59. Feinberg P. Imprinting of a genomic domain of 11p15 and loss of imprinting in cancer: an introduction. Cancer Res. 1999;59(7 Suppl):1743s–6s.
60. Holm TM, Jackson-Grusby L, Brambrink T, et al. Global loss of imprinting leads to widespread tumorigenesis in adult mice. Cancer cell. 2005;8(4):275–85.
61. Ito Y, Koessler T, Ibrahim AEK, et al. Somatically acquired hypomethylation of IGF2 in breast and colorectal cancer. Hum Mol Genet. 2008;17(17):2633–43.
62. Greger V, Passarge E, Höpping W, Messmer E, Horsthemke B. Epigenetic changes may contribute to the formation and spontaneous regression of retinoblastoma. Hum Genet. 1989;83(2):155–8.
63. Wong J, Barrett T, Emond J, Reid J. Promoter is hypermethylated adenocarcinomas at a high frequency in esophageal. Cancer Res. 1997;57:2619–22.
64. Belinsky S, Nikula KJ, Palmisano W, et al. Aberrant methylation of p16(INK4a) is an early event in lung cancer and a potential biomarker for early diagnosis. Proc Natl Acad Sci USA. 1998;95(20):11891–6.
65. Belinsky S. Gene-promoter hypermethylation as a biomarker in lung cancer. Nat Rev Cancer. 2004;4(9):707–17.
66. Barletta JM, Rainier S, Feinberg P. Reversal of loss of imprinting in tumor cells by 5-aza-2¢-deoxycytidine. Cancer Res. 1997;57(1):48–50.
67. Murrell A. Genomic imprinting and cancer: from primordial germ cells to somatic cells. Sci World J. 2006;6:1888–910.
68. Costello JF, Frühwald MC, Smiraglia DJ, et al. Aberrant CpG-island methylation has non-random and tumour-type-specific patterns. Nat Genet. 2000;24(2):132–8.
69. Adorján P. Tumour class prediction and discovery by microarray-based DNA methylation analysis. Nucleic Acids Res. 2002;30(5):e21.
70. Ehrich M, Turner J, Gibbs P, et al. Cytosine methylation profiling of cancer cell lines. Proc Natl Acad Sci USA. 2008;105(12):4844–9.
71. Toyota M. CpG island methylator phenotype in colorectal cancer. Proc Natl Acad Sci USA. 1999;96(15):8681–6.
72. Shen L, Toyota M, Kondo Y, et al. Integrated genetic and epigenetic analysis identifies three different subclasses of colon cancer. Proc Natl Acad Sci USA. 2007;104(47):18654–9.
73. Roman-Gomez J, Jimenez-Velasco A, Agirre X, et al. Lack of CpG island methylator phenotype defines a clinical subtype of T-cell acute lymphoblastic leukemia associated with good prognosis. J Clin Oncol. 2005;23(28):7043–9.
74. Kuang S-Q, Tong W-G, Yang H, et al. Genome-wide identification of aberrantly methylated promoter associated CpG islands in acute lymphocytic leukemia. Leukemia. 2008;22(8):1529–38.
75. J-pierre I. CpG island methylator phenotype in cancer. Nat Rev Cancer. 2004;4(December):988–93.
76. Doi A, Park I-H, Wen B, et al. Differential methylation of tissue- and cancer-specific CpG island shores distinguishes human induced pluripotent stem cells, embryonic stem cells and fibroblasts. Nat Genet. 2009;41(12): 1350–3.
77. Sawan C, Vaissière T, Murr R, Herceg Z. Epigenetic drivers and genetic passengers on the road to cancer. Mutat Res. 2008;642(1–2):1–13.
78. Milani L, Lundmark A, Kiialainen A, et al. DNA methylation for subtype classification and prediction of treatment outcome in patients with childhood acute lymphoblastic leukemia. Blood. 2010;115(6):1214–25.
79. Bennett LB, Schnabel JL, Kelchen JM, et al. DNA hypermethylation accompanied by transcriptional repression in follicular lymphoma. Genes Chromosom Cancer. 2010;48(9):828–41.
80. O'Riain C. UKPMC Funders Group Array-based DNA methylation profiling in follicular lymphoma. Leukemia. 2010;23(10):1858–66.
81. Figueroa ME, Wouters BJ, Skrabanek L, et al. Genome-wide epigenetic analysis delineates a biologically distinct immature acute leukemia with myeloid/T-lymphoid features. Blood. 2009;113(12):2795–804.
82. Figueroa ME, Lugthart S, Li Y, et al. DNA methylation signatures identify biologically distinct subtypes in acute myeloid leukemia. Cancer Cell. 2010;17(1):13–27.
83. Cheung HH, Lee TL, Davis AJ, et al. Genome-wide DNA methylation profiling reveals novel epigenetically regulated genes and non-coding RNAs in human testicular cancer. Br J Cancer. 2010;102(2):419–27.
84. Alaminos M, Davalos V, Cheung N-KV, Gerald WL, Esteller M. Clustering of gene hypermethylation associated with clinical risk groups in neuroblastoma. J Nat Cancer Inst. 2004;96(16):1208–19.
85. Marsit CJ, Christensen BC, Houseman EA, et al. Epigenetic profiling reveals etiologically distinct patterns of DNA methylation in head and neck squamous cell carcinoma. Carcinogen. 2009;30(3):416–22.
86. Martinez R, Martin-Subero JI, Rohde V, et al. A microarray-based DNA methylation study of glioblastoma multiforme. Epigenetics. 2009;4(4):255–64.

87. Wu X, Rauch T, Zhong X, et al. CpG island hypermethylation in human astrocytomas. Cancer Res. 2010;70(7):2718–27.
88. Varier R, Timmers HTM. Histone lysine methylation and demethylation pathways in cancer. Biochem Biophys Acta. 2011;1815:75–89.
89. Fraga MF, Ballestar E, Villar-Garea A, et al. Loss of acetylation at Lys16 and trimethylation at Lys20 of histone H4 is a common hallmark of human cancer. Nat Genet. 2005;37(4):391–400.
90. Seligson DB, Horvath S, Shi T, et al. Global histone modification patterns predict risk of prostate cancer recurrence. Nature. 2005;435(7046):1262–6.
91. Ellinger J, Kahl P, Mertens C, et al. Prognostic relevance of global histone H3 lysine 4 (H3K4) methylation in renal cell carcinoma. Int J Cancer. 2010;127(10):2360–6.
92. Manuyakorn A, Paulus R, Farrell J, Dawson NA, Tze S, Cheung-Lau G, et al. Cellular histone modification patterns predict prognosis and treatment response in resectable pancreatic adenocarcinoma: results from RTOG 9704. J Clin Oncol. 2010;28(8):1358–65.
93. Elsheikh SE, Green AR, Rakha EA, et al. Global histone modifications in breast cancer correlate with tumor phenotypes, prognostic factors, and patient outcome. Cancer Res. 2009;69(9):3802–9.
94. Ellinger J, Kahl P, von der Gathen J, et al. Global levels of histone modifications predict prostate cancer recurrence. Prostate. 2010;70(1):61–9.
95. Bianco-Miotto T, Chiam K, Buchanan G, et al. Global levels of specific histone modifications and an epigenetic gene signature predict prostate cancer progression and development. Cancer Epidemiol Biomarkers Prev. 2010;19(10):2611–22.
96. Wei Y, Xia W, Zhang Z, et al. Loss of trimethylation at lysine 27 of histone H3 is a predictor of poor outcome in breast, ovarian, and pancreatic cancers. Mol Carcinog. 2008;47(9):701–6.
97. Berezovska OP, Glinskii AB, Yang Z, et al. Essential role for activation of the Polycomb group (PcG) protein chromatin silencing pathway in metastatic prostate cancer. Cell Cycle. 2006;5(16):1886–901.
98. Mosashvilli D, Kahl P, Mertens C, et al. Global histone acetylation levels: prognostic relevance in patients with renal cell carcinoma. Cancer Sci. 2010;101(12):2664–9. doi:10.1111/j.1349-7006.2010.01717.x. Epub 2010 Sep 1.
99. I H, Ko E, Kim Y, et al. Association of global levels of histone modifications with recurrence-free survival in stage IIB and III esophageal squamous cell carcinomas. Cancer Epidemiol Biomarkers Prev. 2010;19(2):566–73.
100. Sporn JC, Kustatscher G, Hothorn T, et al. Histone macroH2A isoforms predict the risk of lung cancer recurrence. Oncogene. 2009;28(38):3423–8.
101. Mikesch J-H, Gronemeyer H, So CWE. Discovery of novel transcriptional and epigenetic targets in APL by global ChIP analyses: emerging opportunity and challenge. Cancer Cell. 2010;17(2):112–4.
102. Boukarabila H, Saurin AJ, Batsché E, et al. The PRC1 Polycomb group complex interacts with PLZF/RARA to mediate leukemic transformation. Genes Dev. 2009;23(10):1195–206.
103. Krivtsov AV, Armstrong SA. MLL translocations, histone modifications and leukaemia stem-cell development. Nat Rev Cancer. 2007;7(11):823–33.
104. Guenther MG, Lawton LN, Rozovskaia T, et al. Aberrant chromatin at genes encoding stem cell regulators in human mixed-lineage leukemia. Genes Dev. 2008;22(24):3403–8.
105. Krivtsov AV, Feng Z, Lemieux ME, et al. H3K79 methylation profiles define murine and human MLL-AF4 leukemias. Cancer Cell. 2008;14(5):355–68.
106. Xu S, Powers M. Nuclear pore proteins and cancer. Semin Cell Dev Biol. 2009;20(5):620–30.
107. Wang GG, Cai L, Pasillas MP, Kamps MP. NUP98-NSD1 links H3K36 methylation to Hox-A gene activation and leukaemogenesis. Nat Cell Biol. 2007;9(7):804–12.
108. Wang GG, Song J, Wang Z, et al. Haematopoietic malignancies caused by dysregulation of a chromatin-binding PHD finger. Nature. 2009;459(7248):847–51.
109. Figueroa ME, Reimers M, Thompson RF, et al. An integrative genomic and epigenomic approach for the study of transcriptional regulation. PloS One. 2008;3(3):e1882.
110. Sauvageau M, Sauvageau G. Polycomb group genes: keeping stem cell activity in balance. PLoS Biol. 2008;6(4):e113.
111. Zhang L, Zhong K, Dai Y, Zhou H. Genome-wide analysis of histone H3 lysine 27 trimethylation by ChIP-chip in gastric cancer patients. J Gastroenterol. 2009;44(4):305–12.
112. Yu J, Yu J, Rhodes DR, et al. A polycomb repression signature in metastatic prostate cancer predicts cancer outcome. Cancer Res. 2007;67(22):10657–63.
113. Yu J, Cao Q, Mehra R, et al. Integrative genomics analysis reveals silencing of beta-adrenergic signaling by polycomb in prostate cancer. Cancer Cell. 2007;12(5):419–31.
114. Kondo Y, Shen L, Cheng AS, et al. Gene silencing in cancer by histone H3 lysine 27 trimethylation independent of promoter DNA methylation. Nat Genet. 2008;40(6):741–50.

115. Ke X-S, Qu Y, Rostad K, et al. Genome-wide profiling of histone h3 lysine 4 and lysine 27 trimethylation reveals an epigenetic signature in prostate. Carcinogen. PloS One. 2009;4(3):e4687.
116. Viré E, Brenner C, Deplus R, et al. The Polycomb group protein EZH2 directly controls DNA methylation. Nature. 2006;439(7078):871–4.
117. Widschwendter M, Fiegl H, Egle D, et al. Epigenetic stem cell signature in cancer. Nat Genet. 2007;39(2):157–8.
118. Schlesinger Y, Straussman R, Keshet I, et al. Polycomb-mediated methylation on Lys27 of histone H3 pre-marks genes for de novo methylation in cancer. Nat Genet. 2007;39(2):232–6.
119. Li M, Chen W-D, Papadopoulos N, et al. Sensitive digital quantification of DNA methylation in clinical samples. Nat Biotechnol. 2009;27(9):858–63.
120. Fanelli M, Amatori S, Barozzi I, et al. Pathology tissue-chromatin immunoprecipitation, coupled with high-throughput sequencing, allows the epigenetic profiling of patient samples. Proc Natl Acad Sci USA. 2010;107(50):21535–40.
121. Gupta R, Nagarajan A, Wajapeyee N. Advances in genome-wide DNA methylation analysis. BioTechniques. 2010;49(4):iii–xi.
122. Andersen JB, Factor VM, Marquardt JU, et al. An integrated genomic and epigenomic approach predicts therapeutic response to zebularine in human liver cancer. Sci Transl Med. 2010;2(54):54ra77.
123. Du Z, Song J, Wang Y, et al. DNMT1 stability is regulated by proteins coordinating deubiquitination and acetylation-driven ubiquitination. Sci Signal. 2010;3(146):ra80.
124. Lin YH, Kakadia PM, Chen Y, et al. Global reduction of the epigenetic H3K79 methylation mark and increased chromosomal instability in CALM-AF10-positive leukemias. Blood. 2009;114(3):651–658.

Chapter 10
Primary Epithelial Ovarian Neoplasms: New Concepts Concerning Origin, Pathogenesis and Classification Based on Morphology, Immunomarkers, Molecular Features, and Gene Expression Studies

Bernard Czernobilsky, Leonor Leider-Trejo, Daniele Fanale, and Antonio Russo

Origin, Pathogenesis, Morphology, Classification Immunomarkers

The World Health Organization (WHO) classification of ovarian tumors, which first appeared in 1983 and since then has undergone a number of revisions, [1] is based on morphologic features as well as on the concept that each category of ovarian tumors develops from a specific ovarian cell. According to this histogenetic classification, all the epithelial ovarian neoplasms are derived from the ovarian surface epithelium and/or from ovarian inclusion cysts, which are lined by the above epithelial cells.

In recent years, a new approach to morphologic data, increasing presumptive evidence that the cell of origin of most, if not all, ovarian epithelial tumors may be extraovarian, especially from fallopian tube and uterine endometrial cells, the recognition of precursor lesions, the emergence of certain key immunomarkers as well as molecular and genetic factors [2–4] have brought about a reevaluation of the traditional approach to these tumors. This has resulted in attempts of reclassification or subclassification of ovarian epithelial neoplasms as well as new diagnostic criteria for these tumors [5]. It should also be stressed that in most cases these new concepts correlate with the clinical course of the disease and eventually may also have an impact on the therapeutic approach to these tumors.

Origin and Pathogenesis of Epithelial Ovarian Neoplasms

The long-standing classic theory that all types of ovarian epithelial neoplasm are derived from the ovarian surface epithelium (mesothelium) and/or from its stromal invaginations resulting in ovarian inclusion cysts has been challenged in recent years. Actually, the ovarian surface epithelium bears no resemblance to any of the different epithelial tumors, which involve the ovaries, and a

B. Czernobilsky (✉)
Patho-Lab Diagnostics, Ness Ziona, Israel
e-mail: bc@patho-lab.com

L. Leider-Trejo
Institute of Pathology, Sourasky Medical Center, Tel Aviv, Israel

D. Fanale • A. Russo
Section of Medical Oncology, Department of Surgical and Oncological Sciences,
University of Palermo, Palermo, Italy

A. Russo et al. (eds.), *Diagnostic, Prognostic and Therapeutic Value of Gene Signatures*, Current Clinical Pathology, DOI 10.1007/978-1-61779-358-5_10, © Springer Science+Business Media, LLC 2012

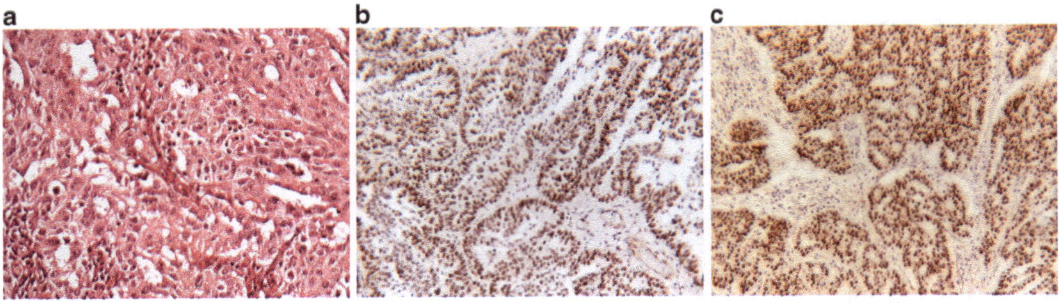

Fig. 10.1 (**a**) Ovarian high-grade serous carcinoma, (**b**) Positive WT1 and (**c**) Positive p53

questionable metaplastic process has been suggested to explain the apparent transformation of these ovarian surface cells to serous, endometrioid, mucinous, clear cell, and other epithelial neoplasm.

In recent years, entirely new concepts concerning the origin and pathogenesis of ovarian epithelial tumors have been advanced and were recently summarized by Kurman and Shih [4].

According to this hypothesis, most ovarian epithelial neoplasms are considered to be of extraovarian origin. Thus, it has been suggested that serous carcinomas develop from the fimbriated portion of the fallopian tube, endometrioid, and clear cell tumors from endometrial tissue passing through the fallopian tube resulting in endometriosis and mucinous as well as Brenner tumors arise from transitional-type epithelial nests at the tubal–mesothelial junction by a process of metaplasia.

Admittedly, much of the above still remains to be proven, but these preliminary data are of great interest and may find support in recent molecular and gene expression studies.

Serous Tumors

The vast majority of high-grade ovarian carcinomas are of the serous type. It is now being recognized that these tumors are morphologically very heterogenous which constitutes most likely an expression of their genetic heterogeneity [3]. Most serous carcinomas not only demonstrate papillary features, but also glandular, cribriform and solid architecture. Squamous metaplasia has also been described in these neoplasms. Thus, it may be difficult in some of these cases to differentiate serous from other ovarian epithelial tumors, such as endometrioid, mucinous, or clear cell carcinoma, when adhering only to the WHO classification and to the traditional morphologic criteria of ovarian tumors [3].

Most useful in the differential diagnosis between ovarian high-grade serous carcinoma and other epithelial neoplasms especially endometrioid carcinoma is widespread WT_1 (Wilms tumor protein, a suppressor gene) expression [6] as well as p53 (tumor suppressor gene) overexpression and mutation which are characteristic for serous high-grade tumors and are absent in most other ovarian carcinomas (Fig. 10.1) [3, 7].

Furthermore, precursor lesions, such as the coexistence of a serous ovarian borderline tumor with an epithelial malignant neoplasm favors the diagnosis of ovarian low-grade serous carcinoma in the latter [8] while tubal intraepithelial carcinoma is characteristically associated with high-grade serous tumor (Fig. 10.2) [9].

Mucinous Tumors

Less than 3% of ovarian mucinous carcinomas of the intestinal type are primary tumors. The vast majority are metastatic from the intestinal tract [10].

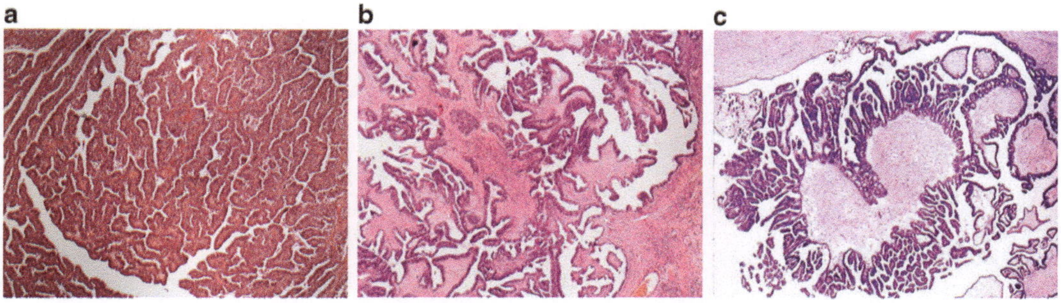

Fig. 10.2 (**a**) Ovarian low-grade serous carcinoma. Precursor lesions: (**b**) Borderline serous tumor (**c**) Micropapillary serous tumor

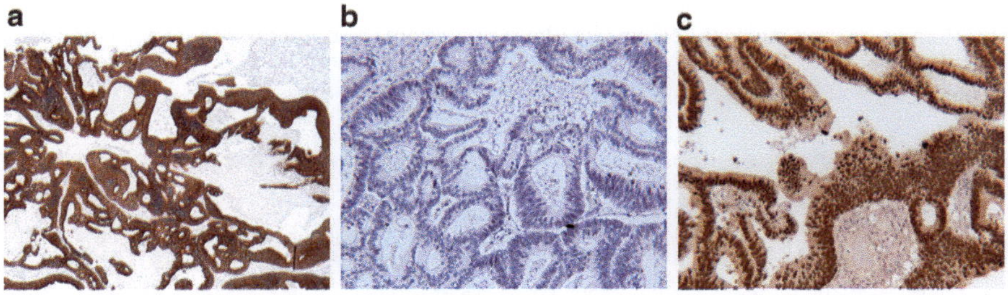

Fig. 10.3 Ovarian mucinous carcinoma metastatic from colonic carcinoma. (**a**) Positive cytokeratin 20, (**b**) Negative cytokeratin 7, (**c**) Positive CDX2

Most ovarian mucinous tumors are of the intestinal type. The so-called müllerian or seromucinous (endocervical) type ovarian tumors are uncommon. They are sometimes included in the group of mixed epithelial ovarian tumors [3].

Since many mucinous tumors lack apical mucin, they may be erroneously diagnosed as other ovarian epithelial neoplasms, especially as endometrioid carcinoma.

Immunohistochemical staining of these tumors is crucial in reaching a correct diagnosis, since primary mucinous tumors show predominantly cytokeratin 7 with lesser expression of cytokeratin 20. Most importantly, CDX2 is negative or occasionally only focally positive in primary mucinous ovarian tumors, while in metastatic carcinomas from the intestinal tract, CDX2 and cytokeratin 20 are strongly and diffusely positive while cytokeratin 7 is absent or only focally positive (Fig. 10.3) [11]. The precursor lesion of ovarian mucinous carcinoma is the mucinous borderline tumor (Fig. 10.4).

Endometrioid Tumors

According to Czernobilsky et al. [12], ovarian endometrioid carcinoma constituted about 23% of all ovarian primary carcinomas in 1970.

With the emergence of immunohistochemical markers, the recognition of precursor lesions and the change in our approach to classic morphologic features endometrioid carcinoma nowadays represents only about 10% of all ovarian carcinomas [13]. Many of erroneously diagnosed endometrioid carcinomas are indeed serous tumors.

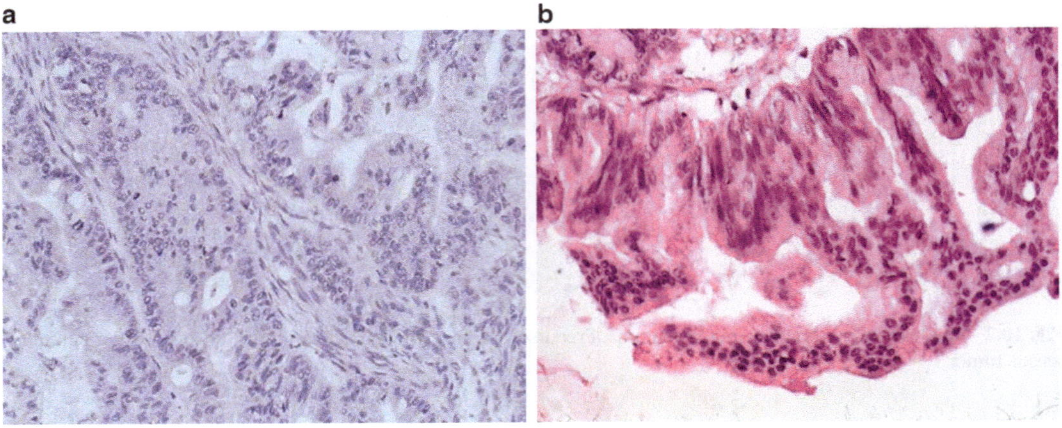

Fig. 10.4 (**a**) Primary ovarian mucinous carcinoma. Precursor lesion: (**b**) Borderline mucinous tumor

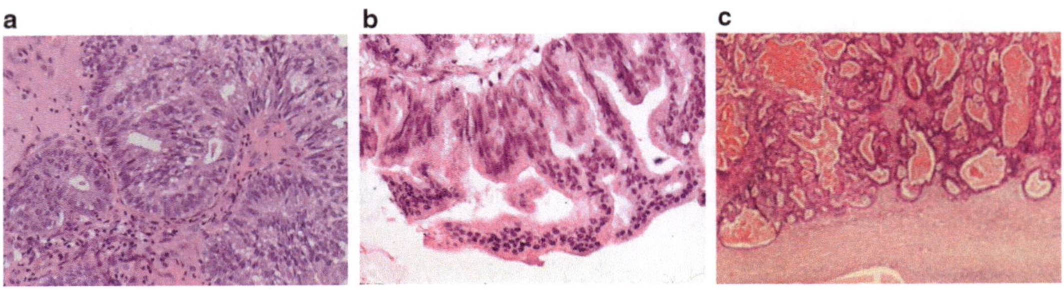

Fig. 10.5 (**a**) Ovarian endometrioid carcinoma. Precursor lesions: (**b**) Endometrioid adenofibroma (**c**) Borderline endometrioid tumor

Thus, endometrioid carcinomas are commonly associated with endometriosis, endometrioid adenofibroma, endometrioid borderline tumors as well as synchronous endometrioid intrauterine carcinoma [2, 3, 14]. These precursor lesions are absent in serous carcinomas (Fig. 10.5). Finally, endometrioid carcinomas in contrast to serous carcinomas, lack WT_1 [15] and p53 overexpression, and usually express estrogen and progesterone receptors [3].

Clear Cell Tumors

This tumor shows a papillary, tubulocystic, and solid architecture with typical hobnail type and clear cells. An oxyphilic cell type has also been described. A typical feature of these tumors is the large, highly atypical nuclei with large nucleoli.

Positive stainings for estrogen and progesterone receptors as well as WT_1 expression exclude a diagnosis of clear cell carcinoma [16]. P53 may be evident but not in the diffuse, prominent way as it is seen in serous tumors [17]. Hepatocyte nuclear factor (HNF) 1b (1beta) has been recently described as a reliable immunomarker in ovarian clear cell tumors [18, 19].

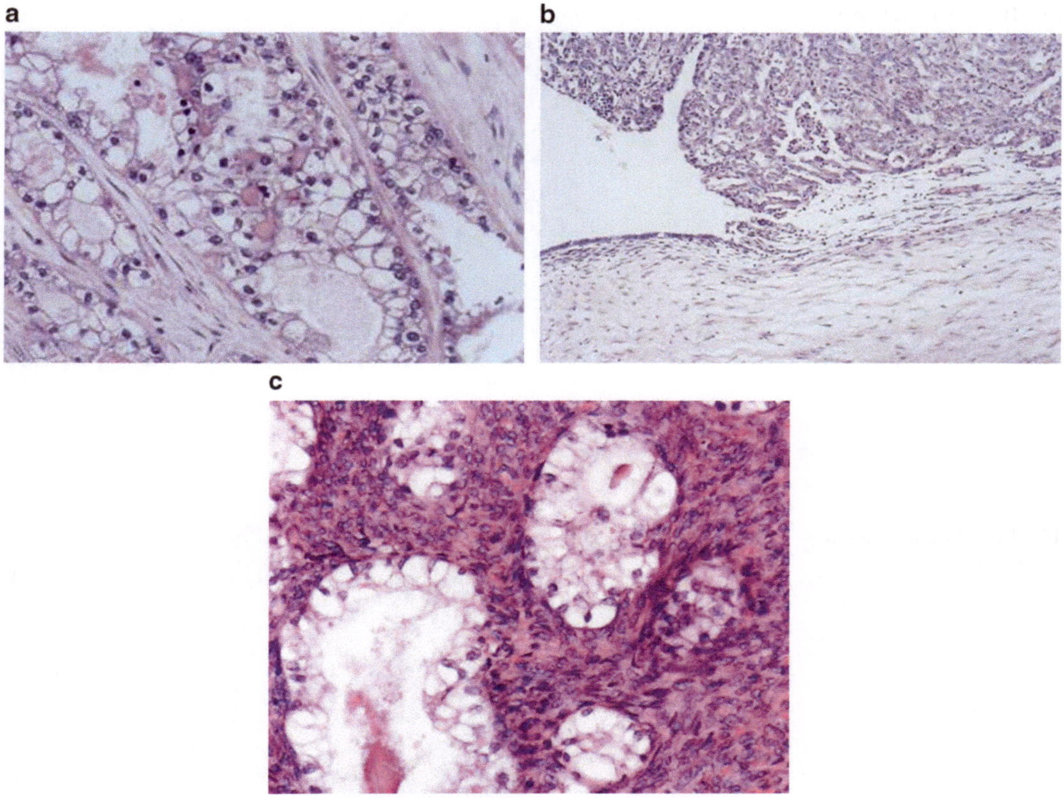

Fig. 10.6 (**a**) Ovarian clear cell carcinoma Precursor lesions: (**b**) Endometriosis with contiguous clear cell carcinoma (**c**) Clear cell adenofibroma

Ovarian carcinomas composed of clear cell elements mixed with other cell types, such as endometrioid or serous carcinoma should not be diagnosed as clear cell carcinoma but rather as serous carcinoma [20]. In other words, the diagnosis of clear cell carcinoma should be reserved for tumors with a homogenous clear cell population and highly atypical nuclei [21].

Precursor lesions of clear cell carcinoma include endometriosis and clear cell adenofibromas (Fig. 10.6) [2].

Transitional Cell Tumors

Most of the tumors classified as transitional cell carcinoma in the WHO classification [1] are actually high-grade serous or high-grade endometrioid carcinomas and frequently express p53 and WT_1, which are typical of serous carcinomas (Fig. 10.7) [22]. The only exception appears to be the benign Brenner tumor which is positive for uroplakin and may thus be of urothelial derivation (Fig. 10.8) [23].

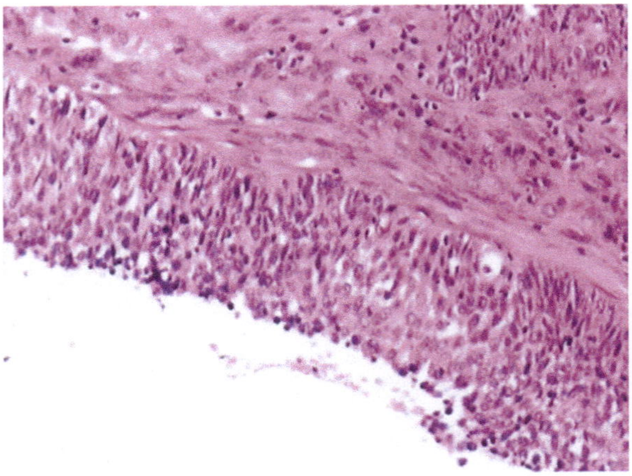

Fig. 10.7 Ovarian transitional cell carcinoma morphologically resembling urothelial cancer but representing high-grade serous carcinoma

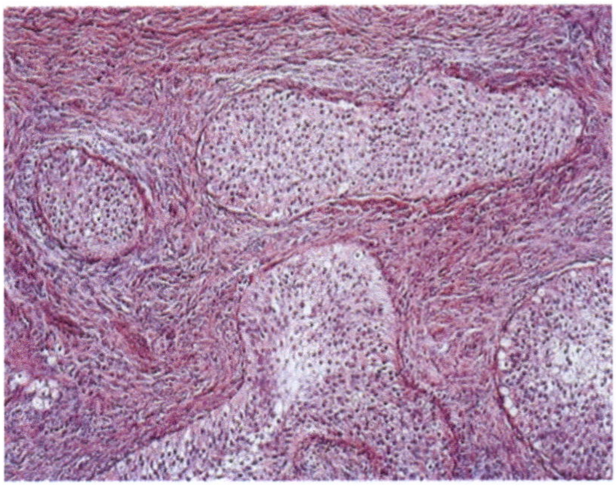

Fig. 10.8 Benign ovarian Brenner tumor

Mixed Epithelial Ovarian Tumors

These are tumors showing two or more distinctive types of neoplasms constituting at least 10% of the tumor. According to Soslow [3], the tumor elements in mixed epithelial tumors should be separable and as such diagnosable as separate neoplasms. Notwithstanding the above definition of mixed epithelial tumors, most of these are considered to be high-grade serous carcinomas, which are supported by diffuse WT_1 staining [3].

Undifferentiated Carcinomas

Most of these tumors which are not histologically classifiable should also be diagnosed as high-grade serous carcinomas especially if the tumor is WT_1 positive [3] and other possibilities, such as metastases to the ovary are excluded (Fig. 10.9). Tables 10.1–10.3 summarize some of the above discussed data.

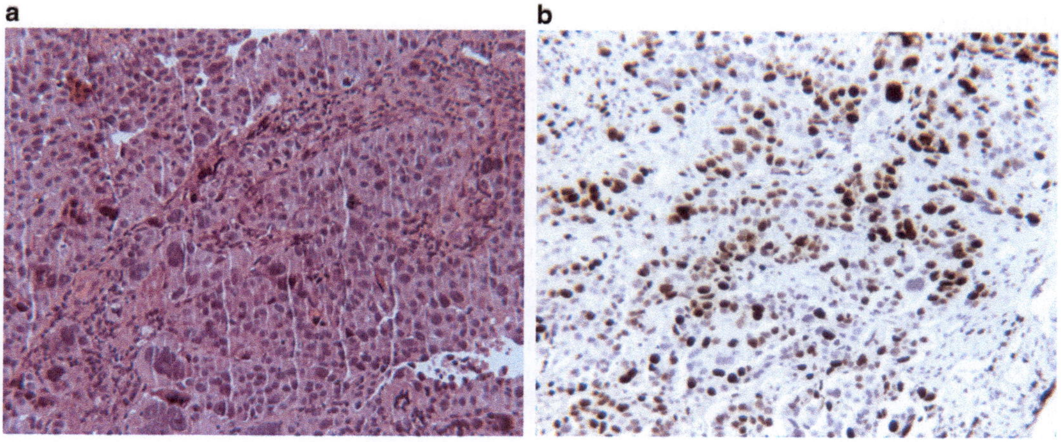

Fig. 10.9 (**a**) Ovarian undifferentiated carcinoma (**b**) Positive WT1

Table 10.1 Proposed subclassification of primary ovarian epithelial tumors [5]

1. High-grade serous, high-grade endometrioid and undifferentiated carcinomas
2. Low-grade serous carcinomas and serous borderline tumors
3. Mucinous carcinomas and mucinous borderline tumors of intestinal type
4. Low-grade endometrioid carcinoma and endometrioid borderline tumors
5. Clear cell carcinomas
6. Transitional cell carcinomas

Table 10.2 Precursor lesions in ovarian carcinomas

Carcinoma	Precursor lesions
Serous (high grade)	Fallopian tube intraepithelial carcinoma
Serous (low grade)	Serous borderline tumor
Mucinous (intestinal type)	Mucinous borderline tumor
Endometrioid	Endometriosis
	Endometrioid borderline tumor
	Synchronous intrauterine endometrioid carcinoma
Clear cell	Endometriosis
	Clear cell adenofibroma
	Clear cell borderline tumor
Undifferentiated	Unknown
Mixed	Unknown

Table 10.3 ER, WT1, P53 in ovarian carcinomas

Ovarian carcinomas	ER	WT1	p53
Serous, high grade	–	+	+
Serous, low grade	+	–	–
Mucinous (intestinal type)	–	–	–
Endometrioid	+	–	–
Clear cell	–	–	+ (variable)
Transitional cell (Brenner tumor excluded)	–	+	+
Undifferentiated[a]	–	+	+
Mixed[b]	–	+	+

ER Estrogen receptor
[a]Usually represents high-grade serous carcinoma
[b]Frequently shows a significant component of high-grade serous carcinoma

Summary

The emergence of new data in recent years concerning the origin and pathogenesis of epithelial ovarian carcinomas culminating in a theory that the latter are of extraovarian origin, the morphologic variability of hitherto well-defined ovarian tumors, the presence of precursor lesions, relatively specific immunohistochemical markers as well as molecular and genetic features, require a reassessment of various aspects of ovarian epithelial tumors.

In addition, it has been proposed by Kurman et al. [2, 4, 24] that ovarian epithelial tumors be divided in type I and type II tumors, similar to the already existing division of endometrial uterine carcinoma.

According to this proposal, type I tumors which are relatively genetically stable consist of micropapillary serous, mucinous, and endometrioid carcinomas. Type II tumors are genetically highly unstable aggressive neoplasms, such as high-grade serous carcinomas, carcinosarcomas, and undifferentiated carcinoma.

The two groups vary in their histologic appearance, precursor lesions, and immunophenotype as well as by molecular and genetic features. The clinical course and prognosis also differ in these two groups. Since clear cell carcinomas show clinical morphological, immunohistochemical, and genetic features which are shared to some degree by both type I and type II ovarian tumors, they cannot at this time be definitely classified [2].

Finally, it is to be expected that a different therapeutic approach to patients with group I and II tumors, including screening and prevention may result in a more favorable response and improve prognosis especially in the patients with high-grade ovarian cancer.

Molecular Features and Gene Expression Studies

Despite the poor prognosis and the importance of early diagnosis, there are no reliable methods for detection of ovarian cancer in the early stages of disease. Since patients diagnosed with stage I epithelial ovarian cancers (EOCs) have a 90% survival rate, it is important to identify novel ovarian cancer biomarkers with potential utility in early stage screening. Little is known of the molecular genetic changes that are associated with the development of invasive ovarian cancer. Cytogenetic analyses of epithelial ovarian tumors have shown frequent structural aberrations of chromosomes 1, 3, 6, and 11, suggesting that inactivation of genes located on these chromosomes may be important in ovarian tumorigenesis. A high frequency of allelic losses (LOH) was observed on 11p, 13q, and 17p. The loss of tumor-suppressor genes on 13q and 17p may be involved in early events of ovarian tumorigenesis and changes on 11p in later events [25].

With regard to the application of new genomic technologies, the gene expression analysis has allowed to identify important differentially expressed genes and molecular pathways that may help to understand the evolution from normal ovarian tissue to ovarian cancer [26]. Several studies have shown distinctive gene expression patterns that can differentiate between histological subtypes of ovarian carcinoma [27–31] or predict response to chemotherapy or survival [32–35]. The different histological subclasses and the clinical phenotypes displayed by EOCs are hypothesized to be driven by specific genes. Differentially expressed genes are included in pathways involved in chromosomal instability, invasion cell, motility, proliferation, and gene silencing and provided new insights into the origin of this cancer [36]. In addition, differences in gene expression patterns may help to characterize ovarian cancer and to identify potential targets for effective prevention and treatment of disease. Normal epithelial ovarian samples have been compared with tumor samples in gene expression profiling studies, generating very distinct groups in hierarchical clustering.

Table 10.4 Gene signatures associated with epithelial ovarian cancers

Histological subtypes	Differentially expressed genes	References
Serous	*FOLR1, PTGS1, WT1, GAS6*	[41]
Endometrioid	*TFF3, SFN, MSX1, CEACAM1b*	[41]
Mucinous	*ERBB3, CCND1, TGF*-a	[42]
	K-ras2, c-JUN, YES1, ECT2	
	CAV-1, SPRY1	
	CDC42, RAC1, IQGAP2, RALA, Cortactin	
	ABCC6, ABCC3	
Clear cell	*SOD2, GPX3, RBP4, UGT1A1, TFPI2, FXYD2, GLRX, ANXA4*	[38] [29]

The choice of a normal control that can be compared to EOC samples in microarray analyses can strongly influence the identification of differentially expressed genes. Seventy-five gene expression profiles of EOC subhistotypes were compared by Zorn et al. to determine the similarities and differences between all samples. The gene signature generated from Zorn et al. identified 166 genes that distinguished the samples into three subtypes: endometrioid, serous, and clear cell [37]. Instead, Schwartz et al. found 158 differentially expressed genes histotype-specific for the endometrioid, serous, clear cell, and mucinous ovarian cancers. Of these 158 genes, 73 genes were clear cell histotype-specific, 64 genes were mucinous histotype-specific, 19 genes were specific for serous, and 2 genes were specific for endometrioid. These data showed that there are large expression differences between the various histological types of ovarian cancer [38]. There is a small set of upregulated genes related to ovarian clear cell tumors including *SOD2* (superoxide dismutase), *GPX3* (glutathione peroxidase 3), *RBP4* (retinol binding protein 4), UGT1A1 (UDP glycosyltransferase1family, polypeptideA1), *TFPI2* (tissue factor pathway inhibitor), *FXYD2* (FXYD domain containing ion transport regulator 2), *GLRX* (Glutaredoxin), and *ANXA4* (Annexin). These genes overlap with genes identified by Zorn et al. [29]. Two of these genes are associated with chemotherapy response: *UGT1A1* detoxifies the active metabolite of irinotecan, whereas *ANXA4* has been associated with paclitaxel resistance [39, 40]. *SOD2*, *GLRX*, and *GPX3* are involved in oxidative stress response and particularly high levels of these in clear cell histotypes may make these tumors more resistant to chemotherapy [38] (Table 10.4).

A microarray analysis of 103 primary ovarian cancers has suggested the contributions of origin and histotype on the tumor gene expression profile [41]. Sixty-two differentially expressed genes were observed between endometrioid versus serous histotypes. Of these 62 genes, endometrioid carcinomas often showed highly expressed *TFF3* (trefoil factor 3), *SFN* (stratifin), *MSX1* (msh homeobox 1), and *CEACAM1b* (carcinoembryonic antigen-related cell adhesion molecule) genes. Serous carcinomas often showed high expression of the following genes: *FOLR1* (folate receptor 1), *PTGS1* (prostaglandin-endoperoxide synthase 1), *WT1* (Wilms tumor 1), and *GAS6* (growth arrest-specific 6). By an extensive Affymetrix microarray analysis, mucinous ovarian cancer was differentiated from the other subtypes of ovarian cancer, due to different gene expression profiles [38, 42, 43]. In mucinos ovarian tumors, Wamunyokoli et al. [42] identified specific differentially expressed genes involved in the following pathways: proliferation and cell cycle regulation (*ERBB3*, *CCND1*, *TGF*-a), transformation (*K-ras2*, *c-JUN*, *YES1*, *ECT2*), signal transduction (*CAV-1* and *SPRY1*), cytoskeleton rearrangement/signal transduction (*CDC42*, *RAC1*, *IQGAP2*, *RALA*, Cortactin), and drug resistance (*ABCC6* and *ABCC3*).

In conclusion, clear cell, serous, endometrioid, and mucinous histotypes appear to show specific gene expression signatures. These differences in gene expression profiles could be useful in the treatment of the different histotypes.

Using a Serial Analysis of Gene Expression (SAGE), David G. Peters et al. [44] have identified several potentially novel biomarkers whose expression is elevated in ovarian cancer. These proteins include CD9, HMGA1, AHCY, GNAI2, CCT3, and TACC3.

Moreover, it has recently been demonstrated by some research groups that the gene expression signatures determined by microarray analysis may act as a prognostic factor in EOC [45]. Several markers with prognostic value have been identified by specific molecular studies. *HER2/neu* and *EGFR* overexpression has been related to poor prognosis [46–48]. Clinical studies have shown that *TP53* overexpression is related with shorter survival in ovarian cancer [49] and resistance to platinum-based chemotherapy [50]. Other molecular studies by microarray identified several genes responsible for platinum resistance involved in the following pathways: proliferation (*FRA1*, *ETV4*, *IGFBP3*, *STAT1*), cell-cycle control (*CDKN1A*, *CDKN1C*, *CDC25C*, *PLK3*), apoptosis (*BAK*, *BAX*, *STAT1*, *c-JUN*, *TP53*), DNA repair (*XRCC9*, *PCNA*, *TP53*, *DDB2*, *GADD45B*, *POLH*), and energy regulation metabolism (STARD4 and FDXR) [51]. Nevertheless, the *NAC-1* expression modulates taxol resistance in ovarian cancer and may provide an effective target for chemotherapeutic intervention in taxol-resistant tumors [52]. The loss of protein p27 is an important prognostic marker for predicting disease recurrence in primary ovarian cancer [53]. Moreover, the amplification of oncogene *MYC* and increased expression levels of cyclin E have been associated with poor prognosis [54, 55]. In a recent work [56], the mammaglobin B (*MGB-2*) gene expression was evaluated in ovarian cancer tissues and in normal ovarian controls by quantitative real-time PCR and then by immunohistochemistry. *MGB-2* expression levels were found increased in EOC compared to normal ovarian controls, both at protein level and mRNA. *MGB-2* expressing tumors were related to clinicopathologic features of the less aggressive tumors. This finding suggest that *MGB-2* is an independent prognostic marker in EOC and its expression is correlated with reduced risk of disease recurrence [56].

Moreover, for the first time, the epithelial cell adhesion molecule (Ep-CAM) overexpression was associated with a decreased overall survival. Ep-CAM represents a novel independent prognostic marker for reduced survival of patient with EOC [57].

In conclusion, whole genome expression profiling has become a vital tool for identifying differentially expressed genes in EOCs, in order to potentially prolong the survival of women diagnosed with this disease [58].

References

1. Tavassoli FA, Devillee P, editors. Pathology and genetics: tumours of the breast and female genital organs. World Health Organization Classification of Tumours. Lyon: IARC; 2003.
2. Kurman RJ, Shih I. Pathogenesis of ovarian cancer: lessons from morphology and molecular biology and their clinical implications. Int J Gynecol Pathol. 2008;27:151–60.
3. Soslow RA. Histologic subtypes of ovarian carcinoma: an overview. Int J Gynecol Pathol. 2008;27:161–74.
4. Kurman RJ, Shih I. The origin and pathogenesis of epithelial ovarian cancer: a proposed unifying theory. Am J Surg Pathol. 2010;34:433–43.
5. Gilks CB. Subclassification of ovarian surface epithelial tumors based on correlation of histologic and molecular pathologic data. Int J Gynecol Pathol. 2004;23:200–5.
6. Hwang H, Quenneville L, Yaziji H, et al. Wilms tumor gene product: sensitive and contextually specific marker of serous carcinomas of ovarian surface epithelial origin. Appl Immunohistochem Mol Morphol. 2004;12:122–6.
7. Leitao MM, Soslow RA, Baergen RN, et al. Mutation and expression of TP53 gene in early stage epithelial ovarian carcinoma. Gynecol Oncol. 2004;93:301–6.
8. Singer G, Kurman RJ, Chang HW, et al. Diverse tumorigenic pathways in ovarian serous carcinoma. Am J Pathol. 2002;160:1223–8.
9. Medeiros F, Muto MG, Lee Y, et al. The tubal fimbria is a preferred site for early adenocarcinoma in women with familial ovarian cancer syndrome. Am J Surg Pathol. 2006;30:230–6.

10. Seidman JD, Kurman RJ, Ronnett BM. Primary and metastatic mucinous adenocarcinomas in the ovaries: incidence in routine practice with a new approach to improve intraoperative diagnosis. Am J Surg Pathol. 2003; 27:985–93.
11. Groisman GM, Meir A, Sabo E. The value of CDx2 immunostaining in distinguishing primary ovarian carcinomas from colonic carcinoma metastatic to the ovaries. Int J Gynecol Pathol. 2004;23:52–7.
12. Czernobilsky B, Silverman BB, Mikuta JJ. Endometrioid carcinoma of the ovary. A clinicopathologic study of 75 cases. Cancer. 1970;26:1141–52.
13. Seidman JD, Horkayne-Szakaly I, Haiba M, et al. The histologic type and stage distribution of ovarian carcinomas of surface epithelial origin. Int J Gynecol Pathol. 2004;23:41–4.
14. Roth LM, Emerson RE, Ulbright TM. Ovarian endometrioid tumors of low malignant potential: a clinicopathologic study of 30 cases with comparison to well-differentiated endometrioid adeno-carcinoma. Am J Surg Pathol. 2003;27:1213–59.
15. Shimizu M, Tok T, Takagi Y, et al. Immunohistochemical detection of the Wilms tumor gene (WT1) in epithelial ovarian tumors. Int J Gynecol Pathol. 2000;19:158–63.
16. Waldstrom M, Grove A. Immunohistochemical expression of Wilms tumor gene protein methylation in ovarian clear cell adenocarcinoma. Arch Pathol Lab Med. 2005;129:85–8.
17. Otis CN, Krebs PA, Quezado MM, et al. Loss of heterozygosity in p53, BRACA1, and estrogen receptor genes and correlation to expression of p53 protein in ovarian epithelial tumors of different cell types and biology behavior. Hum Pathol. 2000;31:233–48.
18. Köbel M, Kallogeer SE, Carrick J, Huntsman D, Asad H, Oliva E, et al. A limited panel of immunomarkers can reliably distinguish between clear cell and high grade serous adenocarcinoma of the ovary. Am J Surg Pathol. 2009;33:14–21.
19. Stadlmann S, Gueth U, Baumhoer D, et al. Glypican-3 expression in primary and recurrent ovarian carcinomas. 2007;26:341–44.
20. Silva EG, Young RH. Endometrioid neoplasms with clear cells: a report of 125 cases in which the alteration is not of typical secretory type. Am J Surg Pathol. 2007;31:1203–8.
21. Han G, Gilks CB, Leung S, et al. Mixed ovarian epithelial carcinomas with clear cell and serous components are variants of high grade serous carcinomas: an interobserver correlative and immunohistochemical study of 32 cases. Am J Surg Pathol. 2008;32:955–64.
22. Logani S, Oliva E, Amin MB, et al. Immunoprofile of ovarian tumor with putative transitional cell (urothelial) differentiation using novel utothelial markers: histogenetic and diagnostic implications. Am J Surg Pathol. 2003;27:1434–41.
23. Riedel L, Czernobilsky B, Lifschitz-Mercer B, et al. Brenner tumors but not transitional cell carcinomas of the ovary show urothelial differentiation: an immunohistochemical study including cytokeratins and uroplakin. Virchows Arch. 2001;438:181–91.
24. Kurman RJ, McConnell TG. Precursors of endometrial and ovarian carcinoma. Arch Virchows. 2010;456:1–12.
25. Gallion HH, Powell DE, Morrow JK, et al. Molecular genetic changes in human epithelial ovarian malignancies. Gynecol Oncol. 1992;47:137–42.
26. Grisaru D, Hauspy J, Prasad M, et al. Microarray expression identification of differentially expressed genes in serous epithelial ovarian cancer compared with bulk normal ovarian tissue and ovarian surface scrapings. Oncol Rep. 2007;18:1347–56.
27. Marquez RT, Baggerly KA, Patterson AP, et al. Patterns of gene expression in different histotypes of epithelial ovarian cancer correlate with those in normal fallopian tube, endometrium, and colon. Clin Cancer Res. 2005;11:6116–26.
28. Meinhold-Heerlein I, Bauerschlag D, Hilpert F, et al. Molecular and prognostic distinction between serous ovarian carcinomas of varying grade and malignant potential. Oncogene. 2005;24:1053–65.
29. Zorn KK, Bonome T, Gangi L, et al. Gene expression profiles of serous, endometrioid, and clear cell subtypes of ovarian and endometrial cancer. Clin Cancer Res. 2005;11:6422–30.
30. Ouellet V, Provencher DM, Maugard CM, et al. Discrimination between serous low malignant potential and invasive epithelial ovarian tumors using molecular profiling. Oncogene. 2005;24:4672–87.
31. Gilks CB, Vanderhyden BC, Zhu S, et al. Distinction between serous tumors of low malignant potential and serous carcinomas based on global mRNA expression profiling. Gynecol Oncol. 2005;96:684–94.
32. Berchuck A, Iversen ES, Lancaster JM, et al. Patterns of gene expression that characterize long-term survival in advanced stage serous ovarian cancers. Clin Cancer Res. 2005;11:3686–96.
33. Bernardini M, Lee CH, Beheshti B, et al. High-resolution mapping of genomic imbalance and identification of gene expression profiles associated with differential chemotherapy response in serous epithelial ovarian cancer. Neoplasia. 2005;7:603–13.
34. Spentzos D, Levine DA, Ramoni MF, et al. Gene expression signature with independent prognostic significance in epithelial ovarian cancer. J Clin Oncol. 2004;22:4700–10.

35. Spentzos D, Levine DA, Kolia S, et al. Unique gene expression profile based on pathologic response in epithelial ovarian cancer. J Clin Oncol. 2005;23:7911–8.
36. Farley J, Ozbun LL, Birrer MJ. Genomic analysis of epithelial ovarian cancer. Cell Res. 2008;18:538–48.
37. Zorn KK, Jazaeri AA, Awtrey CS, et al. Choice of normal ovarian control influences determination of differentially expressed genes in ovarian cancer expression profiling studies. Clin Cancer Res. 2003;9:4811–8.
38. Schwartz DR, Kardia SL, Shedden KA, et al. Gene expression in ovarian cancer reflects both morphology and biological behavior, distinguishing clear cell from other poor-prognosis ovarian carcinomas. Cancer Res. 2002;62:4722–9.
39. Han EK, Tahir SK, Cherian SP, et al. Modulation of paclitaxel resistance by annexin IV in human cancer cell lines. Br J Cancer. 2000;83:83–8.
40. Gagne JF, Montminy V, Belanger P, et al. Common human UGT1A polymorphisms and the altered metabolism of irinotecan active metabolite 7-ethyl-10-hydroxycamptothecin (SN-38). Mol Pharmacol. 2002;62:608–17.
41. Shedden KA, Kshirsagar MP, Schwartz DR, et al. Histologic type, organ of origin, and Wnt pathway status: effect on gene expression in ovarian and uterine carcinomas. Clin Cancer Res. 2005;11:2123–31.
42. Wamunyokoli FW, Bonome T, Lee JY, et al. Expression profiling of mucinous tumors of the ovary identifies genes of clinicopathologic importance. Clin Cancer Res. 2006;12(3 Pt 1):690–700.
43. Heinzelmann-Schwarz VA, Gardiner-Garden M, Henshall SM, et al. A distinct molecular profile associated with mucinous epithelial ovarian cancer. Br J Cancer. 2006;94:904–13.
44. Peters DG, Kudla DM, DeLoia JA, et al. Comparative gene expression analysis of ovarian carcinoma and normal ovarian epithelium by serial analysis of gene expression. Cancer Epidemiol Biomarkers Prev. 2005;14:1717–23.
45. Jochumsen KM, Tan Q, Høgdall EV, et al. Gene expression profiles as prognostic markers in women with ovarian cancer. Int J Gynecol Cancer. 2009;19:1205–13.
46. Kokuho M, Yoshiki T, Hamaguchi A, et al. Immunohistochemical study of c-erbB-2 proto-oncogene product in prostatic cancer. Nippon Hinyokika Gakkai Zasshi. 1993;84:1872–8.
47. Meden H, Kuhn W. Overexpression of the oncogene c-erbB-2 (HER2/neu) in ovarian cancer: a new prognostic factor. Eur J Obstet Gynecol Reprod Biol. 1997;71:173–9.
48. Skirnisdottir I, Sorbe B, Seidal T. The growth factor receptors HER-2/neu and EGFR, their relationship, and their effects on the prognosis in early stage (FIGO I-II) epithelial ovarian carcinoma. Int J Gynecol Cancer. 2001;11:119–29.
49. Ozalp SS, Yalcin OT, Basaran GN. Prognostic significance of deletion and overexpression of the p53 gene in epithelial ovarian cancer. Eur J Gynaecol Oncol. 2000;21:282–6.
50. Reles A, Wen WH, Schmider AG, et al. Correlation of p53 mutations with resistance to platinum-based chemotherapy and shortened survival in ovarian cancer. Clin Cancer Res. 2001;7:2984–97.
51. Konstantinopoulos PA, Spentzos D, Cannistra SA. Gene-expression profiling in epithelial ovarian cancer. Nat Clin Practice Oncol. 2008;5(10):577–87.
52. Ishibashi M, Nakayama K, Yeasmin S, et al. A BTB/POZ gene, NAC-1, a tumor recurrence-associated gene, as a potential target for taxol resistance in ovarian cancer. Clin Cancer Res. 2008;14(10):3149–55.
53. Masciullo V, Sgambato A, Pacilio C, et al. Frequent loss of expression of the cyclin-dependent kinase inhibitor p27 in epithelial ovarian cancer. Cancer Res. 1999;59:3790–4.
54. Diebold J, Suchy B, Baretton GB, et al. DNA ploidy and MYC DNA amplification in ovarian carcinomas. Correlation with p53 and bcl-2 expression, proliferative activity and prognosis. Virchows Arch. 1996;429:221–7.
55. Farley J, Smith LM, Darcy KM, et al. Cyclin E expression is a significant predictor of survival in advanced, suboptimally debulked ovarian epithelial cancers: a Gynecologic Oncology Group study. Cancer Res. 2003;63:1235–41.
56. Tassi RA, Calza S, Ravaggi A, et al. Mammaglobin B is an independent prognostic marker in epithelial ovarian cancer and its expression is associated with reduced risk of disease recurrence. BMC Cancer. 2009;9:253.
57. Spizzo G, Went P, Dirnhofer S, et al. Overexpression of epithelial cell adhesion molecule (Ep-CAM) is an independent prognostic marker for reduced survival of patients with epithelial ovarian cancer. Gynecol Oncol. 2006;103:483–8.
58. Mok SC, Elias KM, Wong K-K, et al. Biomarker discovery in epithelial ovarian cancer by genomic approaches. Cancer Res 2007. doi:10.1016/S0065-230X(06)96001-1.

Chapter 11
Thyroid Carcinoma: Molecular Signature by Histotype-Specific Mutations and Gene Expression Patterns

Umberto Malapelle, Claudio Bellevicine, Lajos Pustzai, and Giancarlo Troncone

Introduction

Thyroid cancer is the most common type of endocrine malignancy [1]. The incidence of thyroid cancer has been steadily rising over the past three decades; since mortality remained unchanged, this increased incidence is more apparent than real and due to changes in diagnostic criteria and to the advent of a more sensitive diagnostic approach based on ultrasound (US) guided fine-needle aspiration (FNA) [2]. Data from the survival, epidemiology, and end results database showed that this cancer constitutes approximately 1.5% of all newly diagnosed cancer cases in the USA [3], its incidence being two to three times higher among women than men and about twice higher in whites than in blacks [1, 2].

Histological Classification

Most of thyroid tumors are primary neoplasms originating from epithelial cells [4]. Except from medullary carcinomas, that originate from thyroid C cells, the vast majority of thyroid neoplasms derives from follicular cells (Table 11.1), including both benign follicular adenoma and malignant neoplasms [5]. These latter are most commonly well-differentiated carcinomas either of papillary or follicular histotype. Oncocytic (Hürthle cell) adenomas and carcinomas are considered as a variant of follicular tumors by the current (2004) WHO classification [5]. Poorly differentiated and anaplastic carcinomas (ATC) also originate from follicular cells, often as the result of dedifferentiation of preexisting papillary or follicular carcinomas [5]. From a prognostic standpoint, well-differentiated papillary and follicular carcinomas have an indolent course, whereas ATC are almost always lethal [5]. Poorly differentiated carcinoma (PDC) falls into a distinctly intermediate prognostic category [6]. Placement into separate classification groups has been proposed for hyalinizing trabecular tumors, which typically have a benign course, despite sharing several histological features with papillary carcinoma [7]. Similarly, the term "well-differentiated tumor of uncertain malignant potential" should be reserved for those encapsulated tumors showing either questionable capsular invasion or only partially developed nuclear features of papillary carcinoma [8].

U. Malapelle • C. Bellevicine • G. Troncone (✉)
Scienze Biomorfologiche e Funzionali, Universita degli Studi di Napoli Federico II, Naples, Italy
e-mail: giancarlo.troncone@unina.it

L. Pustzai
Department of Breast Medical Oncology, MD Anderson Cancer Center, University of Texas, Houston, TX, USA

A. Russo et al. (eds.), *Diagnostic, Prognostic and Therapeutic Value of Gene Signatures*, Current Clinical Pathology, DOI 10.1007/978-1-61779-358-5_11,

Table 11.1 Histological classification of thyroid tumors arising from follicular epithelial cells

A. Benign
Follicular adenoma
Conventional type
Oncocytic type
B. Uncertain Malignant Potential (UMP)
Hyalinizing trabecular tumor
UMP with questionable papillary carcinoma-type nuclear changes
UMP with questionable capsular penetration without nuclear changes
C. Malignant
Papillary carcinoma
Follicular carcinoma
Conventional type
Oncocytic type
Poorly differentiated carcinoma
Anaplastic (undifferentiated) carcinoma

Thyroid cancer is rare in children, but its incidence begins to rise sharply in the second decade of life and peaks in women during the late reproductive age and in men during the sixth decade of life [1]. This age distribution reflects the incidence of the three most common types of thyroid cancer: papillary carcinoma that constitutes approximately 80% of all thyroid cancer cases, follicular carcinoma (15%), and medullary carcinoma (3%) [5]. ATC, which accounts for less than 2% of thyroid tumors, typically occurs in the older age group [5]. The thyroid cancer increased incidence is almost entirely attributed to papillary carcinoma both in its classic and follicular variant (FVPTC) form, due to the detection of cancers smaller than 2 cm [9]. Thus, about half of the increase is due to tumors <1 cm in size and another 40% is due to tumors 1–2 cm. On the contrary, the rates of follicular, medullary, and anaplastic cancer did not change significantly [1].

Molecular Genetics of Thyroid Cancer

The microscopic diagnosis of poorly differentiated and anaplastic thyroid cancer is straightforward; conversely most of the new developments in molecular diagnostics are centered on well differentiated thyroid cancers. In fact on cytology microscopic diagnosis and on histology the prognostication may be challenging [10]. Although several molecular markers are promising, only those molecular tests that have an unquestionable biological and clinical significance and that are sufficiently robust and reproducible in multiple laboratories are suitable for a widespread clinical implementation [11]. In particular testing for somatic mutations and gene rearrangements has already completed most of the steps required to bring a molecular assay from early exploratory studies to widespread implementation [11].

In fact, several genetic abnormalities have been associated to thyroid carcinoma underlying a close correlation between specific genetic lesions and histologic phenotype [12]. Papillary carcinomas show frequently either specific gene rearrangements which gives rise to the formation of the RET/PTC, or more rarely NTRK1, chimeric genes or B-type Raf kinase (BRAF) activating point-mutations [13, 14]. In addition, only in the follicular variant, activating point-mutations in the RAS oncogenes also occur [14]. From a molecular point of view, there are two groups of FTC; one is associated with activating mutations in RAS and the other with gene rearrangements between the peroxisome proliferator-activated receptor g (PPARg) and the PAX8 (or more rarely the CREB3L2) transcription factor [15]. RAS mutations are common in PDTC [16]. ATC is associated to BRAF and RAS mutations [17]; in addition, it often features mutations in TP53, PTEN, PIK3CA and CTNNB1 [10, 18]. Finally, point mutations of the RET gene are found in familial endocrine syndromes (FMTC; MEN2A, and MEN2B), a common feature of which is the medullary thyroid carcinoma, a malignant tumor derived from parafollicular C-cells [19].

BRAF Mutations as Molecular Marker of PTC

The BRAF mutation was recently discovered to be the most common (up to 45%) genetic alteration in PTC [3]. This is a somatic genetic mutation and is not a germline alteration in familial thyroid cancer [20]. The BRAF mutation exclusive occurs in PTC and PTC derived ATC, whereas it does not occur in follicular carcinoma and other types of thyroid tumors [3]. The high prevalence and high specificity of BRAF mutation for PTC underlines the potential clinical utility to diagnosis in those settings, in which morphology is unclear [10].

The presence of only one "hotspot" (codon 600 in exon 15) renders PTC-associated BRAF mutations easily detectable on a technical point of view [4, 5]. Moreover, as a stable DNA molecular marker, BRAF mutation can be easily detected on common DNA specimens, even in low quantities, such as those obtained by FNA needle rinsing [21]. BRAF mutation detection can be achieved by a variety of molecular techniques including real-time PCR amplification and post-PCR melting curve analysis, allele specific PCR, direct nucleotide sequencing, restriction fragment polymorphism analysis, and others [22]. All of these methods demonstrate reliable detection of BRAF mutation in different types of thyroid specimens [23].

The most employed method for BRAF testing remains direct sequencing of PCR amplification products [11, 24]. Direct sequencing relies on the Sanger chain termination method; the incorporation of a chemically modified nucleotide (dideoxynucleotide) terminates extension of the DNA strand at the point of incorporation [24]. This results in a mixture of DNA fragments of different lengths. Each dideoxynucleotide (A, T, C, or G) is labeled with a different fluorescent dye, allowing their individual detection [24]. The newly synthesized and labeled DNA fragments are separated by size through capillary gel electrophoresis. The fluorescence is detected by an automated sequence analyzer and the order of nucleotides in the target DNA illustrated as a sequence electropherogram [24].

For real-time PCR amplification, two probes complementary to wild-type sequences are designed to span the mutation site for the mutational hot spot (exon 15). If no mutation is present, probes will bind perfectly to sample DNA and melt at a higher temperature, showing a single peak on post-PCR melting curve analysis. In contrast, if a heterozygous mutation is present, probes will bind to mutant DNA imperfectly (i.e., with one nucleotide mismatch) and will melt (dissociate) earlier, producing two melting peaks (one for the wild-type allele and one for the mutant allele) or one melting peak at lower temperature if the mutation is homozygous. Each nucleotide substitution produces a melting peak at specific T_m. This method showed similar sensitivity in detection of BRAF mutation in thyroid tumors when compared with direct nucleotide sequencing as the criterion standard [25].

Since the initial discovery of BRAF mutation in human cancers [24], there have been more than 40 mutations identified in the BRAF gene, among which the T1799A point BRAF mutation is the most common and accounts for more than 90% of all the mutations found in the BRAF gene [21]. The T1799A BRAF mutation causes a V600E amino acid change in the BRAF protein, resulting in the constitutive and oncogenic activation of the mutated BRAF kinase [21]. A few other activated BRAF mutants are only rarely found in thyroid cancer, such as the BRAF K601E, resulting in the substitution of lysine with glutamate. This latter was found with a remarkable frequency (up to 9%) in the FVPTC [26].

Molecular Diagnostics and Histological Diagnosis of PTC

Demonstration of the V600E BRAF mutation or of clonal RET/PTC has strong diagnostic value as an indicator of PTC [21, 27]. Unfortunately, searching for these mutations on histological samples has little diagnostic use [12]. In fact, PTC carrying these mutations regardless from displaying classic, tall or oncocytic features is marked by a prevalent papillary pattern of growth [26]. This makes its microscopic recognition straightforward, contributing to diminish the potential diagnostic importance of the detection of BRAF mutations. Conversely, the diagnosis of a follicular patterned tumor, characterized by total encapsulation and no formation of papillae, is challenging [5]. Usually, capsular invasion is evaluated to separate a benign (follicular adenoma/nodular adenomatous) from a minimally invasive follicular carcinoma [5]; similarly, nuclear characteristics separate a benign nodule from an encapsulated FVPTC [5]. However, these morphological criteria are difficult to assess [28, 29]. Recent reports have suggested that the FVPTC subtype is frequently misdiagnosed by pathologists. Even experts in thyroid pathology expressed a fully concordant diagnosis only in 39% of FVPTC demonstrating that in the absence of clear-cut invasive growth this histological diagnosis is often questionable [28]. Once recognized these tumors should undoubtedly be designated as papillary carcinomas, as they represent a distinct variant of papillary carcinoma with a number of characteristic molecular and biological features, some of which are closer to follicular tumors [30]. These PTCs with a follicular architecture (follicular variant of PTC and oxyphilic variant and papillary microcarcinoma cases with a follicular architecture) usually lack the V600E mutation; a different type of activating B-RAF mutation K601E was found in about 7% of the cases [26]. FVPTC also frequently harbor RAS mutations, which are common in follicular tumors and are rare in classic papillary carcinoma [31]. However, RAS mutation cannot be used to define malignancy since it can also be found in follicular adenomas [16].

Thyroid Cytology

In the general population, the incidence of thyroid nodules detected by clinical examination is very high (5–20%) [32]; this figure becomes even higher by ultrasound screening [1]. Most of the thyroid cancers present as a nodule. However, thyroid nodules very rarely represent cancer. The extremely large number of benign thyroid nodules and the small number of admixed malignant ones requires an accurate screening tool. The aim is to determine when surgery is needed and also the correct surgical procedure. FNA cytology efficiently identifies those clinically relevant nodules whose treatment unequivocally requires surgery [33]. The validity of this statement is proven by several large series of concordant matched cytological and histological specimens [34–36]. Thanks to the widespread use of FNA the number of patients requiring thyroid surgery has reduced by more than 50% [34–36]; the rate of surgically resected thyroid nodules found to be malignant has increased by two to three times and the overall cost of managing a thyroid nodule has decreased by more than 25% [36]. However, FNA performance is highly dependent on the operator's experience, on accurate cytopreparation methodology and on effective communication between physicians [37–40].

The Bethesda System for Reporting Thyroid Cytopathology

To avoid confusion among clinicians, special care should be taken to report thyroid FNA in a consistent manner [37–40]. The Bethesda System for Reporting Thyroid Cytopathology (BSRTC) has been proposed at the 2007 National Cancer Institute (NCI) state of art thyroid FNA conference [41].

Table 11.2 Prospective and retrospective studies on the diagnostic value of V600E to refine PTC diagnosis on FNAs

Study	Design	Total series case	*n* Undetermined	Undetermined with V600E (%)	Specificity for PTC
Salvatore et al. [47]	Retrospective	95	11	4 (27)	4/4
Sapio et al. [44]	Prospective	132	16	4 (25)	4/4
Sapio et al. [46]	Prospective	144	94	10 (11)	10/10
Nikiforov et al. [10]	Prospective	328	51	7 (13)	7/7
Kumagai et al. [48]	Prospective	208	22	3 (14)	3/3
Xing et al. [5]	Prospective	45	25	2 (8)	2/2
Pizzolanti et al. [49]	Prospective	156	19	2 (10)	2/2
Jo et al. [50]	Prospective	101	24	7 (29)	7/7
Zatelli et al. [51]	Prospective	469	22	10 (45)	10/10
Cohen et al. [52]	Retrospective	91	55	5 (9)	5/5
Rowe et al. [53]	Retrospective	19	19	3 (16)	3/3
Xing et al. [54]	Retrospective	190	Not specified	Not specified	–
Chung et al. [55]	Prospective	137	25	4 (16)	3/4[a]

[a]The Authors justify the only false-positive FNAB describing a putative PTC precursor lesion "atypical hyperplasia" on matching surgical specimen

This classification scheme divides adequate cases (containing at least six groups of thyreocytes) into a five-tiered risk based system [42]. At the end of this spectrum are those classes termed as "benign" or "malignant"; here, diagnosis is certain and the related post-FNA options are clear. Benign lesions identified by FNAC are generally left untreated, and patients undergo periodic clinical and ultrasound examination, whereas patients with malignant nodules undergo total thyroidectomy [43]. Between "benign" and "malignant" there are three diagnostic categories relative to indeterminate cytology; each features a different probability of malignancy [43]. The associated risk is low (5–10%) for "Follicular lesion of undetermined significance" (FLUS), intermediated for "suspicious for follicular neoplasm" (15–30%), and high for "suspicious for malignancy" (60–75%) [38].

Molecular Testing on Cytology

There is little doubt that patients with indeterminate cytology will benefit from testing for those somatic mutations that occur in about two-thirds of PTC (BRAF mutations and RET/PTC rearrangements) and FTC (RAS mutations and PAX8/PPARg rearrangements) [44]. Thus, some malignant cases that cannot be diagnosed with cytology can be identified and cured [45]. Our group extensively tested BRAF and RET/PTC either on archival cytological slides or prospectively by a FNA dedicated pass [44, 46, 47]; in these studies and in several other series (Table 11.2) all BRAF positive FNA samples studied prospectively and retrospectively were papillary carcinomas. The high specificity of BRAF mutation for PTC underlines its potential clinical utility to refine undetermined cytological diagnosis [21]. However, BRAF testing cost-effectiveness is high only when PTC is strongly suspected on cytology [21, 56]. Deveci et al. in a detailed analysis relative to the histopathologic follow-up of "follicular neoplasm" ($n=339$) and "suspicious for papillary thyroid carcinoma" (PTC) ($n=120$) FNAs showed that these BSRTC classes have a very different malignancy rate (MR) [56]. The MR of "follicular neoplasms" was 22%, whereas that of "suspicious for PTC" was much higher (72%). In both classes most of the malignant cases were FVPTC that represented, respectively, 11% and 54% of cases. Conversely, classic PTC was 2% and 12% of the cases [56].

Since follicular variant PTC do not or only rarely harbor the V600E and only in rare instances the K601E mutation the diagnostic sensitivity of BRAF mutation testing alone is very low for "suspicious of follicular neoplasms" thyroid nodules and only slightly higher for "suspicious for PTC" classes [12]. In an earlier study on archival smears we were able to refine an undetermined diagnosis in 27% of cases [47] and subsequently this rate (25%) was confirmed also in a study of prospectively collected FNAs [44]. Although only a limited portion of patients within the gray zone FNA diagnoses who need surgery may be identified [57], molecular testing has a very high positive predictive value (PPV) [45]. The high specificity of BRAF for PTC has led some Institution to consider BRAF-positive FNA, even in absence of microscopically observed atypical, as the sole indication for initial thyroidectomy [10]. Thus, BRAF is a low sensitive but 100% specific marker; in a recent analysis we used BRAF together with a less specific, but more sensitive, immunocytochemical marker of thyroid cancer – galectin-3 (Gal-3) [58–60]. This is rarely detected in normal thyroid tissue and benign nodules, while its expression has frequently been demonstrated in malignant thyroid tumors. While Gal-3 sensitivity is high, there is some concern about its specificity because Gal-3 expression has been found in adenomatous goiter and Hashimoto's thyroiditis by immunohistochemistry [61]. The BRAF/Gal-3 combined analysis yielded a modest improvement of the positive predictive (from 73.9 to 78.6%), while the negative predictive value increased from 70.8% for Gal-3 alone to 89.5% for Gal-3 and/or BRAFV600E [46]. Thus, a reduction of false-negative cases was achieved [46].

Novel Malignant Thyroid Markers Identified by Gene Profiling

The BSRTC class diagnostic criteria for the "follicular neoplasm" category refer to a monotonous population of three-dimensional groups of follicular cells with scarce colloid; as discussed before, these uncertain features do not correspond to a single entity, but rather to a wide range of inflammatory, hyperplastic, and neoplastic histological lesions [35]. These latter, including follicular adenoma, follicular carcinoma, and follicular variant of PTC, do not or only rarely harbor RET)/PTC rearrangements or BRAF mutations [62]. Thus, panels of mRNA and protein cancer markers are needed to refine indeterminate diagnosis. Microarray studies are widely used to define diagnostic and prognostic signatures in cancers and they have led to the identification of a large list of carcinoma-regulated genes also in thyroid cancers [63]. To look for genes potentially involved in the neoplastic transformation of the thyroid gland, we extracted RNAs from normal human thyroid primary cells and six human thyroid carcinoma cell lines (WRO cell line from FTC, TPC-1 and FB-2 cell lines from PTC, NPA cell line from a poorly differentiated PTC, and ARO and FRO cell lines from ATC) and hybridized them to U95Av2 Affymetrix oligonucleotide arrays (Affymetrix) containing 12,625 transcripts [64, 65]. We looked for genes whose expression was drastically (at least tenfold) up- or downregulated in all the six thyroid carcinoma cell lines versus normal thyroid primary cell culture, on the assumption that genes whose expression was altered in all carcinoma cell lines could be involved in thyroid cell transformation [64, 65].

UbcH10 Deregulation Occurs in Thyroid Cancer

Genes that regulate cell-cycle progression may be differentially expressed in malignant versus benign thyroid nodules [65]. UbcH10 (alias E2C or UBE2C) is a cell cycle-related protein involved in mitosis completion [66]. Its ubiquitin-conjugating enzymatic activity is exerted from G2/M-phase

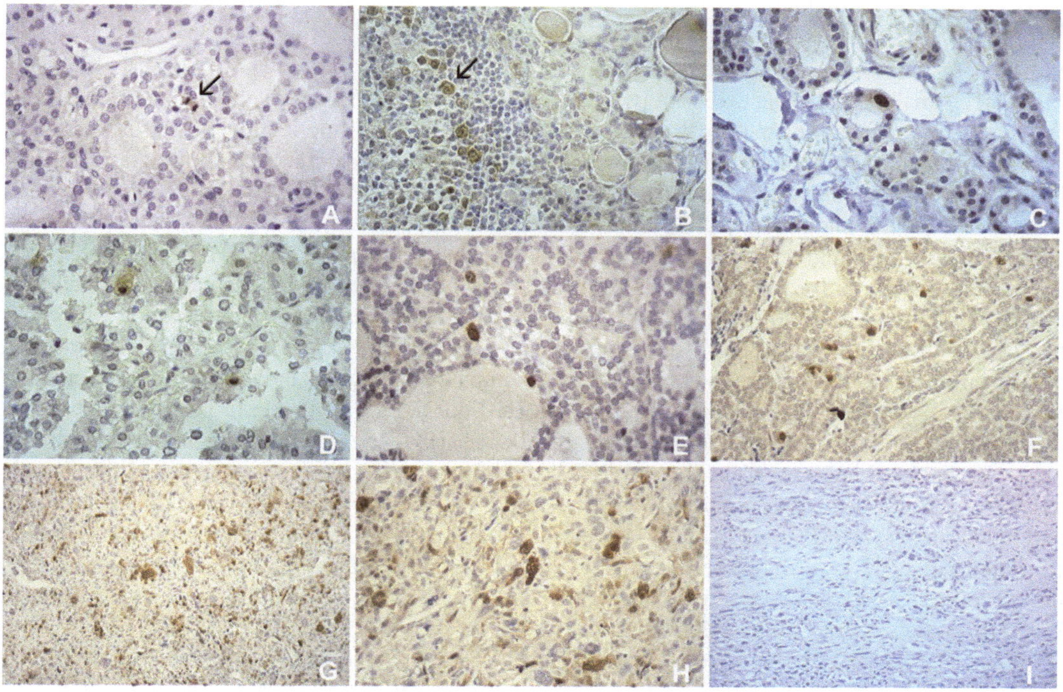

Fig. 11.1 UbcH10 staining pattern in normal, inflammatory, and neoplastic thyroid tissues. Follicular epithelial cells of normal thyroid (**a**) and oxyphilic cells of Hashimoto's thyroiditis (HT) (**b**) do not stain for UbcH10, with occasional mitotic figures (**a**, *arrow*) and lymphoid centroblasts of HT (**b**, *arrow*) providing the appropriate internal positive control. In neoplastic thyroid, UbcH10 staining pattern is strongly related to tumor grade, being weak in follicular adenoma (**c**), slightly more evident in well-differentiated papillary (**d**) and follicular (**e**) carcinomas, whereas stronger in poorly differentiated (**f**) and in anaplastic (**g**) carcinomas. In the latter, most of neoplastic cells show a very intense labeling, with intense nuclear staining (**h**), whereas signal disappeared by antigen incubation (**i**)

to early G1 phase, when UbcH10 together with ubiquitin ligase transfers ubiquitin to the mitotic cyclins, thereby promoting their degradation by the proteosome. Once mitotic cyclins are ubiquinated, UbcH10 triggers its own destruction [67]. This event marks mitotic completion and provides the molecular switch that allows cells to bring cell division to an end and to proceed to the new round of DNA duplication. Thus, UbcH10 is essential in cell cycle progression [68]. Our attention was focused on this gene, since it was upregulated about 150-fold in all of the cell lines tested by the Affymetrix HG_U95Av2 oligonucleotide array cDNA microarray [65]. These results were further confirmed by Western blot analyses, as the expression level of UbcH10 was extremely low in the normal thyroid primary culture cells, but strong in all the cancerous cell lines [65]. Immunohistochemical and RT–PCR analyses on a large panel of thyroid neoplasms of different histotypes revealed that UbcH10 is barely detectable in normal thyroid tissues, goiters, and adenomas, whereas it increases in papillary and follicular tumors, reaching the highest level of expression in ATC (Fig. 11.1) [65]. The block of UbcH10 protein synthesis by RNA interference inhibited the growth of two thyroid carcinoma cell lines. These evidences suggested that UbcH10 deregulation led to the increased thyroid cancer cell proliferation concur with consistent data we generated also in ovarian, breast, and lymphoid neoplasms [66, 68, 69].

UbcH10 on FNAS

Microarray analysis has been extensively used as a first step to identify molecular markers useful in the distinction of malignant from benign tumors [70]. However, application of these diagnostic genes/proteins in combination with routine FNA cytology requires a complex validation process [45]. In this setting, the 3-gene mRNA assay, which included cyclin D2 (CCND2), protein convertase 2 (PCSK2), and prostate differentiation factor, allowed molecular classification of follicular carcinoma and follicular adenoma [10, 71]. In a recent study, we applied UbcH10 to follicular neoplasm and suspicious for malignancy thyroid FNA [72] UbcH10 expression was evaluated at both transcriptional and translational levels. At the mRNA level, its diagnostic performance was compared with those of the best performing components (CCND2, PCSK2) of the 3-gene diagnostic assay; similarly, at the protein level UbcH10 was compared with the standard proliferation marker Ki-67 (Fig. 11.2). Our results suggested that UbcH10 mRNA assessment can be translated into a diagnostic test. FNAs associated to malignancy had UbcH10 mRNA levels higher than those associated with benign histology. CCND2 and PCSK2, the most performing components of the 3-gene assay, were used for comparison. The UbcH10 diagnostic accuracy was similar to that of CCND2 and higher than that of PCSK2 [72]. Moreover, the UbcH10–CCND2 combination further increased the quantitative RT-PCR diagnostic accuracy (Fig. 11.3), and those cytopathology laboratories that have acquired expertise in evaluating thyroid cancer marker at the mRNA level could include UbcH10 in their diagnostic panels [72].

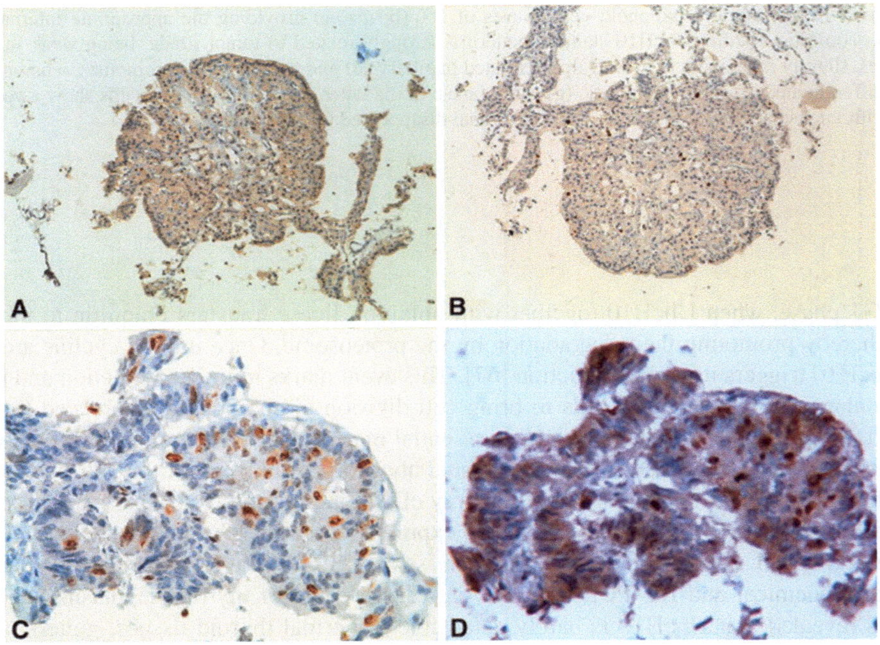

Fig. 11.2 UbcH10 and Ki-67 expression in fine-needle aspiration-derived cell blocks correspond to (**a** and **b**) histologically diagnosed follicular adenoma and (**c** and **d**) papillary thyroid carcinoma (PTC). Both Ki-67 and UbcH10 were found to label more cells in PTC compared with follicular adenoma (H&E counterstain, original magnification, ×10). (**c** and **d**) At higher magnification, it is interesting to note that UbcH10 and Ki-67 shared a similar staining pattern, as shown on parallel cell block sections (H&E counterstain, original magnification, ×20)

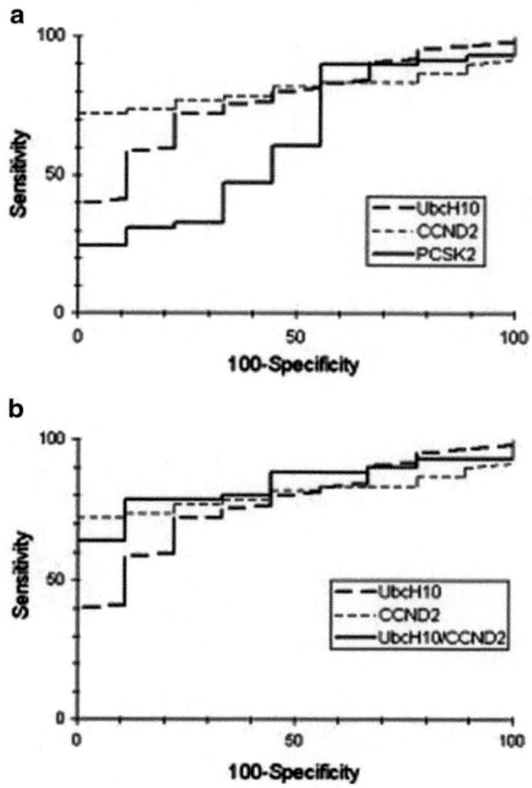

Fig. 11.3 Area under the curve for individual UbcH10, cyclin D2 (CCND2), and protein convertase 2 (PCSK2) expression and their combination using logistic regression analysis is shown. (**a**) The receiver operating characteristic (ROC) curve for any single gene is shown. (**b**) The ROC curves for singly and combined UbcH10 and CCND2 genes are shown

CBX7 Deregulation Occurs in Thyroid Cancer

In addition to UbcH10, among the genes found to be deregulated by our gene profiling screening, we decided to concentrate our studies on a protein belonging to the polycomb group (PcG), the CBX7 protein [73]. The PcG proteins are a class of epigenetic regulators, which always form multiprotein complexes to exert their functions in regulating cell proliferation, senescence, and tumorigenesis via well-known growth regulatory pathways [10, 74]. Several studies have implicated the deregulation of different PcG proteins in carcinogenesis and neoplastic progression [74]. CBX7 is a chromobox family protein and a member of the polycomb repressive complex 1 (PRC1), which together with the PRC2 maintains developmental regulatory genes in a silenced state [75]. Mouse Cbx7 associates with facultative heterochromatin and inactive X chromosome, suggesting a role of the Cbx7 protein in the repression of gene transcription [75]. Data are controversial and the functions and mechanisms of CBX7 in carcinogenesis are still far from clear. Earlier data showed that CBX7 inhibits cellular senescence and extends the lifespan of normal human cells via downregulating the expression of INK4a/ARF locus, and cooperates with c-Myc in lymphomagenesis [76]; more recent studies showed that decrease or loss of CBX7 protein expression correlated with a more aggressive phenotype in pancreatic [64], bladder [77], and colorectal cancer [78], which suggested that CBX7 might act as a potential tumor suppressor. According to this latter possibility, our cDNA microarray analysis data showed that CBX7 is one of the genes with the highest downregulation in all the carcinoma cell lines [73]. In a large panel of cases, CBX7 tissue immunostaining decreased going from benign adenomas to carcinomas (Fig. 11.4). The block of CBX7 expression occurs at

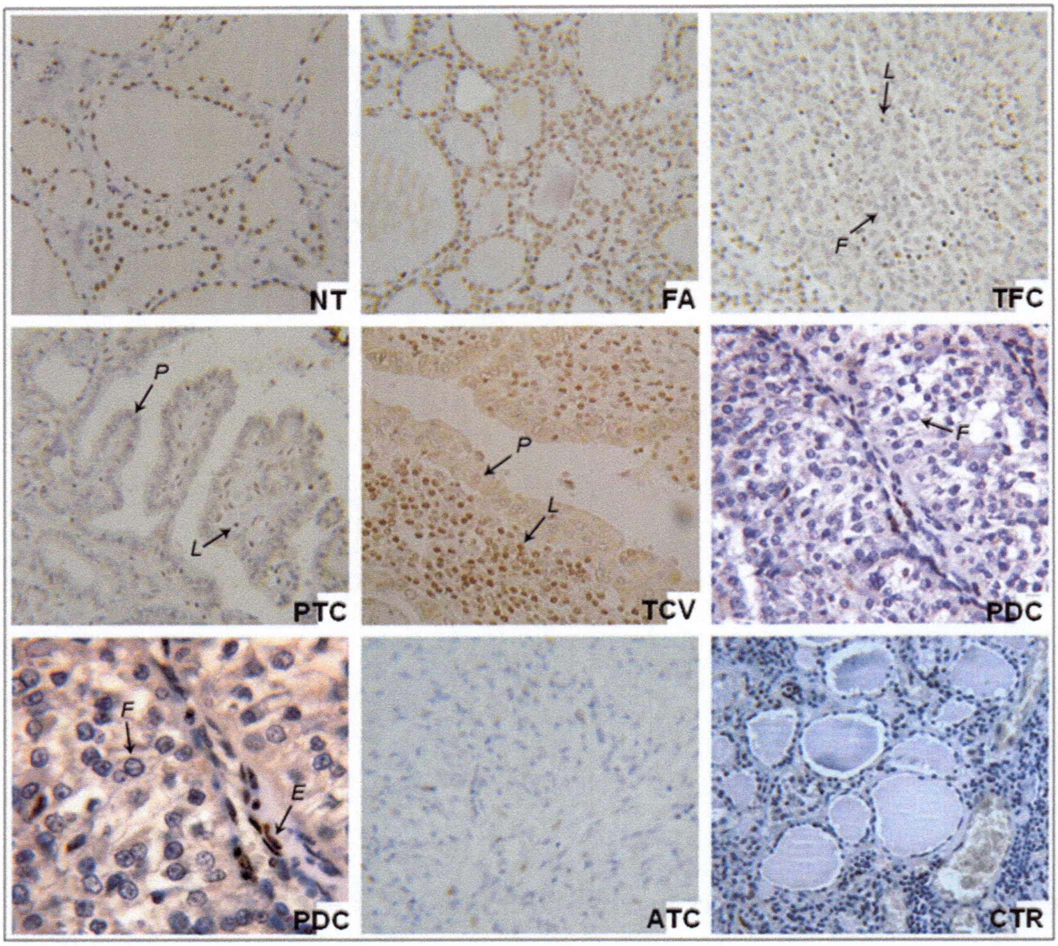

Fig. 11.4 Normal and neoplastic thyroid tissues analyzed for CBX7 protein expression by immunohistochemistry. CBX7 nuclear staining was intense in benign follicular epithelial cells of normal thyroid (1) and follicular adenoma (2), whereas it was weaker in malignant lesions (3), where normal thyroid and papillary carcinoma are adjacent. The decrease of CBX7 in neoplastic lesions was progressive going from well-differentiated cancer, such as minimally invasive follicular carcinoma (4) and "classic variant" papillary carcinomas (PTC; 5) to the "tall cell variant" of PTC (6), whose nuclei are magnified to better show lack of signal, to poorly differentiated (7) and anaplastic (8) carcinomas. The signal disappeared after incubation of the sample with antigen (9). Arrows with letters indicate the following sample features: *P*⟶, nuclei showing cytologic features of PTC negative for CBX7 expression; *N*⟶, normal thyroid adjacent to papillary cancer; *L*⟶, lymphocyte showing CBX7 expression and providing positive internal control

mRNA levels since the qRT-PCR analysis of surgically removed thyroid cancer confirmed the immunohistochemical data. Less-differentiated tumors, namely PDC and TCV PTC and ATC completely lacked CBX7 expression and in most cases LOH at the CBX7 locus (22q13.1) occurred. As the most important application of novel molecular markers is that on FNAs, we applied CBX7 on cytological samples [73].

Immunocytochemistry and qRT-PCR were carried out on cell block specimens obtained from FNC diagnosed as PTC confirmed that CBX7 loss is a reliable novel marker of thyroid cancer also useful in the preoperative setting [74].

From the Bench to the FNA: Sampling and Cytopreparatory Issues

Although the most important application of novel molecular markers is that on cytology, translation from the bench to the FNA is complex [45]. To make DNA- and/or RNA-based testing cost-effective on cytology a close integration with morphology is needed [79]. Thus, the informativeness of the sample for cytology needs to be preserved to keep the accuracy of microscopy high. To this end, each of the steps of traditional cytology, such as preparation of FNA material, search for morphological criteria, assignment to the correct diagnostic class, and suggestion of the appropriate post FNA options should not be altered by molecular analysis. Then, this latter can refine cytology. The final result is to effectively stratify into high-risk and low-risk categories, and the indeterminate cytology classes identified by the BSRTC [45]. Thus, cytological specimens should be properly handled to provide both morphological and molecular information [79]. Our method of preparation of FNA to harvest material sufficient for both tests was recently validated on a series of 128 routinely performed FNA [11]. The rationale behind our sample collection method was to ensure first an adequate cytological diagnosis and, then, to exploit part of the diagnostic material for molecular testing [11]. Thus, two passes from different areas of the lesion are performed. A representative air-dried Diff-Quik stained smear is prepared within few minutes and reviewed on site [11]. In 44 cases, the cytological evidences were sufficient for morphological assessment and the third pass was directly collected in RNA or DNA buffer extraction. Conversely, in 84 cases the specimen was either deemed inadequate by the onsite evaluation or required an additional ethanol-fixed Papanicolaou-stained smear to better evaluate nuclear morphology. Thus, a third pass was dedicated to the preparation of an additional smears and only needle rinsing was collected for BRAF testing. Higher average of extracted DNA concentration was observed in the dedicated pass group (25.9 vs. 7.95 ng/ml). However, the rate of successful exon 15 BRAF amplification was similar with (43/44; 97.7%) or without (79/84; 94%) the dedicated pass. Thus, our protocol is suitable for both tests. When necessary, BRAF testing may also be performed on the residual samples of thyroid nodules, without interfering with routine cytology. Similarly, as far as mRNA markers are concerned, we have shown that in most samples, qRT-PCR analysis does not interfere with cytology [11]. In fact, in a recent study on UbcH10 expression, including 84 cases with a cytological diagnosis of either follicular neoplasm ($n=57$) or suspicious for malignancy ($n=27$), we found that most (73.8%) cases were adequate for both tests [11].

References

1. Sipos JA, Mazzaferri EL. Thyroid cancer epidemiology and prognostic variables. Clin Oncol (R Coll Radiol). 2010;22(6):395–404.
2. Davies L, Welch HG. Increasing incidence of thyroid cancer in the United States, 1973–2002. JAMA. 2006;295(18):2164–7.
3. Aschebrook-Kilfoy B, Ward MH, Sabra MM, Devesa SS. Thyroid cancer incidence patterns in the United States by histologic type, 1992–2006. Thyroid. 2011;21(2):125–34.
4. Volante M, Papotti M. Poorly differentiated thyroid carcinoma: 5 years after the 2004 WHO classification of endocrine tumours. Endocr Pathol. 2010;21(1):1–6.
5. Xing M, Tufano RP, Tufaro AP, Basaria S, Ewertz M, Rosenbaum E, et al. Detection of BRAF mutation on fine needle aspiration biopsy specimens: a new diagnostic tool for papillary thyroid cancer. J Clin Endocrinol Metab. 2004;89(6):2867–72.
6. Volante M, Collini P, Nikiforov YE, Sakamoto A, Kakudo K, Katoh R, et al. Poorly differentiated thyroid carcinoma: the Turin proposal for the use of uniform diagnostic criteria and an algorithmic diagnostic approach. Am J Surg Pathol. 2007;31(8):1256–64.
7. Rothenberg HJ, Goellner JR, Carney JA. Hyalinizing trabecular adenoma of the thyroid gland: recognition and characterization of its cytoplasmic yellow body. Am J Surg Pathol. 1999;23(1):118–25.

8. Papotti M, Rodriguez J, De Pompa R, Bartolazzi A, Rosai J. Galectin-3 and HBME-1 expression in well-differentiated thyroid tumors with follicular architecture of uncertain malignant potential. Mod Pathol. 2005;18(4):541–6.
9. Lin HW, Bhattacharyya N. Clinical behavior of follicular variant of papillary thyroid carcinoma: presentation and survival. Laryngoscope 2010;120(Suppl 4):S163.
10. Nikiforov YE, Steward DL, Robinson-Smith TM, Haugen BR, Klopper JP, Zhu Z, et al. Molecular testing for mutations in improving the fine-needle aspiration diagnosis of thyroid nodules. J Clin Endocrinol Metab. 2009;94(6):2092–8.
11. Troncone G, Cozzolino I, Fedele M, Malapelle U, Palombini L. Preparation of thyroid FNA material for routine cytology and BRAF testing: a validation study. Diagn Cytopathol. 2010;38(3):172–6.
12. Nikiforova MN, Nikiforov YE. Molecular diagnostics and predictors in thyroid cancer. Thyroid. 2009;19(12):1351–61.
13. Kondo T, Ezzat S, Asa SL. Pathogenetic mechanisms in thyroid follicular-cell neoplasia. Nat Rev Cancer. 2006;6(4):292–306.
14. Knauf JA, Fagin JA. Role of MAPK pathway oncoproteins in thyroid cancer pathogenesis and as drug targets. Curr Opin Cell Biol. 2009;21(2):296–303.
15. Lui WO, Zeng L, Rehrmann V, Deshpande S, Tretiakova M, Kaplan EL, et al. CREB3L2-PPARgamma fusion mutation identifies a thyroid signaling pathway regulated by intramembrane proteolysis. Cancer Res. 2008;68(17):7156–64.
16. Volante M, Rapa I, Gandhi M, Bussolati G, Giachino D, Papotti M, et al. RAS mutations are the predominant molecular alteration in poorly differentiated thyroid carcinomas and bear prognostic impact. J Clin Endocrinol Metab. 2009;94(12):4735–41.
17. Salvatore D, Celetti A, Fabien N, Paulin C, Martelli ML, Battaglia C, et al. Low frequency of p53 mutations in human thyroid tumours; p53 and Ras mutation in two out of fifty-six thyroid tumours. Eur J Endocrinol. 1996;134(2):177–83.
18. Smallridge RC, Marlow LA, Copland JA. Anaplastic thyroid cancer: molecular pathogenesis and emerging therapies. Endocr Relat Cancer. 2009;16(1):17–44.
19. Vecchio G, Santoro M. Oncogenes and thyroid cancer. Clin Chem Lab Med. 2000;38(2):113–6.
20. Salvatore G, De Falco V, Salerno P, Nappi TC, Pepe S, Troncone G, et al. BRAF is a therapeutic target in aggressive thyroid carcinoma. Clin Cancer Res. 2006;12(5):1623–9.
21. Xing M. BRAF mutation in papillary thyroid cancer: pathogenic role, molecular bases, and clinical implications. Endocr Rev. 2007;28(7):742–62.
22. Ibrahem S, Seth R, O'Sullivan B, Fadhil W, Taniere P, Ilyas M. Comparative analysis of pyrosequencing and QMC-PCR in conjunction with high resolution melting for KRAS/BRAF mutation detection. Int J Exp Pathol. 2010;91(6):500–5.
23. Kim SK, Kim DL, Han HS, Kim WS, Kim SJ, Moon WJ, et al. Pyrosequencing analysis for detection of a BRAFV600E mutation in an FNAB specimen of thyroid nodules. Diagn Mol Pathol. 2008;17(2):118–25.
24. Davies H, Bignell GR, Cox C, Stephens P, Edkins S, Clegg S, et al. Mutations of the BRAF gene in human cancer. Nature. 2002;417(6892):949–54.
25. Lang AH, Drexel H, Geller-Rhomberg S, Stark N, Winder T, Geiger K, et al. Optimized allele-specific real-time PCR assays for the detection of common mutations in KRAS and BRAF. J Mol Diagn. 2011;13(1):23–8.
26. Trovisco V, Soares P, Sobrinho-Simoes M. B-RAF mutations in the etiopathogenesis, diagnosis, and prognosis of thyroid carcinomas. Hum Pathol. 2006;37(7):781–6.
27. Rhoden KJ, Unger K, Salvatore G, Yilmaz Y, Vovk V, Chiappetta G, et al. RET/papillary thyroid cancer rearrangement in nonneoplastic thyrocytes: follicular cells of Hashimoto's thyroiditis share low-level recombination events with a subset of papillary carcinoma. J Clin Endocrinol Metab. 2006;91(6):2414–23.
28. Lloyd RV, Erickson LA, Casey MB, Lam KY, Lohse CM, Asa SL, et al. Observer variation in the diagnosis of follicular variant of papillary thyroid carcinoma. Am J Surg Pathol. 2004;28(10):1336–40.
29. Mazzaferri EL. Histologic variants of papillary thyroid carcinoma. Endocr Pract. 2001;7(2):139–42.
30. Rosai J, Carcangiu ML, DeLellis RA, Simoes MS. Recommendations for the reporting of thyroid carcinomas. Association of Directors of Anatomic and Surgical Pathology. Hum Pathol. 2000;31(10):1199–201.
31. Castro P, Rebocho AP, Soares RJ, Magalhaes J, Roque L, Trovisco V, et al. PAX8-PPARgamma rearrangement is frequently detected in the follicular variant of papillary thyroid carcinoma. J Clin Endocrinol Metab. 2006;91(1):213–20.
32. Mazzaferri EL. Managing small thyroid cancers. JAMA. 2006;295(18):2179–82.
33. Gharib H. Fine-needle aspiration biopsy of thyroid nodules: advantages, limitations, and effect. Mayo Clin Proc. 1994;69(1):44–9.
34. Sangalli G, Serio G, Zampatti C, Bellotti M, Lomuscio G. Fine needle aspiration cytology of the thyroid: a comparison of 5469 cytological and final histological diagnoses. Cytopathology. 2006;17(5):245–50.
35. Troncone G, Volante M, Iaccarino A, Zeppa P, Cozzolino I, Malapelle U, et al. Cyclin D1 and D3 overexpression predicts malignant behavior in thyroid fine-needle aspirates suspicious for Hurthle cell neoplasms. Cancer. 2009;117(6):522–9.

36. Baloch ZW, Sack MJ, Yu GH, Livolsi VA, Gupta PK. Fine-needle aspiration of thyroid: an institutional experience. Thyroid. 1998;8(7):565–9.
37. Theoharis CG, Schofield KM, Hammers L, Udelsman R, Chhieng DC. The Bethesda thyroid fine-needle aspiration classification system: year 1 at an academic institution. Thyroid. 2009;19(11):1215–23.
38. Cibas ES, Ali SZ. The Bethesda system for reporting thyroid cytopathology. Am J Clin Pathol. 2009;132(5):658–65.
39. Cibas ES, Ali SZ. The Bethesda system for reporting thyroid cytopathology. Thyroid. 2009;19(11):1159–65.
40. Baloch ZW, LiVolsi VA, Asa SL, Rosai J, Merino MJ, Randolph G, et al. Diagnostic terminology and morphologic criteria for cytologic diagnosis of thyroid lesions: a synopsis of the National Cancer Institute Thyroid Fine-Needle Aspiration State of the Science Conference. Diagn Cytopathol. 2008;36(6):425–37.
41. Layfield LJ, Cibas ES, Baloch Z. Thyroid fine needle aspiration cytology: a review of the National Cancer Institute state of the science symposium. Cytopathology. 2010;21(2):75–85.
42. Kocjan G, Cochand-Priollet B, de Agustin PP, Bourgain C, Chandra A, Daneshbod Y, et al. Diagnostic terminology for reporting thyroid fine needle aspiration cytology: European Federation of Cytology Societies thyroid working party symposium, Lisbon 2009. Cytopathology. 2010;21(2):86–92.
43. Layfield LJ, Abrams J, Cochand-Priollet B, Evans D, Gharib H, Greenspan F, et al. Post-thyroid FNA testing and treatment options: a synopsis of the National Cancer Institute Thyroid Fine Needle Aspiration State of the Science Conference. Diagn Cytopathol. 2008;36(6):442–8.
44. Sapio MR, Posca D, Raggioli A, Guerra A, Marotta V, Deandrea M, et al. Detection of RET/PTC, TRK and BRAF mutations in preoperative diagnosis of thyroid nodules with indeterminate cytological findings. Clin Endocrinol (Oxf). 2007;66(5):678–83.
45. Clark DP. Molecular diagnostics on thyroid fine-needle aspirations: the pathway to value creation. Cancer Cytopathol. 2010;118(1):14–6.
46. Sapio MR, Guerra A, Posca D, Limone PP, Deandrea M, Motta M, et al. Combined analysis of galectin-3 and BRAFV600E improves the accuracy of fine-needle aspiration biopsy with cytological findings suspicious for papillary thyroid carcinoma. Endocr Relat Cancer. 2007;14(4):1089–97.
47. Salvatore G, Giannini R, Faviana P, Caleo A, Migliaccio I, Fagin JA, et al. Analysis of BRAF point mutation and RET/PTC rearrangement refines the fine-needle aspiration diagnosis of papillary thyroid carcinoma. J Clin Endocrinol Metab. 2004;89(10):5175–80.
48. Kumagai A, Namba H, Akanov Z, Saenko VA, Meirmanov S, Ohtsuru A, et al. Clinical implications of pre-operative rapid BRAF analysis for papillary thyroid cancer. Endocr J. 2007;54(3):399–405.
49. Pizzolanti G, Russo L, Richiusa P, Bronte V, Nuara RB, Rodolico V, et al. Fine-needle aspiration molecular analysis for the diagnosis of papillary thyroid carcinoma through BRAF V600E mutation and RET/PTC rearrangement. Thyroid. 2007;17(11):1109–15.
50. Jo YS, Huang S, Kim YJ, Lee IS, Kim SS, Kim JR, et al. Diagnostic value of pyrosequencing for the BRAF V600E mutation in ultrasound-guided fine-needle aspiration biopsy samples of thyroid incidentalomas. Clin Endocrinol (Oxf). 2009;70(1):139–44.
51. Zatelli MC, Trasforini G, Leoni S, Frigato G, Buratto M, Tagliati F, et al. BRAF V600E mutation analysis increases diagnostic accuracy for papillary thyroid carcinoma in fine-needle aspiration biopsies. Eur J Endocrinol. 2009;161(3):467–73.
52. Cohen Y, Rosenbaum E, Clark DP, Zeiger MA, Umbricht CB, Tufano RP, et al. Mutational analysis of BRAF in fine needle aspiration biopsies of the thyroid: a potential application for the preoperative assessment of thyroid nodules. Clin Cancer Res. 2004;10(8):2761–5.
53. Rowe LR, Bentz BG, Bentz JS. Utility of BRAF V600E mutation detection in cytologically indeterminate thyroid nodules. Cytojournal. 2006;3:10.
54. Xing M, Clark D, Guan H, Ji M, Dackiw A, Carson KA, et al. BRAF mutation testing of thyroid fine-needle aspiration biopsy specimens for preoperative risk stratification in papillary thyroid cancer. J Clin Oncol. 2009;27(18):2977–82.
55. Chung KW, Yang SK, Lee GK, Kim EY, Kwon S, Lee SH, et al. Detection of BRAFV600E mutation on fine needle aspiration specimens of thyroid nodule refines cyto-pathology diagnosis, especially in BRAF600E mutation-prevalent area. Clin Endocrinol (Oxf). 2006;65(5):660–6.
56. Deveci MS, Deveci G, LiVolsi VA, Baloch ZW. Fine-needle aspiration of follicular lesions of the thyroid. Diagnosis and follow-up. Cytojournal. 2006;3:9.
57. Yip L, Nikiforova MN, Carty SE, Yim JH, Stang MT, Tublin MJ, et al. Optimizing surgical treatment of papillary thyroid carcinoma associated with BRAF mutation. Surgery. 2009;146(6):1215–23.
58. Bartolazzi A, Orlandi F, Saggiorato E, Volante M, Arecco F, Rossetto R, et al. Galectin-3-expression analysis in the surgical selection of follicular thyroid nodules with indeterminate fine-needle aspiration cytology: a prospective multicentre study. Lancet Oncol. 2008;9(6):543–9.
59. Gasbarri A, Marchetti C, Iervasi G, Bottoni A, Nicolini A, Bartolazzi A, et al. From the bench to the bedside. Galectin-3 immunodetection for improving the preoperative diagnosis of the follicular thyroid nodules. Biomed Pharmacother. 2004;58(6–7):356–9.

60. Bartolazzi A, Gasbarri A, Papotti M, Bussolati G, Lucante T, Khan A, et al. Application of an immunodiagnostic method for improving preoperative diagnosis of nodular thyroid lesions. Lancet. 2001;357(9269):1644–50.
61. Mehrotra P, Okpokam A, Bouhaidar R, Johnson SJ, Wilson JA, Davies BR, et al. Galectin-3 does not reliably distinguish benign from malignant thyroid neoplasms. Histopathology. 2004;45(5):493–500.
62. Baloch ZW, Fleisher S, LiVolsi VA, Gupta PK. Diagnosis of "follicular neoplasm": a gray zone in thyroid fine-needle aspiration cytology. Diagn Cytopathol. 2002;26(1):41–4.
63. Segev DL, Clark DP, Zeiger MA, Umbricht C. Beyond the suspicious thyroid fine needle aspirate. A review. Acta Cytol. 2003;47(5):709–22.
64. Pallante P, Terracciano L, Carafa V, et al. The loss of the CBX7 gene expression represents an adverse prognostic marker for survival of colon carcinoma patients. Eur J Cancer. 2010;46(12):2304–13.
65. Pallante P, Berlingieri MT, Troncone G, Kruhoffer M, Orntoft TF, Viglietto G, et al. UbcH10 overexpression may represent a marker of anaplastic thyroid carcinomas. Br J Cancer. 2005;93(4):464–71.
66. Troncone G, Guerriero E, Pallante P, Berlingieri MT, Ferraro A, Del Vecchio L, et al. UbcH10 expression in human lymphomas. Histopathology. 2009;54(6):731–40.
67. Rape M, Kirschner MW. Autonomous regulation of the anaphase-promoting complex couples mitosis to S-phase entry. Nature. 2004;432(7017):588–95.
68. Berlingieri MT, Pallante P, Sboner A, Barbareschi M, Bianco M, Ferraro A, et al. UbcH10 is overexpressed in malignant breast carcinomas. Eur J Cancer. 2007;43(18):2729–35.
69. Berlingieri MT, Pallante P, Guida M, Nappi C, Masciullo V, Scambia G, et al. UbcH10 expression may be a useful tool in the prognosis of ovarian carcinomas. Oncogene. 2007;26(14):2136–40.
70. Pusztai L, Gianni L. Technology insight: emerging techniques to predict response to preoperative chemotherapy in breast cancer. Nat Clin Pract Oncol. 2004;1(1):44–50.
71. Shibru D, Hwang J, Khanafshar E, Duh QY, Clark OH, Kebebew E. Does the 3-gene diagnostic assay accurately distinguish benign from malignant thyroid neoplasms? Cancer. 2008;113(5):930–5.
72. Guerriero E, Ferraro A, Desiderio D, Pallante P, Berlingieri MT, Iaccarino A, et al. UbcH10 expression on thyroid fine-needle aspirates. Cancer Cytopathol. 2010;118(3):157–65.
73. Pallante P, Federico A, Berlingieri MT, Bianco M, Ferraro A, Forzati F, et al. Loss of the CBX7 gene expression correlates with a highly malignant phenotype in thyroid cancer. Cancer Res. 2008;68(16):6770–8.
74. Gil J, Bernard D, Peters G. Role of polycomb group proteins in stem cell self-renewal and cancer. DNA Cell Biol. 2005;24(2):117–25.
75. Bernard D, Martinez-Leal JF, Rizzo S, Martinez D, Hudson D, Visakorpi T, et al. CBX7 controls the growth of normal and tumor-derived prostate cells by repressing the Ink4a/Arf locus. Oncogene. 2005;24(36):5543–51.
76. Scott CL, Gil J, Hernando E, Teruya-Feldstein J, Narita M, Martinez D, et al. Role of the chromobox protein CBX7 in lymphomagenesis. Proc Natl Acad Sci USA. 2007;104(13):5389–94.
77. Hinz S, Kempkensteffen C, Christoph F, Krause H, Schrader M, Schostak M, et al. Expression parameters of the polycomb group proteins BMI1, SUZ12, RING1 and CBX7 in urothelial carcinoma of the bladder and their prognostic relevance. Tumour Biol. 2008;29(5):323–9.
78. Karamitopoulou E, Pallante P, Zlobec I, Tornillo L, Carafa V, Schaffner T, et al. Loss of the CBX7 protein expression correlates with a more aggressive phenotype in pancreatic cancer. Eur J Cancer. 2010;46(8):1438–44.
79. Ohori NP, Nikiforova MN, Schoedel KE, LeBeau SO, Hodak SP, Seethala RR, et al. Contribution of molecular testing to thyroid fine-needle aspiration cytology of "follicular lesion of undetermined significance/atypia of undetermined significance". Cancer Cytopathol. 2010;118(1):17–23.

Index

A. Russo et al. (eds.), *Diagnostic, Prognostic and Therapeutic Value of Gene Signatures*, Current Clinical Pathology, DOI 10.1007/978-1-61779-358-5, © Springer Science+Business Media, LLC 2012

MIX
Papier aus verantwortungsvollen Quellen
Paper from responsible sources
FSC® C105338

If you have any concerns about our products,
you can contact us on
ProductSafety@springernature.com

In case Publisher is established outside the EU,
the EU authorized representative is:
Springer Nature Customer Service Center GmbH
Europaplatz 3, 69115 Heidelberg, Germany

Printed by Libri Plureos GmbH
in Hamburg, Germany